Praise For

The Politically Incorrect Guide® to
Climate Change

"The Politically Incorrect Guide® to Climate Change is a welcome scientific and rational antidote to the liberal news media, the UN, and Al Gore's incessant chattering about climate doom. This book exposes the hypocrisy of Learjet limousine liberals who fly in their own private jets and own multiple homes while preaching to the world about downsizing and energy rationing. Every parent in America should be armed with this book to fight the brainwashing of their kids from kindergarten through college. Marc's book is the ultimate A-Z reference guide that debunks man-made climate change claims using scientific studies and prominent scientists. *The Politically Incorrect Guide® to Climate Change* is the book the UN and Al Gore do not want you to read. The climate scare ends with this book."

> **—SEAN HANNITY**, host of *Hannity* on Fox News and of the nationally
> syndicated radio program *The Sean Hannity Show*

"I have never met Marc Morano, the author of this very interesting book, but I know him well from his excellent blog Climate Depot, which I read regularly. In the book he exposes the climate myths that even scientific organizations like the Physical Society and American Association for the Advancement of Science push. The Earth has existed for maybe 4.5 billion years, and now the alarmists will have us believe that because of the small rise in temperature for roughly 150 years (which, by the way, I believe you cannot really measure) we are doomed unless we stop using fossil fuels. We are now forced to use corn-based ethanol in our gas, subsidized windmills, and solar cells for energy; meanwhile, maybe a billion people worldwide starve and have no access to electricity. You and I breathe out at least thirty tons of CO_2 in a normal life span, but nevertheless the Environmental Protection Agency decided to classify rising carbon-dioxide emissions as a hazard to human health. Marc Morano discusses the reasons and history of all these strange theories in his excellent book *The Politically Incorrect Guide® to Climate Change.* Please read it, you will be amazed!"

> **—IVAR GIAEVER**, Nobel Laureate in physics

"Marc Morano and I have been on the front lines fighting the global warming hoax together. In his book *The Politically Incorrect Guide® to Climate Change*, Marc continues the work we started together—using facts and sound science to show that the radical environmental alarmists' claims are nothing more than climate hysteria. He doesn't just present the truth—he uses open debate to challenge and rebut the claims from the climate extremists."

—**SENATOR JAMES INHOFE OF OKLAHOMA**, member of the U.S. Senate
Environment and Public Works Committee

"This book covers the history of climate, from the global cooling 'coming ice age' scare of the 70s to the 'we have just a few years left to save the planet' that characterizes the current global warming scare. Written in a light reading style, virtually every page is meticulously referenced with sources for the points he makes. Love him or hate him, Morano is very effective in conveying the history and the climate flim-flammery under the guise of science that has been going on the last few decades, mostly thanks to huge government funding of climate science. It reads like a postmortem verification of President Eisenhower's farewell address, which warned of the 'military-industrial complex,' but also said, 'The prospect of domination of the nation's scholars by Federal employment, project allocations, and the power of money is ever present and is gravely to be regarded.'"

—**ANTHONY WATTS**, publisher of WattsUpWithThat, the world's most viewed
climate-themed website

"Marc Morano's remarkable book *The Politically Incorrect Guide® to Climate Change* documents, in their own words, how many honest scientists still insist that hypotheses not confirmed by observation should be rejected. It exposes the pernicious myth that 97 percent of scientists agree that increasing levels of carbon dioxide are an existential threat, one that mandates the surrender of human freedom and wellbeing to an 'enlightened' climate elite. The book documents that many very distinguished scientists do not agree. In fact, more carbon dioxide is already benefitting the world though increased yields for agriculture and forestry, and from shrinking deserts. The hated 'deniers' are right. There is no emergency. When later generations of historians analyze the climate hysteria of our time, this book will be one of their most valuable references."

—**WILL HAPPER**, Cyrus Fogg Brackett Professor of Physics, emeritus,
Princeton University

"*The Politically Incorrect Guide® to Climate Change* is a must read to counter the media's non-stop climate propaganda machine. This book demolishes the star-dazzled media elite claims and exposes the Hollywood climate hypocrites. Morano's new book bypasses the establishment media and brings you the hard data and studies and gives voice to the prominent scientists who declare their dissent from Al Gore's ridiculous claims. A true game-changer in the climate debate."

—**L. BRENT BOZELL**, founder and president of the Media Research Center

"Marc Morano likes to say that he's not a scientist, but he plays one on TV. Well, this book shows that he really is a scientist, in that he follows the scientific method of demanding real-world data to verify hypotheses and predictions. That's more than can be said of lots of Ph.D.s who are still sticking to their story of climate catastrophe, despite its collapse under the solid data of the past few decades. In this book Morano corrects the necessary flaw in his light-hearted movie, which is that as a movie it simply didn't have time for the level of detail and proof needed to resolve a debate. Here Morano provides the details and the footnotes to the studies he and his stable of leading physicists like William Happer and Richard Lindzen and humbler statisticians like me offer to explain why climate hysteria is misguided and why carbon-cutting policies are dangerous to global living standards. From the bogus claim that 97 percent of climate scientists believe that all warming comes from carbon dioxide to the bogus claims about the catastrophic effects of the one degree of global warming since 1880, whether industrial or natural, on polar bear populations, droughts, flooding, storms, and forest fires, you'll find it all documented here. Also, it's all balanced with a full presentation of the views of those who dispute Morano. Unlike the globalarmists, who for decades have declared 'the debate is over,' Morano revels in open debate. He's obviously confident that he'll win it. After reading this book, you will be too."

—**CALEB ROSSITER**, climate statistician, American University

"Marc Morano is to the climate change cult what Galileo was to the believers in a flat Earth. He uses observation, history, and, yes, science to prove the global warming crowd are full of hot air. He is a thorn in the side of those who want to control every aspect of our lives. His fact-based refutations of the secular progressives overcome their hysteria and twisted 'science.' Read this book and then challenge a friend who has drunk the Kool-Aid® to read it."

—**CAL THOMAS**, nationally syndicated columnist

"This book reveals that 'global warming' is not and has never been about the 'science.' *The Politically Incorrect Guide® to Climate Change* reveals the agenda behind the lavishly funded and government-sponsored climate change establishment. Morano unmasks the United Nations' goals of 'global governance,' redistribution of wealth, and global carbon taxes. This book arms every citizen with a comprehensive dossier on just how science, economics, and politics have been distorted and corrupted in the name of saving the planet. Contrary to Al Gore's claims, UN treaties and EPA regulations cannot control the weather or the oceans. A must read."

—**MARK LEVIN**, author of *Men in Black*, *Liberty and Tyranny*, and
Rediscovering Americanism: And the Tyranny of Progressivism

"With his book *The Politically Incorrect Guide to Climate Change*, Marc Morano vies to be the Thomas Paine of the movement to save the world from the tyranny of climate catastrophists. He exposes the seemingly infinite number of absurd claims, and the almost unbounded hypocrisy and venality of the proponents of this clearly inhuman and scientifically implausible attempt to control mankind by controlling and, more importantly, restricting access to, energy. This book is an unrelenting polemic of the best kind."

—**RICHARD LINDZEN**, professor emeritus and former Alfred P. Sloan
Professor of Meteorology, Massachusetts Institute of Techology

"Morano's probably single-handedly, in a civilian sense, the guy (other than me, of course) doing a better job of ringing the bells alarming people of what's going on here."

—**RUSH LIMBAUGH**, nationally syndicated radio host and bestselling author

The Politically Incorrect Guide® to Climate Change

Be sure to check out

The Politically Incorrect Guides® to...

The Politically Incorrect Guide® to
Climate Change

Marc Morano

REGNERY
PUBLISHING
A Division of Salem Media Group

Regnery® is a registered trademark of Salem Communications Holding Corporation

Cataloging-in-Publication data on file with the Library of Congress

ISBN 978-1-62157-676-1
e-book ISBN 978-1-62157-757-7

Published in the United States by
Regnery Publishing
A Division of Salem Media Group
300 New Jersey Ave NW
Washington, DC 20001
www.Regnery.com

Manufactured in the United States of America
10 9 8 7 6

Books are available in quantity for promotional or premium use. For information on discounts and terms, please visit our website: www.Regnery.com.

This book is dedicated to the pioneers who challenged the man-made global warming narrative years before it displaced earlier environmental scares such as the Amazon rainforest deforestation movement—pioneers like atmospheric physicist Fred Singer, MIT's Richard Lindzen, Bill Gray, Pat Michaels, and so many other scientists were first to take on the government-media-academia-UN complex. And one political leader stands out: Oklahoma Senator James Inhofe, author of The Greatest Hoax. Inhofe stood up fearlessly to the alleged "consensus" and the political "solutions" being promoted to "solve" climate change. Senator Inhofe, my former boss, stared down the media and political establishment when other Republicans were intimidated into silence. Others instrumental in paving the way for this book were Ralph Hostetter and my former bosses Rush Limbaugh, Paul Weyrich, and Brent Bozell (and his Media Research Center team), who took on the media and political classes and documented and deflated their claims. In this book I have endeavored to follow in these pioneers' wake and provide the public with a full history and exposé of just what exactly lies behind the man-made "climate change" curtain.

Contents

Foreword

by John Coleman

Don't worry about "climate change," says Marc Morano—there is no significant man-made global warming.

Are you kidding me? We all know that the icecaps are melting, the oceans are about to flood our cities, and more and more superstorms are happening. And the experts are certain that mankind's use of fossil fuels is causing it all. We have all the facts; right?

The truth is that there is a debate about climate change but it has been very one-sided. With the U.S. government, all the scientific organizations, Al Gore, the Science Guy, Hollywood, the Democratic Party, and the United Nations all behind the bad news that our use of fossil fuels is destroying the climate of Earth, anyone on the other side of the debate finds themselves behind the eight ball. Peeking out from behind the eight ball is Marc Morano. In this great book he begins his comprehensive review of the debate about global warming by chatting about the history of climate scares in centuries past—and goes on to decisively debunk the current climate scare. By the time he's done, you will realize you've been hoaxed. Climate change has become a scam.

As the founder of the Weather Channel and a six-decade veteran TV news weatherman, I know a great deal about this topic. We meteorologists are well aware of how limited our ability is to predict the weather. Our predictions become dramatically less reliable as they extend out into the future. When we try to predict just a few weeks into the future our predictions become increasingly inaccurate. Yet the "climate change" establishment that now dominates the UN bureaucracy and our own government science establishment claim that they can predict the temperature of the Earth decades into the future. Their global warming scare is not driven by science; it is now being driven by politics. So today anybody who defies the prevailing "climate change" scare puts his career and his reputation into extreme danger.

That is where we find Marc. He is living life behind the eight ball. He has been there for decades. But whatever you may hear from his enemies in the climate change establishment, he is no crazy denier or shill for Big Oil. The explanation is simple. He is so certain of his data that he is quite comfy there behind the eight ball. When you really study the issue, you realize that Marc Morano is absolutely right. And it turns out he is not alone there behind the eight ball. He has developed relationships with hundreds of brilliant scientists and other experts who are willing to testify, along with Marc, that in fact there is no significant man-made global warming.

This book is exactly what parents need to counter the indoctrination our children are now being subjected to. Starting at a very young age and continuing through their teenage years, American school children are being constantly bombarded with climate change propaganda. This is science gone bad. It has become political. And climate science has been hijacked by the extreme fringe of the environmental movement. The truth is that while climate is naturally changing—as it always has—no crisis is occurring and there is no reason to fear any in the future. This book uses over twelve hundred footnotes to bolster its compelling, scientific, and logical demonstration that Al Gore and the United Nations are dead wrong on climate fears. And maybe even more important, this book uses the climate change establishment's own words to refute their silly claims.

Read this book and Marc will become your hero. Give it to your friends to read. Maybe in the end there will be enough of us who no longer believe the climate change hoax that he and those of us who know he is right can get out from behind the eight ball and enjoy life. Read on, my friend, read on.

CHAPTER 1

The Education of a Climate Denier

I am not a scientist—though I do occasionally play one on TV. Well, actually, I debate scientists there, regularly appearing on television to expose the unscientific claims about catastrophic man-made climate change. My degree is in political science, which happens to be the ideal background for examining man-made global warming claims, which, as ample evidence set forth in this book will demonstrate, are driven more by politics than by science. I have spent the last twenty-five years in a range of disciplines including as a working journalist, documentary maker, radio talk show host, author, and national television correspondent. I have been passionate about environmental issues since I began my career in 1991. I produced a documentary on the myths surrounding the Amazon Rainforest in 2000[1] and have reported extensively—at one time holding both White House and Capitol Hill press credentials—on environmental and energy issues such as deforestation, endangered species, pollution, and climate change. I co-wrote, hosted, and co-produced the 2016 global warming documentary *Climate Hustle*,[2] which was featured in over four hundred theaters in the United States. The film featured current and former UN scientists who have turned against the UN and prominent

Did you know?

★ The purported 97 percent scientific consensus on climate change was "pulled from thin air"

★ Over 750 skeptical scientists were featured in a Senate Environment and Public Works Committee report

★ A former UN IPCC official called global warming "my religion"

1

politically left-wing scientists who have reversed their views and now reject claims of a man-made climate crisis.

I am a climate skeptic, a doubter, a dissenter—and have been smeared as a "denier." But I am not alone in my skepticism. I work regularly with a huge network of internationally renowned scientists, many of them formerly of the United Nations Intergovernmental Panel on Climate Change (UN IPCC). Less charitable names have been used to describe us. The vicious name-calling starts with the "climate denier" epithet that is meant to evoke comparisons to Holocaust deniers. But it doesn't stop there.

"Wanted" poster of Marc Morano outside the 2015 UN climate summit in Paris.

★ ★ ★

Rave Reviews

In 2015, I appeared—as the villain—in the Sony Pictures climate activist documentary *Merchants of Doubt*. The reviews were glowing (as in, I was torched). Reviewers called me "terrifyingly impressive, sadistic"[10] and the "documentary's most engaging character."[11] I was "a magnificent antihero, a cheery, chatty prevaricator,"[12] "slick and scary,"[13] "a loathsome mercenary,"[14] a "sleazy spin doctor,"[15] and a "grinning-skull nihilist."[16] The *Daily Kos* said I was "Evil Personified."[17]

Rolling Stone magazine named me one of the planet's seventeen "climate killers" in its December 2009 cover story. The magazine described me as "the Matt Drudge of climate denial" and "a central cell of the climate-denial machine."[3] In 2009, *Newsweek* magazine declared that "Morano, undeniably, is quickly becoming king of the skeptics."[4] Professor Andrew Watson called me an "asshole" on a live BBC TV broadcast about Climategate.[5] In 2010, *Esquire* magazine profiled my work as a climate skeptic in a sixty-five-hundred-word feature article titled "This Man Wants to Convince You Global Warming Is a Hoax," stating, "He seems to be winning."[6] *Washington Post* climate reporter Juliet Eilperin wrote, "Esquire has pronounced Marc Morano the most important climate skeptic of all."[7]

In 2012, I was (dis)honored by climate activists at Media Matters as the "Climate Misinformer of the Year"—an "award" that former vice president Al Gore helped promote.[8] I went Hollywood in 2014 when *Variety* magazine called me a "charismatic professional climate-change misinformer."[9] While I was attending the UN climate summit in 2015, my face was plastered on posters around the streets of Paris—I was "wanted" as a "climate criminal."

I started out my political life as an eleven-year-old volunteer for Ronald Reagan's 1980 presidential campaign. I always considered myself a Republican—except when it came to environmental issues.

I was not a fan of James Watt, Reagan's Interior secretary, as I was upset by his land-use policies and by what I perceived to be his general anti-environmental stances.

As a kid I loved to spend time in the woods fishing, hiking, and building forts. I even thought my ultimate job would be as a forest ranger, living deep in the wild in the great Pacific Northwest. I was always passionate about animals, having grown up with what can reasonably be called a small zoo in my home, including alligators, snakes, turtles, lizards, frogs, rats, mice, hamsters, and of course dogs. So I was very sensitive to environmentalists' claims of deforestation and species extinction. From the 1980s through the early 1990s, I became increasingly concerned about the destruction of the Amazonian rainforest. I watched documentaries and read up on the issue; vivid images of jungle animals' habitat being cleared, trees cut down by chain saws, made a huge impression on me.

But then, in 1992, I experienced an epiphany. My first doubts on the deforestation issue were raised by physicist Dixy Lee Ray, who attended the Rio Earth Summit and filed reports on Rush Limbaugh's radio show debunking the claims that the Amazonian rainforest was about to disappear. I started investigating and was very surprised to find out that the green movement was in many cases wildly exaggerating deforestation claims.

My investigation of the issue eventually culminated in the production of my own documentary on the Amazon, released in 2000 and titled *Amazon Rainforest: Clear-Cutting the Myths.*[18]

The film made a huge media splash when it was released, helped by my interview with comic actor Chevy Chase, who claimed "socialism works" and pointed to Cuba as the model. My documentary debunked the myth that environmentalists and celebrities are the friends of indigenous people. I interviewed the tribal leaders who have contempt for environmental activists and celebrities because they feel exploited by them. I also spoke in depth with the scientists monitoring the Amazonian rainforests at that time. When I showed them one of the travel guide books on the Amazon claiming that the forests were about to disappear, the scientists threw down the guide book and shouted "Bullshit!" on camera.

My film showed that the Amazon was one of the most intact forests in the world. The greens were using double accounting when they claimed that X number of "football fields" of forest per minute were disappearing—they weren't taking into account the regeneration forests. The film generated huge publicity for taking on Hollywood and the environmentalists. *US Weekly* entertainment magazine did a feature article reporting that celebrities were "infuriated" by my documentary and *Extra*, the nationally syndicated entertainment TV show, featured a segment.[19]

In 2009, the *New York Times* echoed my findings in a feature article on the Amazon, reporting on the growth of the "galloping jungle" as farmlands reverted back to nature. The article noted that for every acre of rainforest cut down each year, more than fifty acres of new forest were growing. Why? Because people were leaving swamps, jungles, and wetlands to move to urban areas.[20]

The *Times* reported, "These new 'secondary' forests are emerging in Latin America, Asia and other tropical regions at such a fast pace that the trend has set off a serious debate about whether saving primeval rain forest—an iconic environmental cause—may be less urgent than once thought." It was during the making of the rainforest documentary that I began my collaboration with Greenpeace founder Patrick Moore, who had left the group in 1986 because he felt it had become too radical. Moore, who holds a Ph.D. in ecology from the University of British Columbia, has been a font of information, experience, and expertise on all things environmental in the years that have followed.

On the rainforest issue, Moore had a simple philosophy: "Save the Trees, Use More Wood." He pointed out, "We should be growing more trees and using more wood. The less wood we use, the more steel and concrete we use."[21] Moore explained that a greater demand for wood products leads to more forested land, noting that 80 percent of the timber produced in the United States comes from private property. He predicted that if "those land owners had no market for wood, they would clear the forest away and grow something else they could make money from instead." Moore explained that, ironically, "When you go into a lumber yard, you are given the impression that by buying wood you are causing the forest to be lost, when in fact what you are doing is sending a signal into the market to plant more trees."

After the success of my Amazon documentary, I went to work for Brent Bozell's Media Research Center in their news division CNSNews.com. Bozell was exposing the mainstream media's gross bias and fake news decades before the term was invented. As an investigative journalist, I specialized in environmental reporting and attended UN Earth summits in South Africa, South America, and other parts of the world. It was during this time that I broadened the focus of my reporting to include climate change as the deforestation issue faded.

In 2006, I went to work for the U.S. Senate Environment and Public Works Committee under then-chairman Senator James Inhofe of Oklahoma. From 2006 to 2009, I was the committee's director of communications and the speechwriter for the senator on green issues, particularly climate change. It was an honor to work for Senator Inhofe, who is absolutely fearless on global warming, being virtually the only Republican in the Senate standing up to the media, politically and financially co-opted academia, and the UN. He has fought them all for decades now. In 2016, Leonardo DiCaprio's National Geographic climate documentary released a top ten list of "climate deniers."[22] I was proud to have made the list at number two, second only to my old boss Senator Inhofe.

In my capacity as communications director, I published the Senate Environment and Public Works Committee's award-winning blog and authored the first ever U.S. government "Skeptic's Guide to Debunking Global Warming Alarmism" in 2006.[23] I also reengineered the media communications of the Senate EPW. EPW press secretary Matt Dempsey and I created an online "war room" to serve as a rapid response team to counter climate change claims by the media and Al Gore. We were so successful that this Senate climate blog crashed the entire U.S. Senate Web server, creating a historic shutdown in 2007 after we were linked by the Drudge Report.[24] The radically reimagined climate strategy got noticed, and in 2008 the EPW Senate blog and website won the coveted Gold Mouse Award, made possible by the John F. Kennedy School of Government at Harvard University and the National Science Foundation.[25]

I travelled the globe attending UN climate summits in such exotic locations as Bali and Kenya and going on a "fact-finding" trek to Greenland.

I also authored the 255-page Senate report of over seven hundred dissenting scientists on man-made global warming, originally published in 2007 and updated in 2008 and 2009.[26] In 2010, the number of dissenting scientists exceeded a thousand.[27] The Gallup polling organization recognized the impact of this U.S. Senate report in a 2008 analysis: "Republican spokespersons and conservative commentators have long challenged IPCC reports as reflecting the 'scientific consensus' on global warming by highlighting the views of a modest number of 'skeptic' or 'contrarian' scientists who question the IPCC conclusions." Gallup concluded, "Growing skepticism about news coverage of global warming clearly goes hand in hand with Republicans' declining belief that it is already occurring."[28]

I am now the publisher of the award-winning Climate Depot website.[29] Climate Depot is a project of the Committee for a Constructive Tomorrow or CFACT. CFACT has been around since 1985 and is run by David Rothbard and Craig Rucker, two guys who understand the modern climate debate and approach it from a free market environmental perspective. CFACT also understands how to have fun poking holes in climate change claims.

At Climate Depot, I work daily with scientists who examine the latest peer-reviewed studies and data on the climate.

Nationally syndicated radio talk show host Rush Limbaugh has praised Climate Depot. "It's a great place to keep up on the global warming debate," Limbaugh said on air in 2009.[30] "Morano's probably single-handedly, in a civilian sense, the guy (other than me, of course) doing a better job of ringing the bells alarming people of what's going on here," Limbaugh explained. I worked for Limbaugh's nationally syndicated *Rush Limbaugh* television show during its four-year run from 1992 to 1996. I served as his onscreen television reporter and producer, and Limbaugh referred to me as "Our Man in Washington." I had the dubious distinction of being the first journalist in history to have his television camera seized at the Clinton White House in 1993 while on assignment with the Limbaugh TV show.

This book will serve as a reference guide for readers who know they are not getting the full story from the *New York Times*, CBS News, and the UK *Guardian*. The case against man-made climate change fears has only strengthened in recent years.

Why should people care about whether man-made "global warming" is a threat and whether the proposed "solutions" are really necessary? The fact is, the climate agenda literally impacts every aspect of your life. The purported solutions to this non-problem will affect what kind of lightbulbs and appliances you are allowed to buy, the size of your home and how it is heated and air conditioned, how you travel, the food you eat, the clothes you wear, and how many kids you can have. They will have enormous impacts on land use, jobs, prices, the world's economy, and even our national sovereignty. Some scientists are even advocating for shrinking the human race to decrease our carbon footprint—and for medicating us to make us care more about the climate.

United Nations officials and the media claim that the world must act on climate change or else a calamity faces humanity. But left out of this discussion are the many scientists skeptical of the scientific claims and goals behind the United Nations climate agenda. A top UN IPCC official has stated openly that "we redistribute de facto the world's wealth by climate policy."[31] In fact, the UN climate panel is nothing more than a political lobbying organization masquerading as a scientific body.

The motivations of many climate "authorities" are anything but scientific. Rajendra Pachauri, former head of the UN IPCC panel, has announced that global warming "is my religion."[32]

Pachauri also admitted that the purpose of the UN IPCC climate reports was to make the case so that "rational people across the globe will see that action is needed on climate change."[33] And he conceded that the UN climate reports were tailored to meet the political needs of governments: "We are an intergovernmental body and we do what the governments of the world want us to do. If the governments decide we should do things differently and come up with a vastly different set of products we would be at their beck and call," Pachauri said in 2013.[34]

And for those who aren't true believers, we'll see that there are enormous financial and career incentives to stay on the global warming bandwagon— plus intimidation and drastic disincentives to punish anyone who thinks about hopping off.

In my investigations into the climate change issue, I've uncovered shameless scare-mongering, scientific fraud, and countless instances of the misappropriation of the reputation and authority of science to prop up what is essentially a political campaign to put the world's economy under the management of "experts" at the United Nations. Along the way, every argument advanced to support the claim of catastrophic man-made global warming has proven to be severely lacking.

The repeated claim of a 97 percent "consensus" in support of catastrophic man-made climate change? It's nothing more than a talking point designed to silence anyone who dares to question the very dubious "science" on global warming. As we shall see, UN IPCC lead author Richard Tol, a professor of the economics of climate change at Vrije University in Amsterdam, has examined the 97 percent claim and found that it was simply "pulled from thin air."[35]

And no wonder scientists don't really agree. The evidence that human activity is causing a dangerous rise in temperatures is thin to nonexistent.

Carbon dioxide causes warming? As we'll see, CO_2 can have a warming effect, but you cannot distinguish the impact from rising CO_2 from natural variability.

This year is the hottest on record? That's what we keep being told—when, in fact, the claimed "record heat" is within the margin of error between so-called hottest years—a fancy way of saying the temperature standstill is continuing.

Extreme weather events are getting worse? That's what they say, but we'll see data showing that there's no such trend.

Global warming threatens the polar bears with extinction? As we'll see, polar bear populations are hitting record highs.

Climate change will create instability and increase the risk of war? This book will show you research demonstrating that there is less conflict during warmer eras than in cold ones.

And even if we really were facing catastrophic global warming as a result of human activity, the proposed solutions still wouldn't make any sense.

To fight climate change, former UN climate chief Christiana Figueres has called for a "centralized transformation" that is "going to make the life of everyone on the planet very different."[36]

But will UN treaties, EPA regulations, and personal sacrifices for the planet actually impact the climate? As we'll see, even the Obama administration EPA chief admitted there would be no measurable climate impact from these proposed remedies.[37]

The notion that a UN agreement to limit emissions will somehow alter the Earth's temperature or storminess borders on belief in witchcraft. Philip Stott, professor emeritus of Biogeography at the University of London, points out that "climate change is governed by hundreds of factors, or variables, and the very idea that we can manage climate change predictably by understanding and manipulating at the margins one politically-selected factor [CO_2], is as misguided as it gets."[38]

If we had to rely on the UN or EPA to save us from global warming, we would all be doomed. Nothing the UN is proposing to solve climate change would have any significant impact on temperatures or extreme weather events even if you accept the climate warmists' scientific claims.

In any case, the UN's agenda of limiting the Earth's temperature to a rise of no more than two degrees Celsius is a target "plucked out of thin air,"[39] according to an email exposed in the Climategate scandal—about which, much more later.

And the climate agenda by which the UN aims to meet its arbitrary target will cripple the economies of the industrialized nations while doing nothing for the climate. Worse, it will stop desperately needed economic growth in the developing world, where over a billion people still lack life-saving access to electricity. The UN agenda seeks to replace the proven success of coal, oil, and gas with not-ready-for-prime-time "green" energy. Wind and solar are simply not capable of powering the world economy at this time. And, as we shall see, the world's poor are not oblivious to the irony of the prosperous nations' plans keep them from leveraging their way out of poverty with the fossil fuels that made us rich.

On climate change, the science is not "settled." The debate is not "over." This book is designed to provide you with the facts you need to understand and resist a political agenda that has no real basis in science, that threatens our very sovereignty and prosperity, and that promises to trap millions in grinding poverty.

CHAPTER 2

Climate Change Déjà Vu

Everyone seems to be warning about climate change and its dire consequences.

"Here is the takeaway: Unless the world changes course quickly and dramatically, the fundamental systems that support human civilization are at risk," declared Brian Williams, summarizing a 2014 UN report on NBC News.[1]

NBC newsman Tom Brokaw also warned about the perils of global warming in 2006: "In the coming centuries, New York could be abandoned, its famous landmarks lost to the sea." NASA scientist James Hansen told Brokaw, "Boston, Philadelphia, Washington, Miami.... They would all be underwater."[2]

John Holdren, the Obama administration's science czar, painted a frightening picture: "As global temperatures rise, they may cause the massive West Antarctic Ice Sheet to slip more rapidly. Then we'll be facing a sea-level rise not of one to three feet in a century, but of 10 or 20 feet in a much shorter time. The Supreme Court would be flooded. You could tie your boat to the Washington Monument. Storm surges would make the Capitol unusable."[3]

ABC News reporter Anne Thompson noted that "coastal communities...are at risk and in the worst case scenario could disappear altogether."[4]

Did you know?

★ An analysis found that more than 250 scientific papers about global *cooling* were published from the 1960s to the 1980s

★ Scientists who now warn of global warming previously warned of global cooling

★ The congressional testimony that launched the global warming panic used "stagecraft" to increase the dramatic impact on the public

The UN IPCC's Michael Oppenheimer warned, "If the sea level rise occurred fast enough, some major cities might have to be abandoned—like, for instance, London."[5]

The world is running a fever, and the effects will be dire. As another commentator observed,

> Snows are less frequent and less deep. They do not often lie, below the mountains, more than one, two or three days, and very rarely a week. They are remembered to have been formerly frequent, deep and of long continuance. The elderly inform me the earth used to be covered with snow about three months in every year. The rivers, which then seldom failed to freeze over in the course of the winter, scarcely ever do now. This change has produced an unfortunate fluctuation between heat and cold in the spring of the year which is very fatal to fruits.

This same observer also noted, "I remember that when I was a small boy, say 60 years ago, snows were frequent and deep in every winter."

Who said that? Al Gore? Leonardo DiCaprio? Nope. That's Thomas Jefferson, in his 1799 book *Notes on the State of Virginia*."[6]

Two Centuries of Climate Scares

But here's a more modern quote: "A considerable change of climate, inexplicable at present to us, must have taken place in the polar regions...the Greenland seas...have been hitherto covered [in ice], have in the last two years entirely disappeared."

That's the president of the UK Royal Society. Not the current president, but Joseph Dalton Hooker, the president of the Society on November 20, 1817.[7]

Let's fast forward in time: "A great climatic change was now carrying the world, slowly and irresistibly, towards world-wide drought."

That was from a professor at the Royal Geographical Society—in 1914.[8]

But here's a much more recent quote from the *New York Times*: "Some scientists estimate that the polar ice pack is 40% thinner and 12% less in area than it was a half century ago, and that even within the lifetime of our children the Arctic Ocean may open, enabling ships to sail over the North Pole.... "

That quotation is from an October 19, 1958, article titled "The Changing Face of the Arctic."[9]

People were worrying about a changing climate in the eighteenth century.

And for more than a hundred years, skeptics have been calling them out on the evidence.

A skeptical editorial in the January 10, 1871, *Brisbane Courier* could be weighing in on our current climate debate: "Every season is sure to be 'extraordinary,' almost every month one of the driest or wettest, or windiest, coldest or hottest, ever known.... THREE consecutive years of drought.... Others speculating quite as conjecturally and even more absurdly, seem to attribute the impending change of climate—of which they assume the reality—to the operation of men."[10]

The 1871 editorial writer even foreshadowed how the promoters of the climate change panic would not allow actual observational evidence to sway their claims: "Much observation, which ought to correct a tendency to exaggerate, seems in some minds to have rather a tendency to increase it."

Sixty-two years later, Australia's chief weather expert called the belief in unprecedented or unusual climate change an "error of human memory." As the *Adelaide News* reported on February 1, 1933, "'When people compare the present with the past, they remember only the abnormal.' The Commonwealth Meteorologist (Mr. Watt) smiles when he is asked about what is wrong with the weather, because all the records show that it is normal."[11]

Mr. Watt would have a field day with the claims that our extreme weather events today are now somehow unprecedented.

The Inconvenient Global Cooling Scare of the 1970s

It seems that people in every era believe their time on Earth features unprecedented weather.

But they haven't always blamed it on warming.

"There are strong signs that these recent climate disasters were not random deviations from the usual weather, but instead signals of the emergence of a new normal for world climates." That was written in 1974 by Walter Orr Roberts, the founder of the National Center for Atmospheric Research (NCAR), who was warning of—global cooling.[12]

The next year, *Newsweek* also raised the alarm about a cooler climate: "There are ominous signs that the Earth's weather patterns have begun to change dramatically and that these changes may portend a drastic decline in food production—with serious political implications for just about every nation on Earth." There was an urgent need for government action: "Climatologists are pessimistic that political leaders will take any positive action to compensate for the climatic change, or even to allay its effects.... The longer the planners delay, the more difficult will they find it to cope with climatic change once the results become grim reality."[13]

The same tipping-point rhetoric we're used to hearing from Al Gore and Co. on global warming permeated the 1970s global cooling scare. And of course no climate hype would be complete without appeals to a supposed scientific consensus. *Newsweek*'s 1975 article noted that scientists "are almost unanimous in the view that the [cooling] trend will reduce agricultural productivity for the rest of the century."

A cooler Earth was supposedly causing more extreme weather. The *Newsweek* article argued, "Last April, in the most devastating outbreak of tornadoes ever recorded, 148 twisters killed more than 300 people and caused half a billion dollars' worth of damage in 13 U.S. states.... To scientists, these seemingly disparate incidents represent the advance signs of fundamental changes in the world's weather."[14]

The *New York Times* reported, "Many weather scientists expect greater variability in the earth's weather and, consequently, greater risk of local disasters.... "[15]

In recent years, there have been attempts to downplay and outright deny the global cooling fears of the 1970s. A 2008 paper that purported to erase history and claimed to reveal "The Myth of the 1970s Global Cooling Consensus"[16] was one such attempt. Journalist James Delingpole poked fun at this historical revisionism.

Delingpole wrote, "The full extent of these activists' skullduggery has been uncovered by researcher Kenneth Richard, writing at NoTricksZone. Richard shows that during the 1960s and 1970s, there was an 86 percent scientific consensus that the planet was on a cooling path. But this was airbrushed out of history so successfully that even now if you do a Google search on '70s global cooling scare' the top results claim it never really happened."[17]

The 2016 survey by Kenneth Richard on the NoTricksZone website found 285 scientific papers on global cooling published from the 1960s to the '80s. The panic about a coming ice age was pushed by many scientists, including some who would later champion global warming.

"The cooling trend heralds the start of another ice age, of a duration that could last from 200 years to several millennia," reported the *Ukiah Daily Journal* on November 20, 1974.

★ ★ ★
Global Cooling Flashback

John Holdren, chief science advisor to President Obama, is a leading proponent of the global warming scare who claimed in 2014 that "the thing that keeps me up at night the most is the climate change issue." But back in 1971, he was warning that pollution and volcanic ash could "start a new ice age."[18]

"Sixty theories have been advanced," it added "to explain the global cooling period."[19]

"Climate experts believe the next ice age is on its way," Leonard Nimoy intoned ominously, in a 1978 episode of *In Search Of*.... The man who had played Spock warned, "If we are unprepared for the next advance [in ice] it could mean hunger and death on a scale unprecedented in all of history. What scientists are telling us now is that the threat of an ice age is not as remote as once thought...[and it] could turn most habitable portions of planet into frozen desert."

The late Stanford University professor Stephen Schneider of the Center for Environment Science and Policy was featured on the show discussing with Nimoy whether or not attempts to "loosen ice caps" with nukes and to melt sea ice by blanketing it in soot would help alleviate the expected global cooling: "Could we do things, yes. But I am not sure that would make things better."[20]

By 1977, Schneider was less certain about global cooling. "We just don't know enough to choose definitely at this stage whether we are in for warming or cooling—or when," he wrote.[21]

As Tony Heller, using the pen name Steven Goddard, has reported at Real Science, "Every major climate organization endorsed the ice age scare, including NCAR, CRU, NAS, NASA—as did the CIA."[23] The CIA issued reports warning about the dire implications of the imminent cooling. In 1974, the Office of Research and Development of the Central Intelligence Agency published "A Study of Climatological Research as it Pertains to Intelligence Problems,"[24] warning that the world's food supply was in danger:

★ ★ ★

Air Travel Was So Cool

Some scientists blamed the supposed global cooling trend on airplanes. As the *New York Times* reported in 1975, "A federally sponsored inquiry into the effects of possible climate changes caused by heavy supersonic traffic in the stratosphere has concluded that even a slight cooling could cost the world from $200 billion to 500 times that much in damage done to agriculture, public health and other effects."[22]

★ ★ ★

Global Cooling Heats Up the Headlines

The media went wild promoting the global cooling panic. Here's just a sampling of the headlines:

"There's a New Ice Age Coming!"
— *Windsor Star*, September 9, 1972[28]

"International Team of Specialists Finds No End in Sight to 30-Year Cooling Trend in Northern Hemisphere"
— *New York Times*, January 5, 1978[29]

"Air Pollution May Trigger Ice Age, Scientists Feel,"
— *Telegraph*, December 5, 1974[30]

"Climate Changes Called Ominous; Scientists Warn Predictions Must Be Made Precise to Avoid Catastrophe"
— *New York Times*, January 19, 1975[31]

"Geologist Says Winters Getting Colder"
— *Middlesboro Daily News*, January 16, 1978[32]

"Worrying about a New Ice Age"
— *New York Times*, February 23, 1969[33]

"Colder Winters Held Dawn of New Ice Age"
— *Washington Post*, January 11, 1970[34]

"Climate: Chilling Possibilities"
— *Science News*, March 1, 1975[35]

"Scientists Agree World Is Colder"
— *New York Times*, January 30, 1961[36]

"British Climate Scientist Predicts New Ice Age"
— *Christian Science Monitor*, September 23, 1972[37]

"New Ice Age Coming—It's Already Getting Colder"
— *Los Angeles Times*, October 24, 1971[38]

"The Ice Age Cometh"
— *Canberra Times*, May 31, 1975[39]

"The western world's leading climatologists have confirmed recent reports of a detrimental global climate change. The stability of most nations is based upon a dependable source of food, but this stability will not be possible under the new climatic era.... The University of Wisconsin was the first accredited academic center to forecast that a major global climatic change was underway."[25] Two years later, the CIA issued another report, this time warning that global cooling was bringing "a promise of famine and starvation to many areas of the world."[26]

In 1977, ABC News featured warnings of global cooling on the evening news program. ABC's Harry Reasoner introduced the segment, noting,

"An editorial in the Hartford Courant in 1897 actually said it, although Mark Twain usually gets the credit, 'Everybody talks about the weather, but nobody does anything about it.' In his comment tonight, Howard K. Smith talks about the weather and suggests that we better do something about it. Howard?"[27]

Smith warned viewers, "We are over-ready for a return of the ice. Experts like Reid Bryson, the head of the biggest meteorological department in the world, in Wisconsin, believe that since 1945 that has been in progress. The returning to an ice age."

"If they're right, the present wave is no temporary blip but a long-term trend that will go on and get worse," he added. "The argument that we face some long cold years is pretty convincing," Smith concluded.

There is one key difference, though, between the global cooling scare of the 1970s and our current global warming panic. While the media did hype the scary predictions of a new ice age, they weren't completely in the tank for climate change the way they are today. Take for example a 1976 *New York Times* article that featured climatologists predicting global cooling and warned of "highly erratic weather for decades to come." The *Times* reported that climatologists "believe that the earth's climate has moved into a cooling cycle.... And that, they say, has profound implications—most of them bad—for world food production, economic stability and social order. With the world's population now so high, the results of even minor year-to-year shifts in climate could be catastrophic, they say." But the paper of record also let "skeptical scientists" who mocked such catastrophic global cooling predictions have their say: "Skeptical Scientists: Some scientists think all

---

★ ★ ★

Global Cooling Flashback

In 2014, *Time* magazine explained that "a polar vortex" was an example of how "global warming could be making the occasional bout of extreme cold weather in the U.S. even more likely." In 1974, *Time* listed the same "circumpolar vortex" as one of the "indications of global cooling."[40]

of that is nonsense, mainly because climatologists can offer no scientific proof to back up their theories. If meteorologists, using sophisticated computers, can forecast weather only a day or two in advance, they ask, how can climatologists project climate years ahead?" One of the skeptics quoted by the *Times* said some of the climatologists' predictions were "right out of fantasy land."[41]

The next year another *New York Times* piece reiterated the meteorologists' criticism: "Meteorologists, or short-term weather forecasters, dispute the scientific credentials of climatologists, saying they are working in a new area without much base data and with no 'proof' to back up their assertions," the 1977 article explained.[42]

It's downright refreshing seeing a *Times* piece on climate change with this kind of balance, one that allows scientists to dispute each other's claims. Unfortunately, these examples of balanced journalism are from forty years ago.

The Big Switch

The early 1980s was a hiatus between climate change panics. It was not until the decade neared its close that global warming became the cause du jour.

Nineteen eighty-eight is the year when global warming dawned on the world's consciousness. It was in that year that the UN IPCC—the United Nations climate panel that would become the central hub for global warming science, issuing high-profile reports every five years or so—was formed.

The UN Intergovernmental Panel on Climate Change (IPCC) was formed in 1988 to examine how CO_2 and other greenhouse gases impact the climate. It had every incentive to declare a "crisis" because the UN was also going to be in charge of coming up with a "solution." If the UN failed to find CO_2 was a problem, it would also deny itself the opportunity to be in charge of regulating the world's economies and planning the energy mixes for the next hundred years and beyond. This conflict of interest has been inherent in every action the UN climate panel has taken. UN IPCC scientist Vincent Gray, an expert reviewer, has called the IPCC process a "perversion of science."[43]

Previous environmental scares were never this centrally organized and lavishly funded. That may help explain why none of them was bolstered by a one-sided narrative fully supported by academia and the media. The UN's involvement in climate change ensured that there would be only one acceptable view on the subject—that a man-made climate crisis was at hand.

David Kear, geoscientist and former UN consultant, witnessed how the IPCC corrupted science. "A huge international bureaucratic industry was born—with Cabinet Ministers, government departments, company sections, travel, conferences, treaties, carbon credits, and carbon trading, and very much more. The challenge was often heard that we must curb our carbon emissions or sacrifice our grandchildren's well-being. In truth, those children were being saddled with a gigantic debt to pay for everything encompassed by the Warmers' 'carbon footprints,' including the salaries and expenses of the loudest proponents," Kear wrote in 2014.[44]

The UN IPCC is stacked with environmental activists. Climate policy expert Donna Laframboise has documented the green activists who serve on the IPCC. Ove Hoegh-Guldberg, for example, who has served as a lead author of the IPCC reports is "a full-blown environmental activist" with a "long history of employment with both the WWF and Greenpeace," Laframboise's research revealed.[45]

"The IPCC has a very long, very sordid history of recruiting personnel linked to activist organizations," Laframboise explained.[46] "The World Wildlife Fund is an NGO. Greenpeace is an NGO. The people who work for those organizations are not scientific experts. They are advocates, activists, and partisans. They have an agenda. They are paid a salary to advance that agenda," she added.

UN IPCC lead author Richard Tol has pointed out, "Governments nominate academics to the IPCC—but we should be clear that it is often the environment agencies that do the nominating."

As Tol explained, "It is rare that a government agency with a purely scientific agenda takes the lead on IPCC matters. As a result, certain researchers are promoted at the expense of more qualified colleagues. Other competent people are excluded because their views do not match those of their government. Some authors do not have the right skills or expertise, and are nominated on the strength of their connections only."

And the IPCC's motivation is to generate media impact for their reports. "An ambitious team wants to make a bigger splash than last time. It's worse than we thought. We're all gonna die an even more horrible death than we thought six years ago. Launching a big report in one go also means that IPCC authors will compete with one another on whose chapter foresees the most terrible things," Tol said.[47]

Nineteen-eighty-eight was also the year of NASA scientist James Hansen's dramatic testimony to Congress about the urgency of global warming. This was the key moment when Hansen warned, "the greenhouse effect is here." As the *New York Times* reported at the time, "'Global warming has reached a level such that we can ascribe with a high degree of confidence a cause and effect relationship between the greenhouse effect and observed warming,' Hansen said at the hearing today, adding, 'It is already happening now.'"[48]

Hansen's testimony as NASA's lead global warming scientist was a huge media sensation.

But one of Hansen's former supervisors explained that Hansen's dramatic testimony was not well received at NASA.

Retired senior NASA atmospheric scientist Dr. John S. Theon explained in 2009, "We were somewhat appalled. We were certainly embarrassed," by Hansen's testimony.[49]

Global Cooling Flashback

In 1988, NASA's James Hansen was the authority trotted out before a U.S. Senate committee to launch the global warming panic. But sixteen years earlier, in the global cooling scare era, "a computer program developed by Dr. James Hansen" had been applied by other scientists to the question of "the carbon-dioxide [that] fuel-burning puts in the atmosphere." They found "no need to worry" about the carbon dioxide, but warned that other particles from burning fossil fuels "could screen out so much sunlight that the average temperature could drop by six degrees" in fifty years, bringing on a new ice age by 2021.[51]

"Hansen was never muzzled even though he violated NASA's official agency position on climate forecasting (i.e., we did not know enough to forecast climate change or mankind's effect on it)," Theon wrote. "Hansen thus embarrassed NASA by coming out with his claims of global warming in 1988 in his testimony before Congress," Theon added. "I probably would have been removed if I had tried to cut off Jim Hansen's funding, after all he had Al Gore...on his team."

It wasn't until nearly two decades later that we learned the extent to which that testimony was staged to manipulate the public—exactly like so much of the climate change "science" we have been bombarded with ever since.

The theatrical hearing was orchestrated in part by Senators Al Gore and Timothy Wirth, as was revealed in a 2007 PBS *Frontline* special, which reported on the "stagecraft" employed at the Hansen hearing to facilitate the transition from cooling fears to warming fears.

PBS correspondent Deborah Amos reported that "Sen. Timothy Wirth was one of the few politicians already concerned about global warming, and he was not above using a little stagecraft for Hansen's testimony":

TIMOTHY WIRTH: We called the Weather Bureau and found
out what historically was the hottest day of the summer. Well,
it was June 6th or June 9th or whatever it was. So we scheduled

the hearing that day, and bingo, it was the hottest day on record in Washington, or close to it.

DEBORAH AMOS: [on camera] Did you also alter the temperature in the hearing room that day?

TIMOTHY WIRTH: What we did is that we went in the night before and opened all the windows, I will admit, right, so that the air conditioning wasn't working inside the room. And so when the—when the hearing occurred, there was not only bliss, which is television cameras and double figures, but it was really hot.

★ ★ ★

Whither the Humble Armadillo?

The armadillo holds the distinction of being used as a mascot for both the global cooling scare in the 1970s and the global warming scare today. ABC News has cited the expansion of the animal's range—in two different directions—as evidence for both scares.

"I've noticed in my driving now that I've seen armadillos in places where I've never seen them before. Much farther north than I've ever seen them in the past," said a truck driver featured in a 2007 ABC News segment on climate change and its impacts.[53]

But go back to 1977 and the same ABC News was touting the little armadillo as proof of—yes, you guessed it—global cooling!

"The signs of cooling have already begun. They began about 1945. Homely things like the flight of the heat loving armadillos from Nebraska to Mexico," Howard K. Smith intoned ominously during a segment on how the evidence was piling up for a coming ice age.[54]

And it isn't just ABC News. The *New Castle News* reported in 1973 that "warmth-loving" armadillos were "retreating southward" to escape global cooling. It was "seen as a sign of a cooling climatic trend."[55] But a 2011 *Scientific American* article reported, "The armadillo is moving north thanks to climate change.... Armadillos have settled into southern Illinois, Indiana, Kansas and Missouri—all areas that were 'totally unexpected.'"[56]

> [from the 1988 hearing] WIRTH: Dr. Hansen, if you'd start us off, we'd appreciate it.
>
> TIMOTHY WIRTH: The wonderful Jim Hansen was wiping his brow at the table at the hearing, at the witness table, and giving this remarkable testimony.[50]

But in 2015, Senator Wirth was seemingly pressured to recant his version of events. Hansen accused Wirth of fibbing about the "stagecraft." "He just made these up later to make it seem interesting," Hansen said.[52] Wirth suddenly remembered that his stories about the behind-the-scenes machinations of the hearing were totally made up!

Wirth now says he was in error: "Some myths about the hearing also have circulated over the years, including the idea that windows were left open or the air conditioning was not working. While I've heard that version of events in the past, and repeated it myself, I've since learned it didn't happen."

It's hard to believe that global warming science can be "settled" when major players behind promoting it can't agree on basic events surrounding their biggest day.

CHAPTER 3

"Pulled from Thin Air":
The 97 Percent "Consensus"

One of the warmists' most persistent talking points is the climate change "consensus." We are told ad nauseam that 97 percent of scientists all agree.

Early global warming "consensus" claims came from Vice President Al Gore. In 1992, Gore reassured the public that "only an insignificant fraction of scientists deny the global warming crisis. The time for debate is over. The science is settled."[1]

Former secretary of state John Kerry claimed in 2014 that "Ninety-seven percent of the world's scientists" agree and "tell us this is urgent."[2] President Obama touted the figure as well.[3]

CNN claims "between 95–97% of scientists agree that climate change is happening now, that it's damaging the planet and that it's manmade."[4]

Tom Friedman of the *New York Times* tries to intimidate anyone who does not accept the consensus claims. "97% of experts say this—3% say that and conservatives are saying 'I will go with the 3%', that's not conservative, that's Trotskyite radical," Friedman has said.[5]

In almost every single climate debate I have been involved in, the consensus claims have been made repeatedly.

Did you know?

★ Science groups that are nearly 100 percent dependent on government funding endorse the "consensus"

★ In one UN meeting, two scientists were outnumbered by more than forty bureaucrats

★ Numerous scientific organizations have signed on to the "consensus" without ever polling their membership

Fooling Even the Scientists

The 97 percent "consensus" talking point has taken in even climate scientists.

American University climate statistics professor Caleb Rossiter talks about how he used to accept the alleged "consensus": "If we had this interview, ten years ago, I would have said I never thought about climate and I assumed all the scientists who are reporting and telling the president and the prime minister in England are right."[7]

Climatologist Judith Curry has explained, "I didn't have any reason to not accept the judgment of my colleagues. You know the consensus and the IPCC process and you know I bought into it. You know 'don't trust what one scientist says. Trust what these hundreds, thousands of international scientists have come up with with years of deliberation'.... So I bought into that and supported...the consensus."[8]

"When somebody asked me eight or ten years ago, 'Leighton, what's causing Global Warming?' I said, 'Well, I guess it's carbon dioxide.' That's all I'd ever heard," noted geologist Leighton Steward.[9]

But UK scientist Philip Stott has explained why consensus claims are harmful to science: "Science does not function by consensus and most certainly not by politically driven consensus. In fact, the history of consensus in science is terrible from Galileo right the way through the beginning of the 20th century when 95% of scientist, for goodness sake, believed in eugenics. Science has to by its very nature be skeptical."[10]

Groupthink prevails in climate science today, as climatologist Dr. Roy Spencer explains. "So, basically what you get is you get hundreds of scientists that just repeat what they've heard. You know, in the medical community it might have been years ago, you know, that all medical experts, all doctors agreed that stomach ulcers were caused by stress and spicy food," Spencer has said.[11]

MIT climate scientist Richard Lindzen, now retired, ripped the consensus claims as being a "propaganda" tool to help fund scientists. "It was the narrative from the beginning. In 1998, [NASA's James] Hansen made some vague remarks. *Newsweek* ran a cover that says all scientists agree. They never really tell you what they agree on. It is propaganda. So all scientists agree it's probably warmer now than it was at the end of the Little Ice Age. Almost all scientists agree that if you add CO_2 you will have some warming. Maybe very little warming. But it is propaganda to translate that into it is dangerous and we must reduce CO_2," Lindzen explained. "If you can make an ambiguous remark and you have people who will amplify it 'they said it not me' and the response of the political system is to increase your funding, what's not to like?"[12]

Climate campaigners base their consensus claims on several factors, including the UN Intergovernmental Panel on Climate Change (IPCC) and a few key surveys.

Former secretary of state John Kerry refers to the UN IPCC as "the gold standard" of science.

"The gold standard for research on this subject is provided by the Intergovernmental Panel on Climate Change, a collaboration of more than 2,000 scientists from 130 countries," Kerry has claimed.[14]

During a contentious live televised global warming debate during the 2009 Copenhagen

Economists versus Climatologists

"You take 400 economists and put them in the room and give them exactly the same data and you will get 400 different answers as to what is going to happen in the economic future. I find that refreshing because it tells me that these guys don't have an agenda. But if you take 400 climatologists and put them in the same room and give them some data about a system which they understand very imperfectly, you are going to get a lot of agreement and that disturbs me. I think that's arguing with an agenda."
—geologist Robert Giegengack of the University of Pennsylvania.[13]

UN climate summit on UK's Sky News, Professor Mark Maslin of the University College London asserted there are "5000 leading climate scientists" with the UN IPCC.

Maslin was debating me, and I countered, "Your idea that there are 5000 UN scientists—you need to apologize and retract that immediately. The biggest number you can come up with if you include [UN bureaucrats] and delegates is 2800."

I said again, "The professor needs to retract it. There is no 5000 [UN IPCC climate scientists]. And interestingly a few days ago [Professor Maslin] said 4000 [UN scientists]. Why not just say 100,000?"

Maslin counter-claimed that "every single intelligent person" listens to UN scientists and accepts that man-made global warming is a serious problem.[15]

★ ★ ★

Why Warmists Duck Debates

In 2010, rhetoric professor Jean Goodwin of North Carolina State University did a very complimentary nine-part analysis of my debate with warmist Professor Mark Maslin. Goodwin's conclusion? "Morano looks like he's having fun, while [debate opponent] Maslin appears increasingly irritated." Goodwin also noted that even warmist Randy Olson had concluded the Maslin debate was a "'K.O.'—in Morano's favor, of course. [Olson's] right, and I want to use the first series of posts to examine why."

As Goodwin explained, "Our educational system no longer includes training in the use of topoi, but skilled debaters reinvent the practice for themselves. Morano certainly has." Goodwin's bottom line on the debate? She found that I engaged in "a remarkable bit of oratory.... A similarly fine display of skill occurs towards the end of the debate, where Morano starts off by saying 'Let's go one at a time,' and then picks off the three arguments Maslin had made in a minute-long speaking term, in precise reverse order: arctic sea ice; temperature increases; and 5000 climate scientists.... Remarkably, Morano seems to have been prepared to attack.... How many scientist-spokespeople are willing to do this sort of prep?"[16]

Whole Dozens of Scientists!

The notion that "hundreds" or "thousands" of UN scientists agree does not hold up to scrutiny. Fifty-two scientists participated in the much ballyhooed 2007 IPCC Summary for Policymakers.[17] The 2013 5th Assessment Report by the UN IPCC increased the number of participating scientists by fourteen to just sixty-six scientists.[18]

The *Guardian* reported on how the sausage is made for the UN IPCC reports: "Nearly 500 people must sign off on the exact wording of the summary, including the 66 expert authors, 271 officials from 115 countries, and 57 observers."

Remember, this is allegedly a scientific process. And yet it somehow features "government officials" having a say in each line of the report's summary.[19]

Climate scientist Mike Hulme took apart the claim that the UN speaks for the world's scientists. Hulme noted, "Claims such as '2,500 of the world's leading scientists have reached a consensus that human activities are having a significant influence on the climate' are disingenuous." In fact the key scientific case for CO_2 driving global warming was reached by a very small gaggle of people. "That particular consensus judgement, as are many others in the IPCC reports, is reached by only a few dozen experts in the specific field of detection and attribution studies; other IPCC authors are experts in other fields," Hulme explained.[22]

UN climate panel lead author William Schlesinger freely admits that very few UN

★ ★ ★

Scientists Outnumbered by Bureaucrats

In April 2014 Harvard professor Robert Stavins revealed his disgust with the UN IPCC process for which he was a lead author: "It has been an intense and exceptionally time-consuming process, which recently culminated in a grueling week...some 195 country delegations discussed, revised, and ultimately approved (line-by-line) the "Summary for Policymakers" (SPM)...the resulting document should probably be called the Summary by Policymakers, rather than the Summary for Policymakers.[20] During one session, Stavins said he was one of only two IPCC authors present, surrounded by "45 or 50" government officials.[21]

scientists are climate experts. "Actually there's a huge range of disciplines represented there. I'm going to have to give you a guess. That's something on order of 20% have some dealing with climate," Schlesinger conceded in 2009.[23]

UN IPCC lead author Dr. Richard Tol revealed how business at the UN climate panel, the IPCC, is really conducted. "The fact that there are people, sort of, who are nominally there does not really mean that they support what is going on. I mean, [IPCC] working group two was essentially run by a small clique of people," Tol said after testifying to the U.S. Congress.

"Ultimately a small group forms, and it runs the thing. And unfortunately, those—those—that small group, I would think, are not the most representative or the most balanced or the most unbiased of people."[24]

UN IPCC expert reviewer John McLean agrees. "The reality is that the UN IPCC is in effect little more than a UN-sponsored lobby group, created specifically to investigate and push the 'man-made warming' line."[25]

Proponents of man-made global warming like to note how the National Academy of Sciences (NAS), the American Geophysical Union (AGU) and the American Meteorological Society (AMS) have issued statements endorsing the so-called "consensus" view that man is driving global warming. But this claim that the alleged "consensus" is true because the governing boards of politically savvy groups like the National Academy of Sciences endorse it is pure politics.

The AMS, for example, has issued consensus statements, but surveys of rank-and-file meteorologists found that up to 75 percent of them did not agree with the UN climate panel claims—though their dissent did not stop the governing board from joining in the alleged "consensus."[26]

Neither the NAS nor the AMS ever allowed member scientists to vote directly on these climate statements. Essentially, only two dozen or so members on the governing boards of these funding-dependent institutions produced the "consensus" statements. In many cases, member scientists are completely unaware that any such consensus climate statement has been released to the public by the governing board.

"Not a single one of those scientist organizations that have issued these very dramatic statements agreeing with IPCC and the UK Royal Society— actually polled their scientist members and showed that a majority of their members agree.... Many scientists who do not agree with statement attributed to all of them," noted Tom Harris, executive director of the International Climate Science Coalition. The organizations "never poll their rank and file" members before issuing the consensus statements, Harris explained.

In some cases, even the boards of the science organizations have not even cast votes on the statements, as was the case with Royal Society of Canada. "A fellow of the Royal Society of Canada learned about his group having signed from the newspaper. The president signed it because he assumed it was consistent with the consensus of world scientists," Harris's research revealed. "The [consensus] statement is just political."

As Harris explained, "The hierarchy of the Royal Society of Canada signed this declaration which is now listed as one of the groups that agree with the dangerous anthropogenic global warming cause, without even consulting their science academy, let alone their rank and file scientists," Harris added.[27]

Many science organizations have faced open rebellion on the part of their skeptical member scientists for these politically inspired actions. Groups such as the AMS, the American Chemical Society, and the American Physical Society have seen blowback from rank-and-file scientists.[28]

The National Academy of Sciences came under fire for its lobbying when a $5.8 million NAS study was used to lobby for a climate change cap-and-trade bill in 2010.[29] The *Washington Times* reported that the federally funded NAS "report urges that a cap-and-trade taxing system be implemented to reduce so-called greenhouse gas (GHG) emissions." The science

★ ★ ★

A Harvard Consensus

In 2017 Princeton Professor Emeritus of Physics William Happer drew parallels to today's man-made climate change claims. "I don't see a whole lot of difference between the consensus on climate change and the consensus on witches. At the witch trials in Salem the judges were educated at Harvard. This was supposedly 100 per cent science. The one or two people who said there were no witches were immediately hung. Not much has changed," Happer quipped.[34]

group also urged passage of a carbon tax that same year,[30] completing its transformation to an advocacy group.

NAS is virtually 100 percent dependent on government funding. According to the NAS website: "About 85 percent of funding comes from the federal government through contracts and grants from agencies and 15% from state governments, private foundations, industrial organizations and funds provide by the Academies member organizations."[31]

MIT climate scientist Richard Lindzen harshly rebuked then-NAS president Ralph Cicerone in his congressional testimony in November 2010. "Cicerone [of NAS] is saying that regardless of evidence the answer is predetermined. If government wants carbon control, that is the answer that the Academies will provide," Lindzen testitifed.[32]

Dubious Evidence for a Ubiquitous Number

The alleged "consensus" in climate science does not hold up to scrutiny. But what about the specific claim that 97 percent of scientists agree?

MIT's Richard Lindzen has explained the "psychological need" for the 97 percent claims. "The claim is meant to satisfy the non-expert that he or she has no need to understand the science. Mere agreement with the 97 percent will indicate that one is a supporter of science and superior to anyone denying disaster. This actually satisfies a psychological need for many people," Lindzen said in 2017.[33]

But what is the basis for this specific number, and what exactly is this overwhelming majority of scientists supposed to be agreeing on?

In 2014, UN lead author Richard Tol explained his devastating research into the 97 percent claim. One of the most cited sources for the claim was a study by Australian researcher John Cook, who analyzed the abstracts of 11,944 peer-reviewed papers on climate change published between 1991 and 2011.[35] Cook and his team evaluated what positions the papers took on mankind's influence on the climate and claimed "among abstracts expressing a position on AGW, 97.1% endorsed the consensus position that humans are causing global warming." The 97 percent number took off. This 97 percent claim was despite the fact that 66.4 percent of the studies' abstracts "expressed no position on AGW" at all.

"The 97% estimate is bandied about by basically everybody. I had a close look at what this study really did. As far as I can see, this estimate just crumbles when you touch it. None of the statements in the papers are supported by the data that is actually in the paper," Tol said. "But this 97% is essentially pulled from thin air, it is not based on any credible research whatsoever." Tol's research found that only sixty-four papers out of nearly twelve thousand actually supported the alleged "consensus." Tol published his research debunking the 97 percent claim in the journal *Energy Policy*.[36]

Meteorologist Anthony Watts summed up Tol's research debunking Cook's claims. The "97% consensus among scientists is not just impossible to reproduce (since Cook is withholding data) but a veritable statistical train wreck rife with bias, classification errors, poor data quality, and inconsistency in the ratings process," Watts wrote.[37]

Andrew Montford of the Global Warming Policy Foundation had authored a critique of Cook's claim the previous year. "The consensus as described by the survey is virtually meaningless and tells us nothing about the current state of scientific opinion beyond the trivial observation that carbon dioxide is a greenhouse gas and that human activities have warmed

★ ★ ★

Butterflies Reveal:
How Climate "Consensus" Is Crafted

"If a scientist studies butterflies, he may choose to do a modeling 'if/then' study on how warmer temps 100 years from now may impact butterflies. His results would most likely have a range of projected temperatures and thus impacts. This butterfly scientist may never even look at the probability temps may rise a certain amount, only on how rising temperatures could theoretically impact butterflies. The high end of the butterfly impacts would be heralded by his university and a press release may proclaim: 'Butterflies to face doom in 100 years due to global warming.' The media would lead with stories about how butterflies are seriously threat- ened and his University would seek out more grant funding for these types of high profile media enhanced studies. Now the butterfly scientist did nothing wrong. He did not cook his book, he did not alter data, but he merely engaged in speculation of the climate 100 years out. He took advantage of the state sponsored science of his day, man-made global warming. This butterfly scientist would then be heralded as another member of the alleged 'consensus' of scientists despite the fact that he never once would have looked at how CO_2 impacts temps."

—my explanation to *Yale Alumni Magazine* for a profile of Climategate professor Michael Mann[40]

the planet to some unspecified extent," Montford found. "The survey methodology therefore fails to address the key points that are in dispute in the global warming debate."[38]

Climatologist Roy Spencer and Heartland Institute's Joe Bast noted that even if a certain study accepts the premise of man-made global warming, that paper may not even study how CO_2 impacts temperatures: The methodology is "flawed," noted Spencer, adding, "a study published earlier this year in Nature noted that abstracts of academic papers often contain claims that aren't substantiated in the papers."[39]

In 2015, former Margaret Thatcher advisor Christopher Monckton also examined the 97 percent claim. Monckton's analysis found that "only 41 papers—0.3% of all 11,944 abstracts or 1.0% of the 4014 expressing an opinion, and not 97.1%" had actually endorsed the claim that "more than half of recent global warming was anthropogenic."[41]

As Monckton explained, "They had themselves only marked 64 out of 11,944 of the papers as representing that view of the consensus, and that is not 97.1% that's 0.5%.... There is no consensus." The 97 percent claim is "fiction. '97 percent' was a figure that was arrived at many years ago by the people who've pushed this 'agenda,'" Monckton noted. "They then realized that they needed some sort of support for it, so they did a couple of very dopey papers."[42]

In 2013, climatologist David Legates from the University of Delaware and his team of researchers had also challenged Cook's 97 percent claims. "The entire exercise was a clever sleight-of-hand trick," Legates explained. "What is the real figure? We may never know. Scientists who disagree with the supposed consensus—that climate change is man-made and dangerous— find themselves under constant attack."[43]

Another survey that claimed 97 percent of scientists agreed was based not on thousands of scientists or even hundreds of scientists... or even ninety-seven scientists, but only seventy-seven. And of those seventy-seven scientists, seventy-five formed the mythical 97 percent consensus. In other words, in this instance the 97 percent of scientist wasn't even ninety-seven scientists. This was a 2009 study published in in *Eos, Transactions American Geophysical Union* by Maggie Kendall Zimmerman, a student at the University of Illinois, and her master's thesis advisor Peter Doran.[44]

As Lawrence Solomon revealed in the *National Post*,

> The number stems from a 2009 online survey of 10,257 earth scientists, conducted by two researchers at the University of Illinois. The survey results must have deeply disappointed the researchers—in the end, they chose to highlight the views of a subgroup of just 77 scientists, 75 of whom thought humans

contributed to climate change. The ratio 75/77 produces the 97% figure that pundits now tout.

The two researchers started by altogether excluding from their survey the thousands of scientists most likely to think that the Sun, or planetary movements, might have something to do with climate on Earth—out were the solar scientists, space scientists, cosmologists, physicists, meteorologists and astronomers. That left the 10,257 scientists in disciplines like geology, oceanography, paleontology, and geochemistry that were somehow deemed more worthy of being included in the consensus.

This was "a quickie survey that would take less than two minutes to complete, and would be done online." And still less than a third of those surveyed even sent in an answer!

The questions, as Solomon noted, "were actually non-questions":

1. When compared with pre-1800s levels, do you think that mean global temperatures have generally risen, fallen, or remained relatively constant?
2. Do you think human activity is a significant contributing factor in changing mean global temperatures?

As Solomon explained, those two points do not give a complete picture of what's at issue. They don't even mention carbon dioxide—which, as we'll explore at length in the next chapter, is the heart of the climate change debate. "From my discussions with literally hundreds of skeptical scientists over the past few years, I know of none who claims that the planet hasn't warmed since the 1700s, and almost none who think that humans haven't contributed in some way to the recent warming—quite apart from carbon dioxide emissions, few would doubt that the creation of cities and the clearing of forests for agricultural lands have affected the climate," Solomon pointed out.[45]

The Fake Science Is Not Fooling the Public

But the relentless campaign to convince Americans that all scientists agree appears to have failed. A 2016 Pew Research survey found Americans don't believe the "97 percent consensus" of climate scientists claim. "Just 27% of Americans say that almost all climate scientists agree human behavior is mostly responsible for climate change," Pew found.[46]

Climatologist Dr. Curry worries that the consensus claims are anti-science.

"To me these kinds of claims of 'settled science'—it's really antithetical to the scientific process. It reflects 'confirmation bias,' 'groupthink,'" she said.[47]

Climate statistics professor Caleb Rossiter of American University noted that claims that all scientists agree are "nothing new." As Rossiter explained, "If we were here 100 years ago and I was in the psychology department, I'd be telling you that by the science of craniology—black people are stupider than white people, West Europeans are smarter and more creative than Eastern Europeans—and this was called phrenology. And all the data and statistics that they could line up and shuffle supported it, and everybody believed it."[48]

CHAPTER 4

The Tail Does Not Wag the Dog

At this point, "climate change" is supposed to be "settled science." And yet the actual scientific data simply doesn't support the claim that there *is* catastrophic global warming—let alone the notion that human activity is causing it. Whether we are talking about global temperatures, sea levels, polar ice, polar bears, or extreme weather, global warming proponents' claims are demonstrably false. And it's no wonder that the higher levels of CO_2 in the atmosphere have not actually resulted in the warming that the climate activists are always predicting. There is very little evidence to support their dire claims—and copious data showing that a changing climate is actually driven by many other factors.

The basic theory of man-made climate change is that human activity is increasing the amounts of "greenhouse gases" in Earth's atmosphere, trapping more heat and warming the Earth to potentially harmful levels.

NASA's website describes the atmosphere as "a jacket for the planet": "Earth is a great planet to live on because it has a wonderful atmosphere around it. This jacket of gases does a lot for us. It keeps us warm, it gives us oxygen to breathe, and it's where our weather happens. The atmosphere surrounds our planet like the peel of an orange."[1]

Did you know?

★ The scientist who originated the greenhouse gas theory of global warming also believed that "exposing children to electricity would make them smarter"

★ Ice ages have occurred when carbon dioxide levels were up to ten times as high as they are today

★ Rising temperatures *precede* rising CO_2 levels in ice core data

The concern is that man-made emissions of certain gases are altering that atmosphere in dangerous ways.

Those greenhouse gases include carbon dioxide (CO_2) from the burning of fossil fuels such as coal, oil, and natural gas. Natural sources of CO_2 include decomposition, ocean outgassing, and plant respiration. Besides carbon dioxide, other greenhouse gases include methane (CH_4) and nitrous oxide (N_2O) and water vapor. Methane is a byproduct of the production of and transport of coal, natural gas, and oil, as well as being emitted by livestock. Nitrous oxide is released from agricultural and industrial processes and as a result of the burning of fossil fuels and solid waste.

But of these gases, carbon dioxide is the main focus of the climate change debate. The warmists' theory is that the additional CO_2 that we have pumped into the air since the beginning of the Industrial Revolution is trapping so much more heat that it is giving the earth a fever.

As Tom Brokaw has explained, "Most scientists agree that rising temperatures are caused by an increase of greenhouse gases in the atmosphere, primarily carbon dioxide, fueled by mankind's consumption of fossil fuels."[2]

"The CO_2 we are putting into the atmosphere right now is going to add to warming for decades into the future," claims Penn State professor Michael Mann.[3]

The Origins of the Global Warming Scare

The greenhouse effect theory originated in the nineteenth century with Nobel Prize–winning

★ ★ ★
A Tiny Fraction of a Tiny Fraction

CO_2 makes up 0.04 percent of the atmosphere. Although human emissions of CO_2 from fossil fuels and other activities can build up in the atmosphere, they only make up 3.5 percent of all the CO_2 emitted each year, with the rest occurring naturally. Water vapor makes up 95 percent of all greenhouse gases in the atmosphere. Greenhouse gases make up only 2 percent of the total atmosphere. Does this small human contribution to the small amount of CO_2 in the atmosphere really drive the climate?[4]

Swedish physicist Svante Arrhenius. In 1896, Arrhenius calculated the potential warming impacts of rising CO_2 on the earth's surface temperature. Arrhenius's book *Worlds in the Making* laid out what he termed the "hothouse" theory of the earth's atmosphere.[5]

According to climate historian Spencer Weart, scientists "were aware" of Arrhenius's theory by 1903, but at first it "was regarded as speculative, and it had no policy implications since warming was not expected until centuries later, if at all, and was assumed to be benign."[6] A 1910 newspaper article noted the positive impacts Arrhenius expected from rising CO_2: "the consumption of coal at present is returning to the atmosphere the carbon dioxide of which it was robbed when the deposits of carbon were stored away in the coal beds during the carboniferous period...a doubling of the quantity in the atmosphere would more than double the rate of growth of plant."[7]

According to a 1913 article in the *Pueblo Leader*, Arrhenius had predicted that doubling and tripling the levels of carbon dioxide "will make the climate warmer, by acting like the glass roof of a green house. With the carbon dioxide increased from two and one-half to three times, the temperature of the whole world will be raised 8 to 9 degrees centigrade.... All the good soil of Canada will be in as temperate a climate as that now enjoyed by Missouri." The so-called father of global warming

★ ★ ★

Another Arrhenius (or Should That Be Erroneous?) Theory

Not all of Arrhenius's views caught on. In 1911, he proposed that "exposing children to electricity would make them smarter." As the *Rodney and Otamatea Times* reported in 1911, "At the suggestion of Prof. Svante Arrhenius, an experiment is being tried in Stockholm on fifty school children. The children are divided into two groups identical in point of health, height, weight, etc. and are placed in two classrooms of the same dimensions, and similarly situated as regards exposure of light. In each class-room, exactly the same teaching is given, but one of the classrooms is subjected to electricity, while the other is not. As yet, the experiment has not drawn to a close, but it is reported that the 'electrified children' have shown a greater mental and physical development than those in the other classroom."[9]

theory forecasted that Greenland would have "a good climate for farming."[8] His predictions have not yet been borne out by reality.

The next big development, after Arrhenius had laid the foundation for concern about the greenhouse effect, came a half a century later, with the monitoring and measuring of carbon dioxide in the atmosphere.

In 2014, Weather Channel founder John Coleman, a meteorologist, looked into the origins of CO_2 monitoring and traced it back at least in part to government funding:

> The story begins with an Oceanographer named Roger Revelle. He served with the Navy in World War II. After the war he became the Director of the Scripps Oceanographic Institute in La Jolla in San Diego, California. Revelle saw the opportunity to obtain major funding from the Navy for doing measurements and research on the ocean around the Pacific Atolls where the US military was conducting atomic bomb tests. He greatly expanded the Institute's areas of interest and among others hired Hans Suess, a noted Chemist from the University of Chicago, who was very interested in the traces of carbon in the environment from the burning of fossil fuels. Revelle tagged onto Suess' studies and co-authored a paper with him in 1957. The paper raises the possibility that the carbon dioxide might be creating a greenhouse effect and causing atmospheric warming. It seems to be a plea for funding for more studies....
>
> Next Revelle hired a Geochemist named David Keeling to devise a way to measure the atmospheric content of carbon dioxide. In 1960 Keeling published his first paper showing the increase in carbon dioxide in the atmosphere and linking the increase to the burning of fossil fuels. These two research papers became the bedrock of the science of global warming, even

though they offered no proof that carbon dioxide was in fact a greenhouse gas.[10]

Human Beings Don't Breathe Out Poison

Is carbon dioxide in fact the key driver of global temperatures?

Carbon dioxide is a trace gas in the atmosphere. We exhale it from our mouths and noses, and it is necessary to all plant life on the planet. In 2013, it officially exceeded the 400 parts per million (ppm) threshold in the atmosphere—to the dismay of climate change activists. The rising level of CO_2 has prompted multiple efforts at cap-and-trade climate bills, EPA regulations, UN climate pacts, all in an attempt to avert the allegedly terrible consequences.

Former vice president Al Gore declared the 400 ppm level "a sad milestone. A call to action."[12] *New York Times* reporter Justin Gillis compared trace amounts of CO_2 to "a tiny bit of arsenic or cobra venom" and warned that rising CO_2 means that "the fate of the earth hangs in the balance."[13] The *New Yorker* magazine declared, "Everything we use that emits carbon dioxide needs to be replaced with something that doesn't."[14] And a UK *Guardian* editorial declared, "Swift political action can avert a carbon dioxide crisis."[15]

But despite the man-made global warming fear movement's panic, prominent scientists, especially experts who study the earth's geologic history, dismissed the 400ppm level of carbon dioxide as a non-event. Scientists pointed out that there are literally hundreds of factors that govern Earth's climate and temperature—not just CO_2. Renowned climatologists have declared that a doubling or even tripling of CO_2 would not have major impacts on the earth's climate or temperature.

Einstein's Successor Touts the Virtues of Carbon Dioxide

Renowned physicist Freeman Dyson of Princeton's Institute for Advanced Study, who has been called Einstein's successor,[16] says, "I like carbon dioxide, it's very good for plants. It's good for the vegetation, the farms, essentially carbon dioxide is vital for food production, vital for wildlife.

"The effects of CO_2 on climate are really very poorly understood.... The experts all seem to think they understand it, I don't think they do...Climate is a very complicated story. And we may or may not understand it better (in the future). The main thing that is lacking at the moment is humility. The climate experts have set themselves up as being the guardians of the truth and they think they have the truth and that is a dangerous situation."[17]

Geologically speaking, the earth is in a "CO_2 famine" today, as William Happer, Cyrus Fogg Brackett professor emeritus at Princeton University, the author of two hundred peer-reviewed scientific papers, explained in a Senate testimony in 2009. "Many people don't realize that over geological time, we're really in a CO_2 famine now. Almost never [have] CO_2 levels been as low...280 (parts per million—ppm)—that's unheard of. Most of the time [CO_2 levels] have been at least 1000 (ppm) and it's been quite higher than that," Happer told the Senate committee.

"Earth was just fine in those times," Happer added. "The oceans were fine, plants grew, animals grew fine. So it's baffling to me that we're so frightened of getting nowhere close to where we started," Happer explained.

"I believe that the increase of is not a cause for alarm and will be good for mankind," he added.

"What about the frightening consequences of increasing levels of CO_2 that we keep hearing about? In a word, they are wildly exaggerated, just as the purported benefits of prohibition were wildly exaggerated," Happer said.

"At least 90% of greenhouse warming is due to water vapor and clouds. Carbon dioxide is a bit player," he added. "But the climate is warming and CO_2 is increasing. Doesn't this prove that CO_2 is causing global warming through the greenhouse effect? No, the current warming period began about 1800 at the end of the little ice age, long before there was an appreciable increase of CO_2. There have been similar and even larger warmings several times in the 10,000 years since the end of the last ice age. These earlier warmings clearly had nothing to do with the combustion of fossil fuels. The current warming also seems to be due mostly to natural causes, not to increasing levels of carbon dioxide."[18]

> ★ ★ ★
> # Corrupting the Language
> "Warming and increased CO_2 will be good for mankind...CO_2 is not a pollutant and it is not a poison and we should not corrupt the English language by depriving 'pollutant' and 'poison' of their original meaning."
> —Princeton professor William Happer to Congress[22]

Peer-reviewed studies have documented that there have been temperatures similar to our temperatures when carbon dioxide was five times higher than today's levels.[19] And a study in 2013 found that the present-day carbon dioxide level of 400 ppm was exceeded 12,750 years ago—not as a result of human activity—when CO_2 may have reached up to 425 ppm.[20]

But climate activists still declare carbon dioxide, which we and the other animals exhale from our mouths (we inhale oxygen) to be "poison." A Leonardo DiCaprio video features climate activist Thom Hartmann calling for "putting a price tag on each ton of CO_2 poison."[21]

Professor Happer and NASA moonwalker and geologist Harrison H. Schmitt pointed out in the May 8, 2013, *Wall Street Journal*, "Thanks to the single-minded demonization of this natural and essential atmospheric gas by advocates of government control of energy production, the conventional wisdom about carbon dioxide is that it is a dangerous pollutant. That's simply not the case."[23]

Don't Spit into the Wind

"You can go outside and spit and have the same effect as doubling carbon dioxide."

—renowned atmospheric scientist Reid Bryson, founding director of the Institute for Environmental Studies[27]

MIT climate scientist Richard Lindzen has mocked claims that carbon dioxide is dangerous.

"CO_2, it should be noted, is hardly poisonous. On the contrary, it is essential for life on our planet and levels as high as 5000 ppm are considered safe on our submarines and on the space station (current atmospheric levels are around 400 ppm, while, due to our breathing, indoor levels can be much higher)," he said in 2017.[24]

What Really Causes Climate Change

The claim by global warming activists that CO_2 is the global temperature control knob has been challenged in the peer-reviewed literature. That's simply not what the earth's geologic history shows.

As many scientists have pointed out, variations in global temperature correlate much better with solar activity and with complicated cycles of the oceans and atmosphere than with CO_2. "There isn't the slightest evidence that more carbon dioxide has caused more extreme weather," Happer and Schmitt wrote.[25]

One peer-reviewed study found the climate of the "ancient" Earth similar to ours—despite CO_2 levels five times higher than those today. Geologists reconstructed Earth's climate belts between 460 and 445 million years ago and found "ancient climate belts were surprisingly like those of the present."[26]

Geoffrey G. Duffy, an award-winning professor at the University of Auckland in New Zealand, who has authored hundreds of scientific studies, pointed out, "Even doubling or tripling the amount of carbon dioxide will virtually have little impact, as water vapor and water condensed on particles as clouds dominate the worldwide scene and always will."[28]

In fact, climate change is governed by hundreds of factors, not just CO_2.

University of London Professor Emeritus of Biogeography Philip Stott, whom we have already met, rebuts the notion that CO_2 is the main climate change driver. "As I have said, over and over again, the fundamental point has always been this: climate change is governed by hundreds of factors, or variables, and the very idea that we can manage climate change predictably by understanding and manipulating at the margins one politically-selected factor (CO_2), is as misguided as it gets," Stott wrote in 2008.[29]

Atmospheric scientist Robert L. Scotto, past member of the American Meteorological Society (AMS) who has authored or co-authored numerous technical publications and reports, has said, "Based on the laws of physics, the effect on temperature of man's contribution to atmospheric CO_2 levels is minuscule and indiscernible from the natural variability caused in large part by changes in solar energy output."[31]

CO_2 may have a small effect on global temperatures, but the sun has a much larger one.

Award-winning Israeli Astrophysicist Dr. Nir Shaviv critiqued the UN IPCC report in 2013 for failing to recognize the power of the sun. "The IPCC and alike are captives of a wrong conception," Shaviv wrote. "The IPCC is still doing its best to avoid the evidence that the sun has a large effect on climate. They of course will never admit this quantifiable effect because it would completely tear down the line of argumentation for a mostly man-made global warming of a very sensitive climate."[32]

And it is not simply the sun that can cause climate change. It is the sun, volcanoes, the tilt of the earth's axis, water vapor, methane, clouds, ocean

★ ★ ★

If You Believe in Magic

Atmospheric physicist Richard Lindzen, retired Alfred P. Sloane professor at the Massachusetts Institute of Technology, blasted the notion that CO_2 is the control knob for the climate. Believing that, he said, "is pretty close to believing in magic. Instead you are told that it is believing in 'science.' Such a claim should be a tip-off that something is amiss. After all, science is a mode of inquiry rather than a belief system."[30]

★ ★ ★
An Admission against Interest

In a 2008 article, even the climate activists at Real Climate let slip, "The actual temperature rise is an emergent property resulting from interactions among hundreds of factors."[36]

cycles, plate tectonics, albedo, atmospheric dust, atmospheric circulation, cosmic rays, particulates like carbon soot, forests, and land use, and more. Ecologist Patrick Moore, the co-founder of Greenpeace who left the organization in 1986 because he felt it was too radical, has argued, "Water is the most important greenhouse gas by far. And it's at a far higher concentration in the air than carbon dioxide."[33]

A paper by Danish physicist Henrik Svensmark, published by the Royal Astronomical Society in the UK, suggested that changes driven by the stars help regulate the amount of carbon dioxide in the air and that cosmic rays from exploded stars create more cloud cover which then cools the Earth.[34]

"Climate is the most complex coupled nonlinear chaotic system known to man. Of course there are human influences in it, nobody denies that. But what outcome will they get by fiddling with one variable (CO_2) at the margins? I'm sorry, it's scientific nonsense," Stott wrote.[35]

"The energy mankind generates is so small compared to [the] overall energy budget that it simply cannot affect the climate," noted Anatoly Levitin, the head of geomagnetic variations laboratory at the Institute of Terrestrial Magnetism, Ionosphere, and Radiowave Propagation of the Russian Academy of Sciences. "The planet's climate is doing its own thing, but we cannot pinpoint significant trends in changes to it because it dates back millions of years while the study of it began only recently. We are children of the Sun; we simply lack data to draw the proper conclusions."[37]

In an April 2017 presentation, atmospheric physicist Richard Lindzen said, "Doubling CO_2 involves a 2% perturbation to this budget. So do minor changes in clouds and other features, and such changes are common. In this complex multifactor system, what is the likelihood of the climate (which, itself, consists in many variables and not just globally averaged temperature anomaly) is controlled by this 2% perturbation in a single variable?"[38]

Takeda Kunihiko, vice-chancellor of the Institute of Science and Technology Research at Chubu University in Japan, has gone even further: "CO_2 emissions make absolutely no difference one way or another.... Every scientist knows this, but it doesn't pay to say so."[39]

The Positive Benefits of CO_2

Physicist Lubos Motl, formerly of Harvard University, dismissed concerns about the rise of atmospheric CO_2 to 400 ppm in a May 12, 2013, essay titled "Why We Should Work Hard to Raise the CO_2 Concentration." As Motl explained, "CO_2 is primarily plant food while its other implications for Nature are negligible in comparison. Humanitarian orgs should work hard to help mankind to increase the CO_2 concentration.... CO_2 is the key compound that plants need to grow—and, indirectly, that every organism needs to get the food at the end."[40]

The late geologist Bob Carter, a professor emeritus at James Cook University in Australia who authored over a hundred scientific papers, pointed out in 2015, "We are currently living on a carbon dioxide starved planet. And were we to double carbon dioxide in the atmosphere, which is the figure everybody fears, that would be a small step back towards restoring the amount of carbon dioxide in the atmosphere."[41] He called "the idea that doubling carbon dioxide is going to be environmentally catastrophic...just a silly idea," and added, "We're not dealing with a scientific issue, we haven't been dealing with a scientific issue now for 15 years. We're dealing with the determined political issue. It's a campaign cause."

Greenpeace co-founder Patrick Moore testified in the U.S. Senate, "Today, we live in an unusually cold period in the history of life on earth and there is no reason to believe that a warmer climate would be anything but beneficial for humans and the majority of other species.... It is extremely likely that a warmer temperature than today's would be far better than a cooler one."[42]

Analyses have shown that CO_2 loses its warming impact as its levels increase. Geologist Leighton Steward, winner of EPA Administrator's Award, noted that "carbon dioxide is a greenhouse gas—it does trap some heat, but its ability to trap more heat declines logarithmically. You can't use carbon dioxide to control the climate. The plants are growing more robustly, food crops, the trees, the forest. Earth has been getting greener, and greener and greener! We're just fertilizing the plants."[43]

In 2013, prominent award-winning Swedish climate scientist Lennart Bengtsson, declared CO_2's "heating effect is logarithmic: the higher the concentration is, the smaller the effect of a further increase."[44]

Bengtsson noted that global warming would not even be noticeable without modern instruments. "The warming we have had last a 100 years is so small that if we didn't have climatologists to measure it we wouldn't have noticed it at all."[45]

Global warming activists were stunned by this public about-face by one of the "consensus" scientists, and Bengtsson was subjected to harassment by the climate change community.

New Zealand climate scientist Chris de Freitas pointed out on May 1, 2009, that "warming and CO_2 are not well correlated.... the effect of CO_2 on global temperature is already close to its maximum."[46] He explained, "Adding more has a decreasing effect," and also stressed that the "[c]urrent warm phase...is not unprecedented.... From the results of research to date, it appears the influence of increasing CO_2 on global warming is almost indiscernible. Future warming could occur, but there is no evidence to suggest it will amount to much."

Meteorologist Tom Wysmuller, formerly of NASA, has pointed out "total disconnects" between temperature and CO_2 "going back ten centuries." As Wysmuller wrote, "From 1000 AD to 1800, over a period of relatively stable

CO_2 values that bounced around the 280 ppm level, temperatures plummeted in the Little Ice Age (LIA) and then rebounded over a century later. CO_2 values neither led nor followed the temperature declines and recoveries.... CO_2 seems to have had little impact in EITHER direction on the observed temperatures over that 10,000 year period.... If CO_2 is to be considered a major driver of temperatures, it is doing a counterintuitive dance around the numbers."[47]

A Book You're Not Supposed to Read

Heaven and Earth: Global Warming, the Missing Science by Ian Plimer (Taylor Trade Publishing, 2009).

Australian Geologist Ian Plimer wrote on August 8, 2009, "At present, the Earth's atmosphere is starved of CO_2."[48] Plimer added, "On all time scales, there is no correlation between temps and CO_2. If there is no correlation, then there can be no causation."[49]

If Anything, It's the Other Way Around

Ivy League geologist Robert Giegengack, former chair of the Department of Earth and Environmental Science at the University of Pennsylvania, has spoken about the "natural interplay" between temperature. Giegengack noted that "for most of Earth's history, the globe has been warmer than it has been for the last 200 years. It has rarely been cooler."[50]

In my interview with him for my film *Climate Hustle*, he said, "I'm impressed by the fact that the present climate, from the perspective of a geologist, is very close to the coldest it's ever been." He also said, "The concentration CO_2 in the atmosphere today is the close to the lowest it has ever been."[51]

Giegengack has authored two hundred peer-reviewed studies and spent much of his academic career in the doing field research on the history of climate on almost every continent:

[Gore] claims that temperature increases solely because more CO_2 in the atmosphere traps the sun's heat. That's just wrong.... It's a natural interplay. As temperature rises, CO_2 rises, and vice versa. Variations in planetary alignment are most likely responsible... When gravitation from other planets' orbits causes the Earth to move closer to or further from the sun, temperatures increase or decrease. When we find that CO_2 levels follow that directly, it's hard for us to say that CO_2 drives temperature. It's easier to say temperature drives CO_2.[52]

"The driving mechanism is exactly the opposite of what Al Gore claims, both in his film and in that book," Giegengack explained. "It's the temperature that, through those 650,000 years, controlled the CO_2; not the CO_2 that controlled the temperature."

I asked Giegengack about CO_2:

Is carbon dioxide the control knob?

Giegengack: I don't see anything in the long term geologic record to support that conclusion. CO_2 is one of many, many, many variables that influence the Earth's temperature. There may be variables we don't know about, that we haven't yet discovered.

Are you afraid of rising CO_2 concentrations?

Giegengack: No, no I'm not. CO_2 is not the villain that it has been portrayed.

Giegengack explained to me that "natural processes close to the earth's surface move CO_2 around in quantities that dwarf the amount that we are generating."

"The record that shows how much higher CO_2 has been in the past under circumstances when life on earth as we know it continued to thrive," he explained.

"I haven't been impressed by the kinds of climate change that I have observed in my lifetime," he added.[53]

Greenpeace co-founder and ecologist Patrick Moore has also pointed out that rising temperatures *preceded* rising CO_2 levels.

Moore told the U.S. Senate that Earth's geologic history "fundamentally contradicts" CO_2 climate fears:

> We had both higher temps and an ice age at a time when CO_2 emissions were 10 times higher than they are today. There is no scientific proof that human emissions of carbon dioxide (CO_2) are the dominant cause of the minor warming of the Earth's atmosphere over the past 100 years....
>
> The fact that we had both higher temperatures and an ice age at a time when CO_2 emissions were 10 times higher than they are today fundamentally contradicts the certainty that human-caused CO_2 emissions are the main cause of global warming....
>
> Well, in fact, in *An Inconvenient Truth*, Al Gore used the graphs of the Vostok ice course.... He said this proves that CO_2 and temperature are directly caused related. But he didn't show that actually the temperature goes up first, usually by 800 years before the CO_2 goes up.[54]

Andrei Kapitsa, a Russian geographer and Antarctic ice core researcher, has argued, "The Kyoto theorists have put the cart before the horse. It is global warming that triggers higher levels of carbon dioxide in the atmosphere, not the other way round."[55]

★ ★ ★

It's Not the Crime, It's the Cover-Up

Was Gore's omission of the fact that temperatures rise before CO_2 in the atmosphere does intentional or accidental? For a clue, we can look at a controversy involving Laurie David, Gore's co-producer on *An Inconvenient Truth*. When David co-authored a children's book titled *The Down-to-Earth Guide to Global Warming*,[56] she was accused by the Science and Public Policy Institute of including "an altered temperature and CO_2 graph that falsely reverses the relationship found in the scientific literature." David and her co-author "mislabeled the blue curve as temperature and mislabeled the red curve as CO_2 concentration," according to the Science and Public Policy Institute.[57] "The peer-reviewed literature is unanimous in finding that in climate records CO_2 changes have historically followed temperature changes," the group noted. After it was publicly exposed, David admitted the key scientific error in the book, terming it a "minor error."[58]

Climatologist Roy Spencer, a former NASA scientist who co-developed the monitoring of global temperatures with satellites and now a principal research scientist at the University of Alabama in Huntsville, explained, "As a climate scientist, I will say that I do indeed believe that increasing CO_2 should cause some amount of warming; the theory on that is reasonably sound. But the relationship between temperatures and CO_2 inferred from the Antarctic ice core record is, to me, not very convincing."

According to Spencer,

In Gore's first movie, a huge stage prop was constructed that showed on a giant graph the relationship between atmospheric CO_2 and estimated global temperatures over hundreds of thousands of years. Then Gore ascended in a man-lift to show where atmospheric CO_2 concentrations were recently rising well above what had been experienced previously. The implication was that skyrocketing CO_2 levels would lead to skyrocketing temperatures.

What wasn't revealed, however, was that the direction of causation between CO_2 and temperature might well have been opposite that implied by Gore. During the interglacial warm periods, life flourished, and there were increases in atmospheric CO_2 which lagged the temperature increases by hundreds of years. Furthermore, warmer ocean waters release more CO_2 into the atmosphere. In other words, the changing temperatures might well have caused the changing CO_2 levels in the atmosphere, rather than the other way around as is needed for global warming theory. In that case, the CO_2 might have had little causal influence on temperature.

We need to keep reminding ourselves just how small the influence of increasing CO_2 is on the global energy balance. The energy flows that naturally occur in the climate system are huge compared to the small, ~1% perturbation in those flows that we have theoretically imposed upon the system.

A Book You're Not Supposed to Read

An Inconvenient Deception: How Al Gore Distorts Climate Science and Energy Policy by Roy Spencer (Amazon Digital Services, 2017).

Feedback in the climate system plays a big role in the climate. "The climate system already has a built-in mechanism for cooling itself against the biggest warming influence, the sun. Clouds reject about 30% of the incoming sunlight and reflect it back out to space. Like the thermostat in your home, there are natural mechanisms for how warm temperatures on Earth can get, just as greenhouse gases limit how cold temperatures can get," Spencer explained. "There is no way to know by just how much of recent warming is human-caused because we don't know whether that direct warming is being reduced or enhanced by changes in clouds, precipitation etc. These so-called 'feedbacks,' which can either add to or subtract from

★ ★ ★

Bill Nye and Al Gore Accused of Faking CO_2 Experiment

In 2011, Gore and Bill Nye the Science Guy produced a video purporting to demonstrate with a simple experiment how carbon dioxide causes warming.

As Roy Spencer explained, "They used pure CO_2 put in a glass bowl, and shined a light on it. They claimed to measure a higher temperature in the bowl than if the bowl didn't have pure CO_2 in it. But as shown by Anthony Watts in a detailed analysis of the video on his popular award-winning website wattsupwiththat.com, the experiment

didn't actually work, and they had to fake the results in post-production to complete the video. For example, the thermometers before and after adding CO_2 weren't even in the bowls."[60]

Watts said, "The experiment as presented by Al Gore and Bill Nye 'the science guy' is a failure, and not representative of the greenhouse effect related to CO_2 in our atmosphere. The video as presented, is not only faked in post production, the premise is also false and could never work with the equipment they demonstrated."[61]

the direct warming, are very uncertain. So, statements by scientists that 'most' or 'all' of the warming is due to humans are statements of faith, not science. No one really knows, and like Al Gore, they rely on you not knowing the details so you cannot easily disagree with them."[59]

Geologist Robert Giegengack was bitterly disappointed by Gore's 2006 film, *An Inconvenient Truth*: "I voted for Gore in 2000. I think that if he ran again, depending on who he ran against, I might vote for him. He's a smart man," he said.

But after viewing Gore's film, Giegengack had this reaction: "I was appalled. I was appalled because he showed this wonderful correlation between the CO_2 and the temperature. And he either deliberately misrepresented the point he was making or didn't understand it. So it was irresponsible of Al Gore not to say 'look at this, the CO_2 and the temperature follow each other for 450,000 years before humans burned their first ton of coal.' He had to say that because that was the important lesson of that curve, not what happened out the other end, which is what he focused on it."

Giegengack now declares that his undergraduate students at University of Pennsylvania are more informed about climate than the former vice president.

"Every kid in that room knew more about the climate than Al Gore and I say that with some confidence," he said. "Gore wasn't poor before he made the film, but that is not why he made it, I think he was a true-believer, he was a zealot. And he disappointed me because he did not give his audience credit for enough intelligence."

Giegengack noted that the earth is not waiting for humans to come to the rescue. "The Earth is fine. It has been around for four and half billion years. It was here before we were here. We can't save the Earth." [62]

CHAPTER 5

The Ice Caps Are Melting!

Once again, it's the hottest year on record! The polar ice caps are melting! Seas will rise! New York City is going to flood! The polar bears will die!

The original "global warming" scare scenario, introduced to the public at large by the notorious unairconditioned Congressional hearings in 1988 and hyped by Al Gore in *An Inconvenient Truth*, was that rising carbon dioxide levels were already causing hotter weather, which would soon have all kinds of dire consequences. Melting polar ice would flood coastal regions, and higher temperatures would damage agriculture and wildlife habitats. As we shall see, the failure of these dire consequences to materialize has forced the warmists to modify their claims—or at least to change which horrors they emphasize, and to alter the language they use to describe them. For reasons that we will explore in more detail in chapter eleven, they have even resorted to changing the name of their panic from "global warming" to "climate change." Nevertheless, we continue to hear hysterical claims about the warmest year ever and the melting of the ice caps. So let's take the activists' warming claims one by one and compare them to the scientific data about what is actually happening.

Did you know?

★ Polar bears are doing so well that climate campaigners have dropped them as their icon

★ Sea level rise rates have not accelerated for over a century

★ Antarctica is gaining ice

The Antarctic

Professor Sir David King, the UK government's chief scientist, claimed in 2004 that Antarctica could soon be the only habitable continent. His prediction—based the idea that global warming would melt the South Pole's ice and render Antarctica hospitable territory for a human race forced out of the now-temperate zones by rising temperatures—still hasn't come true.[1]

Antarctica is failing to follow the predictions of man-made global warming activists. In recent years, Antarctic sea ice has been at or near all-time record high extent in recent years. "Antarctic sea ice yearly wintertime maximum extent hit record highs from 2012 to 2014 before returning to average levels in 2015," NASA reported. In 2016, Antarctic ice dropped to record low levels in the past forty years of satellite monitoring due "in part due to a unique one-two punch from atmospheric conditions."[2]

Antarctic sea ice grew 1.43 percent per year from 2000 until 2008, according to a study published in the *Journal of Climate*, and then in 2014 it broke all-time records for sea ice expansion since satellite monitoring began over three decades before growing to a record high extent.[3]

In that same year climatologist Judith Curry noted that "we're seeing records set for Antarctic sea ice extent but the climate models predict that the Antarctic should be losing sea ice and that's exactly the opposite of what's happening."

A 2013 paper in the peer-reviewed *American Meteorological Journal* found climate models failing to predict the sea ice expansion that was actually happening. "The negative SIE [sea ice extent] trends in most of model runs over 1979–2005 are a continuation of an earlier decline, suggesting that the processes responsible for the observed increase over last 30 years are not being simulated correctly," the study noted.[4]

Land-based Antarctic glacier ice has been growing, on net. A 2015 NASA study found that Antarctica "is not currently contributing to sea

★ ★ ★
Scientists Trapped in Disappearing Ice

It seems that every year a global warming expedition gets trapped in ice at one of the poles. Here is an excerpt of my 2014 Climate Depot report about the "Clitantic" ship:

"The leader of the Antarctic global warming expedition that has been mired in thick Antarctic ice for more than a week and is finally facing a helicopter rescue attempt is claiming that expanding sea ice is consistent with man-made global warming theory. Chris Turney, a professor of climate change at Australia's University of New South Wales, 'remained adamant that sea ice is melting, even as the boat remained trapped in frozen seas,' according to a Fox News interview."[5]

level rise" and noted that the "mass gains of the Antarctic ice sheet are greater than losses."

The NASA team "calculated that the mass gain from the thickening of East Antarctica remained steady from 1992 to 2008 at 200 billion tons per year, while the ice losses from the coastal regions of West Antarctica and the Antarctic Peninsula increased by 65 billion tons per year." Glaciologist Jay Zwally of the NASA Goddard Space Flight Center explained, "The good news is that Antarctica is not currently contributing to sea level rise, but is taking 0.23 millimeters per year away."[6]

"It's hard to see a global warming signal from the mainland of Antarctica right now," David Bromwich, a researcher with the Byrd Polar Research Center at Ohio State University, noted in 2007. "Part of the reason is that there is a lot of variability there."[7]

A 2013 study published in the journal *Nature* found the majority of East Antarctic glaciers have advanced in size since 1990.[8]

Veteran polar scientist Heinrich Miller noted in 2012, "If anything over last 30 years we have a slight cooling trend" in Antarctica.[9]

Despite the lack of alarming news about ice in Antarctica—and despite research showing that Antarctica was as warm or warmer during the

★ ★ ★

Everything Old Is New Again

The same panicked claims about Antarctic ice melt and D.C. landmarks were made in a 1990 *Today Show* segment featuring Paul Ehrlich, who warned that because of potential melting Antarctic ice, "You Could Tie Your Boat to the Washington Monument."

In fact, fears about melting Antarctic ice were also hyped in 1922 and 1901. One media report in 1922 warned that an Antarctic ice sheet collapse could lead to a "Biblical deluge." Another 1922 article noted, "Mountain after mountain of [Antarctic] ice will fall into the sea, be swept northwards by the currents, and melt, thus bringing about, but at a much more rapid rate, the threatened inundation of the land by the rising of the sea to its ancient level."

A 1901 article raised the specter of "London on the Border of Destruction." The British capital was "to be wiped out by a huge wave," the story explained. "Geologists believe that this great ice sucker has reached the stage of perfection when it [Antarctica] will break up again, letting loose all the waters of its suction over the two hemispheres, and completely flooding the low-lying lands of Europe, Asia, and North America."[13]

Medieval Warm Period than today[10]—news outlets and climate activists keep warning of the dire consequences of future Antarctic glacier ice melt. Melting Antarctic ice "will raise global sea levels by 10 feet or more in the centuries to come," WUSA TV in Washington, D.C., warned ominously in 2014.[11] "That would submerge tunnels and subways in Manhattan and would put much of south Florida underwater." The report warned that many parts of D.C. will one day be under water because of projected Antarctic melting.

Also in 2014, an article by Seth Borenstein and Luis Andres Henao of the Associated Press titled "Glacial Melting in Antarctica Makes Continent the 'Ground Zero of Global Climate Change'" made similar claims. The AP reported, "The world's fate hangs on the question of how fast the ice melts."[12]

The 2014 AP article claimed that "scientists in two different studies use the words 'irreversible' and 'unstoppable' to talk about the melting in West Antarctica."[14] But the focus on West Antarctica angers scientists, who see it as cherry picking. Al Gore, for example, has warned, "The West Antarctic Peninsula is warming about four times faster than the global average."[15] Climate scientist Ben Herman, past director of the Institute of Atmospheric Physics and former head of the Department of Atmospheric Sciences at the University of Arizona, explained, "It is interesting that all of the AGW [anthropogenic global warming] stories concerning Antarctica are always about what's happening around the [western] peninsula, which seems to be the only place on Antarctica that has shown warming. How about the net 'no change' or 'cooling' over the rest of the continent, which is probably about 95% of the land mass."[16]

A 2014 study published in the journal *Earth and Planetary Science Letters* suggested that the West Antarctic glacier may be melting from geothermal heat from volcanoes below.[17] A 2017 study published in the Geological Society Special Publications series conducted "a survey of the region of the West Antarctic Rift System [and] revealed 91 new volcanoes hidden within the ice."[18]

And scientists have known as far back as 1977 that the melting of West Antarctica has "nothing to do with climate." Richard Cameron, Glaciology Program Manager at the National Science Foundation, explained at that time, "We're seeing the west [Antarctic] ice sheet on its way out. It seems to be doing something completely different than the east ice sheet. It has nothing to do with climate, just the dynamics of unstable ice."

And scientists have long known that Antarctica was subject to massive melting in the past. In 1932, a report in the *Queenslander* newspaper noted that a "great world change is taking place on the Antarctic Continent. Its glaciers are shrinking." The article quoted a Commander L. A. Bernacchi, who noted that ice had receded "at least 30 miles since it was first seen and surveyed" and added that the "process may have been going on for centuries."[19]

More recently, even mainstream reporters have called out the hype about Antarctica. *New York Times* climate reporter Andrew Revkin pushed back on claims of West Antarctica ice melt fears, calling the media reporting an "awful misuse of 'collapse' in headlines on centuries-long ice loss in W. Antarctica."[20]

And a 2017 *Los Angeles Times* report hyping how "a chunk of ice the size of Delaware broke off from the [West] Antarctic Peninsula" did quote glaciologist Martin O'Leary of the British Antarctic research organization Project MIDAS explaining that the calving of the ice was natural: "We're not aware of any link to human-induced climate change."[21]

Geologist Don Easterbrook has ripped media fears about sea level and potential Antarctica melt. "It's absolutely absurd. And it is not based on good science at all. The Western Antarctic ice sheet has not collapsed in the past that we know of. There's no history of this having happened ever before. So the ocean is warming. The question is can it get under the entire Western Antarctic ice sheet and float it and cause it to collapse. And the answer is no." In addition, Easterbrook says that the bulk of the Antarctic continent is not showing any global warming signal. "The main part, 90% of the Antarctic ice sheet is in the East Antarctic ice sheet and it is not melting, it's growing."[22]

A 2017 study published in the journal *Science of the Total Environment* found that the Antarctic Peninsula has been cooling and that the previous warm period during the end of the twentieth century was "an extreme case." The earlier warm period has now shifted to a cooling period on the peninsula between 1999 through 2014. The Antarctic peninsula had been touted by climate activists as one of the fastest-warming places on the planet.[23]

Another 2017 study, published in *Earth and Planetary Science Letters*, found that the Antarctic ice sheet has been stable for millions of years, during warmer temperatures than the continent is currently experiencing.[24]

Greenland and the Arctic

It's pretty much the same story of hype in the Arctic.

Fears that ice at the North Pole and Greenland will melt have been around for decades. A November 2, 1922, *Washington Post* article was headlined, "Arctic Ocean Getting Warm: Seals Vanish and Icebergs Melt."[25]

Global warming activists have long hyped Arctic satellite data, which began in 1978, to claim record low Arctic sea ice—while ignoring the satellite data that has shown record or near record sea ice expansion in the Antarctic in previous years.[26] Moreover, the satellite monitoring of Arctic ice began at the end of a forty-year cold cycle (remember the 1970s fears of a coming ice age?),[27] when ice was most likely at its highest extent in the modern era.

In 2012, after weeks of media hype blaming global warming for a satellite-era record low summer-sea ice extent in the Arctic, NASA finally admitted that an Arctic cyclone, which "broke up" and "wreaked havoc" on sea ice in August of that year, had "played a key role." According to NASA, "The cyclone remained stalled over the arctic for several days...pushing [sea ice] south to warmer waters, where it melted."[28] Recent Arctic ice changes are not proof of man-made global warming, nor are they unprecedented, unusual, or cause for alarm, according to experts and multiple peer-reviewed studies.

The studies have shown that the previous low Arctic sea ice hyped by the media in 2007 was due to high pressure days, unusual winds, and ocean currents. A NASA study published in the peer-reviewed journal *Geophysical Research Letters* in that year found that "unusual winds" in the Arctic blew "older thicker" ice to warmer southern waters. Another peer-reviewed study the same year, from NASA, found that cyclical changes in ocean currents were impacting Arctic sea ice. Another NASA study, in 2012, blamed an "unusual" high pressure system that led to more sunny days and a subsequent reduction in Arctic sea ice.[29]

Arctic Model Fudging "Gives Wrong Answers"

"The guys who are running the long-term climate models have a tough problem.... In order to get the results before you die, you have to fudge some things. And what they fudge is the small-scale stuff. But it turns out that probably the small-scale stuff is important and fudging it gives you wrong answers."

—**Oceanographer Jane Eert, science coordinator of the Three Oceans Project, a federal study of Canada's Arctic, Atlantic, and Pacific oceans**[30]

Arctic sea ice was 22 percent greater in 2016 than in 2012, a record low year in the satellite-monitoring era.[31] The 2016 Arctic sea ice minimum was part of a ten-year "hiatus" in the melting of sea ice with "no significant change in the past decade," according to climate analyst David Whitehouse of the UK Global Warming Policy Forum. "There is no general decrease in minimal ice area, by this measure, between 2007–2016—ten years! The case can be made that the behavior of the Arctic ice cover has changed from the declining years of 1998–2007."[32]

According to NASA, the 2017 "Arctic sea ice minimum extent is the eighth lowest in the consistent long-term satellite record, which began in 1978."[33] A 2017 analysis by Kenneth Richard found that "since mid-2005, the Arctic temperature trend has stabilized, with no significantly detectable warming.... When viewed from a longer-term context, the current Arctic temperature trends are not unusual. In fact, the warmth of the last 12 years was matched during the 1920s to 1940s, with about 50 years of Arctic cooling in between."[34]

In September 2017, the Arctic sea ice extent monthly average was "193,000 sq miles higher than 10 years ago," and it has been very stable since 2007. "Earlier observations showed that Arctic ice extents were low in the 1940s, grew thereafter up to a peak in 1977, before declining...and the decline ceased in 2007, 30 years after the previous peak. Now we have a plateau in ice extents, which could be the precursor of a growing phase of the quasi-60

★ ★ ★

Ice-Free Poles?

One of the most inexplicable claims about Arctic ice came from Obama science czar John Holdren, who in 2009 claimed that winter sea ice could soon disappear: "if you lose the summer sea ice, there are phenomena that could lead you not so very long thereafter to lose the winter sea ice as well. And if you lose that sea ice year round, it's going to mean drastic climatic change all over the hemisphere."[36]

Holdren's claim of a year-round ice-free Arctic did not sit well with scientists. "Oh my, Unless the continents really diverge away so that the Arctic is no longer enclosed you will have winter sea ice. So that's not going to happen," noted climatologist Judith Curry.

I confronted Holdren on the subject of his Arctic sea ice claims in 2014 after he testified at a House science hearing. "Mr. Holdren, you said there'd be ice-free Arctic in the winter. Did you still stand by that prediction? Did you want to retract that . . . any comment on that?"

Holdren replied: "I am late for a meeting."[37]

In 2017 a study in the journal *Nature* predicting that winter ice in the Arctic Sea would "increase towards 2020" cast further doubt on Holdren's claim. The study, appearing in the journal *Nature*, predicted winter Arctic sea ice will "increase towards 2020."[38]

year Arctic ice oscillation," according to an October 2017 analysis by Ron Clutz of *Science Matters*.[35]

Greenland has long been a focus of climate fears. Tony Heller of Real Climate Science explains, "There were reports in 1940 from leading scientists that the glaciers of NE Greenland melting so fast it was nearing a catastrophe. They were worried about sea ports being drowned. Exactly the same rhetoric as now."[39] The May 6, 1940, *Brisbane Courier-Mail* reported, "It would not be exaggerating to say that these [Greenland] glaciers were nearing a catastrophe." The paper cited "famous Swedish authority, Professor A. W. Ahlmann in a lecture to the Swedish Geographical Society" and warned, "The melting had increased rapidly. By far the largest number of local glaciers in north-east Greenland had receded very greatly during recent decades, and it would not be exaggerating to say that these glaciers were nearing a catastrophe."[40]

But wait, in 1904, Greenland glaciers were already retreating "very considerably," according to an article in *Scientific American*.[41] "Dr. M. C. Engell of Copenhagen, who visited the Jakobshavn glacier last summer, had made a collection of facts which seems to show conclusively that the glaciers of Greenland are also receding," reported the August 13, 1904, *Scientific American*. "In the past fifty-three years the face of the glacier has retreated about eight miles. Not only is it shorter than it formerly was, but its mass has otherwise been reduced to a very considerable extent...It is found also that the other glaciers in that neighborhood are in the process of retreat, and the evidence collected by Dr. Engell shows that this process has been going on for a long series of years."

And perhaps you remember Joseph Dalton Hooker from chapter two, the president of the UK Royal Society who warned of a melting Greenland—in 1817.

Geologist Don Easterbrook dismisses modern claims of a Greenland melt: "It's clear that it was warmer in Greenland in the 1930s than it is right now. And so this is nothing unusual and so what you can say is along with the various other glaciers of the world that they advance and retreat, advance and retreat as the climate warms and cools and warms."[42]

A 2006 peer-reviewed study published in the *Journal of Geophysical Research* concluded, "The warmest year in the extended Greenland temperature record is 1941, while the 1930s and 1940s are the warmest decades." The paper, authored by B. Vinther, K. Andersen, P. Jones, K. Briffa, and J. Cappelen and titled "Extending Greenland Temperature Records into the Late 18th Century," examined temperature data from Greenland going back to 1784.[43]

A study by Danish researchers from Aarhus University in the same year found that "Greenland's glaciers have been shrinking for the past century, suggesting that the ice melt is not a recent phenomenon caused by global warming." Glaciologist Jacob Clement Yde was quoted in an August 21, 2006, *Agence France-Presse* report explaining that the study was "the most comprehensive ever conducted on the movements of Greenland's glaciers." As Yde explained, "Seventy percent of the glaciers have been shrinking regularly since the end of the 1880's."[44]

Another study in 2006, by a team of scientists led by Petr Chylek of Space and Remote Sensing Sciences of the Los Alamos National Laboratory, found that "although there has been a considerable temperature increase during the last decade (1995 to 2005) a similar increase and at a faster rate occurred during the early part of the 20th century (1920 to 1930) when carbon dioxide or other greenhouse gases could not be a cause."[45]

A 2016 study published in journal *Science* found that ice extent is not being lost from Greenland's interior.[46]

That same year, climatologist Pat Michaels declared the "death of the Greenland disaster story": "History shows the current goings-on in Greenland to be irrelevant, because humans just can't make it warm enough up there to melt all that much ice. For example, in 2013, Dorthe Dahl-Jensen and her colleagues published a paper in *Nature* detailing the history of the ice in Northwest Greenland during the beginning of the last interglacial, which included a 6,000 year period in which her ice core data showed averaged a whopping 6 C warmer in summer than the 20th century average," Michaels and his colleague Chip Knappenberger wrote. "Greenland only lost around 30% of its ice with a heat load of (6 X 6000) 36,000 degree-summers. The best humans could ever hope to do with greenhouse gases is—very liberally—about 5 degrees for 500 summers, or (5 X 500) 2,500 degree-summers. In other words, the best we can do is 500/6000 times 30%,

★ ★ ★

Eyewitness Account

In 2007, I traveled to Greenland on a fact-finding trip with U.S. senators. Here is an excerpt of my Senate report: "The July 27–29 2007 U.S. Senate trip to Greenland to investigate fears of a glacier meltdown revealed an Arctic land where current climatic conditions are neither alarming nor linked to a rise in man-made carbon dioxide emissions, according to many of the latest peer-reviewed scientific findings.... As a representative of Environment & Public Works Committee Ranking Member, Senator James Inhofe (R-Okla.), I made the trek to the Arctic Circle with the Senate delegation to the land the Vikings once farmed during the Medieval Warm Period. Senators and their staff viewed majestic giant glaciers and icebergs in the Kangia Ice Fjord and in Disko Bay via helicopter, boat and on foot, during the three day 24 hours of daylight trip which began in the Arctic city of Kangerlussuaq, Greenland."[48]

or a 2.5% of the ice, resulting in a grand total of seven inches of sea level rise over 500 years. That's pretty much the death of the Greenland disaster story, despite every lame press release and hyped 'news' article on it."[47]

Sea Level

"It's our choice how fast the seas rise," declared Brenda Ekwurzel of the environmental pressure group the Union of Concerned Scientists in 2014.[49]

And UN IPCC lead author Michael Oppenheimer claimed in 2014, "There's no question that sea level rise, on the whole over the last few decades, has accelerated compared to what it was in the past."[50] Geologist Bob Carter offered this rebuttal: "That statement is wrong. The sea level is not accelerating. It is if anything diminishing. There's no evidence that the modern rates of sea level change."[51]

The data shows that sea level rise rates have been essentially steady for over a century, with no recent acceleration. In fact, for perspective, sea levels have been rising since the last ice age ended more than ten thousand years ago.[52]

Former NASA Climatologist Roy Spencer explained in 2016, "Sea level rise, which was occurring long before humans could be blamed, has not accelerated and still amounts to only 1 inch every 10 years. If a major hurricane is approaching with a predicted storm surge of 10–14 feet, are you really going to worry about a sea level rise of 1 inch per decade?"[53]

A 2013 study in the journal *Global and Planetary Change* found that global sea level rise decelerated 44 percent since 2004 to a rate of only 7 inches per century.[55]

★ ★ ★

El Niño Ate My Homework

A study in the peer-reviewed journal *Nature Climate Change* in 2014 found that global sea level rise had actually decelerated 31 percent since 2002. The authors did speculate that the decelerated sea level rise might be due to an increase in rain over land due to the El Niño Southern Oscillation. Or, put simply, the land ate the ever-threatened acceleration in sea level rise.[54]

"If you look at the total global sea level from about 1850 until the present time it's been rising at a fairly constant rate, rather slow—about 7 inches a century.... It's about 1 to 2 mm a year so if you're 50 years old you experienced a sea level rise about 3 ½ inches and you probably didn't even notice it," geologist Easterbrook has explained.[56]

"Sea has risen four hundred twenty feet since the end of last of glaciation period. And none of that had anything to do with people," Greenpeace co-founder Patrick Moore points out.

"Today it's only rising by few millimeters a year. I believe almost entirely due to natural causes which mean going on throughout the history of the earth."[57]

Sea level expert Nils-Axel Mörner, a geologist who headed the Department of Paleogeophysics & Geodynamics at Stockholm University, ridiculed the claim in Al Gore's *An Inconvenient Truth* that Florida would be half covered by rising seas. "These are models. They are doing it wrongly, and this is lobbying. Geologic facts are on one side, lobbying and models are on

★ ★ ★

When Reality Fails to Alarm, Make Scary Predictions

In 2015 NASA's former lead climate scientists James Hansen released a study projecting sea level rise of up to ten feet in the next fifty years and warned that unchecked climate change is shaping up to be "highly dangerous" for the world.[59] "Science by press release: Journalists received 'summary' of Hansen's paper via PR firm," noted Meteorologist Ryan Maue.[60] Even Hansen's fellow climate change activist scientists found Hansen's skewing of the science over the top. UN IPCC lead author Kevin Trenberth said Hansen's study was "rife with speculation . . . many conjectures and huge extrapolation based on quite flimsy evidence." Climategate's Michael Mann admitted Hansen's estimates were "prone to a very large 'extrapolation error.'"[61]

the other side." Morner added, "The rapid rise in sea levels predicted by computer models simply cannot happen."[58]

The climate change debate is complicated by the existence of dueling data sets. Global temperature, for example, can be measured by satellites, or by weather balloons, or by surface thermometers. And those different measures may not agree. In the case of sea level, you can pick tide gauges or satellite altimeter measurements. According to NASA, regular record keeping of tide gauge data did not start until the late eighteenth century in the Northern Hemisphere and the late nineteenth century in the Southern Hemisphere.[62]

The tide gauges show sea level rising at a rate of less than the thickness of one nickel per year. If you want to show more pronounced sea level rise, you can use the adjusted satellite altimeter data, which began in 1992. The satellite data show sea level rising at twice the rate of tide gauges or slightly more than two pennies a year.

German meteorologist Klaus-Eckart Puls analyzed the sea level data in 2014: "Numerous evaluations of coastal-level measurements over 200 years, and more recently by gravity measurements of the GRACE satellites, demonstrate again and again a sea rise of about 1.6 mm/yr [note: A U.S. penny is 1.52 mm thick]. In contrast, the published—since 1992 altimeter measurements with the satellite systems TOPEX / POSEIDON / JASON—have twice as high values of 3.2 mm/yr. The significant discrepancy is still unclear."[63]

★ ★ ★

Comforting Hurricane Sandy Survivors

UN IPCC lead author Michael Oppenheimer shamelessly exploited sea level rise fears to a roomful of Union Beach, New Jersey, residents devastated by Hurricane Sandy on the Showtime TV series *Years of Living Dangerously*. "We are unlucky. In this area of the world, the sea is projected to rise faster than the global average. If it's four feet higher—which is what we think will happen in this area—if sea level rise is not slowed down by rather stark emissions cuts—then the 100 year storm happens every five years," Oppenheimer declared, to audible gasps from the audience full of people who had recently lost their homes to the storm. One woman in the audience called his remarks "very frightening." Another was brought to tears, announcing to the room, "I'm not rebuilding. After listening to you, I am kinda glad I am not rebuilding. It scares me a little bit more." Oppenheimer closed his speech by urging all those impacted by Sandy to pressure Republican Governor Chris Christie to take action on climate change. "I hope you take this not as a partisan remark," he added.[64]

I confronted Oppenheimer about his remarks to the New Jersey residents.

Morano: "Do you have any ethical responsibility to Sandy victims when you did that and they gasped in the room? It seems to me irresponsible because it is just a prediction or a scenario of the future, and many scientists would disagree with you."

Oppenheimer: "My predictions on sea level rise were meant for the really long term…. What I said to the Sandy victims was completely consistent with the consensus in the scientific community; it's information that they wanted to have…. The projections are not encouraging for that neighborhood of the world."[65]

A 2017 study by Nils-Axel Mörner analyzed the discrepancy and found that while the tide gauges show no acceleration in sea level rise, the satellite measurements had been "manipulated" to enhance sea level rise acceleration.

According to Mörner, the satellite altimetry measurements are "very questionable as they seem to overestimate observed sea level changes by 100–400%. It seems quite weird to claim that it would be the satellite altimetry that is right and that the true observations in the field are wrong (still this is

★ ★ ★

Shrinking or Growing?

The media and climate activists like to claim that Pacific islands are being inundated by global warming–induced sea level rise. But statistician Bjorn Lomborg cited the latest science on "those non-disappearing Pacific Islands." "Once a year or so, journalists from major news outlets travel to the Marshall Islands, a remote chain of volcanic islands and coral atolls in the Pacific Ocean, to report in panicked tones that the island nation is vanishing because of climate change…. Yet new research shows that this is not the entire—or even an accurate—picture," Lomborg wrote in 2016.

"Using historic aerial photographs and high-resolution satellite imagery, Auckland University scientists Murray Ford and Paul Kench recently analyzed shoreline changes on six atolls and two mid-ocean reef islands in the Marshall Islands. Their peer-reviewed study, published in the September 2015 issue of *Anthropocene*, revealed that since the middle of the 20th century the total land area of the islands has actually grown…. Telling viewers in the U.S. starkly that they're 'making this island disappear,' as a report from CNN's John Sutter did in June 2015, makes for good, blame-laden television."[66]

what the people around the IPCC and the Paris agreement at COP21 [the 2015 United Nations Climate Change Conference] continue to claim)…. The satellite altimetry values provided by NOAA and University of Colorado do not agree with tide gage data…. It is the satellite altimetry data which have been 'corrected' to give a rise in the order of 3.0 mm/yr. This 'correction' may, of course, be classified as a 'manipulation' of facts, like the manipulation temperature measurements recently revealed."[67]

Mörner concluded, "Up to the present, there has been no convincing recording of any acceleration in sea level, rather the opposite: a total lack of any sign of an accelerating trend."

Klaus-Eckart Puls agrees, explaining, "The sea-rise is linear for at least 100 years, there is no acceleration of the increase. A signal due to anthropogenic CO_2 (AGW) is nowhere visible."[68]

Geologist Robert Giegengack explains, "At the present rate of sea-level rise it's going to take 3,500 years to get up there [to Gore's predicted rise of 20 feet]. So if for some reason this warming process that melts ice is cutting loose and accelerating, sea level doesn't know it. And sea level, we think, is the best indicator of global warming."[69]

Climatologist Judith Curry, formerly of Georgia Institute of Technology, said, "Sea level will continue to rise, no matter what we do about CO_2 emissions."[70]

Meteorologist Tom Wysmuller has noted: "For the past 130 years there has been ZERO acceleration in sea-level rise as directly measured by tide gauges in tectonically inert areas (land neither moving up nor down), even as CO_2 has risen almost 40% in the same period."[72]

Other studies of the Pacific Islands have found similar results. A 2010 study found the "Pacific islands growing, not sinking."

A report in Australia's ABC News noted "Climate scientists have expressed surprise at findings that many low-lying Pacific islands are growing, not sinking."[73]

"Islands in Tuvalu, Kiribati and the Federated States of Micronesia are among those which have grown, largely due to coral debris, land reclamation and sediment. The findings, published in the magazine New Scientist, were gathered by comparing changes to 27 Pacific islands over the last 20 to 60 years using historical aerial photos and satellite images," the newspaper reported. "Eighty per cent of the islands we've looked at have either remained about the same or, in fact, gotten larger," the study's author, Auckland University's Associate Professor Paul Kench, found. "Some of those

"Entirely without Merit"

"I find the Doomsday picture Al Gore is painting—a six-meter sea level rise, fifteen times the IPCC number—entirely without merit."

—atmospheric scientist Dr. Hendrik Tennekes, former director of research at The Netherlands' Royal National Meteorological Institute[71]

islands have gotten dramatically larger, by 20 or 30 per cent.... We've now got evidence the physical foundations of these islands will still be there in 100 years.'"

The Polar Bears

The photogenic polar bear has been the icon for the modern global warming movement. "They are looking for poster children," explains geologist Bob Carter. "It suits that advertising purpose. It has nothing to do with science."[74] The fact is that polar bear populations are at or near historic highs. Scientists point out that the computer models predicting polar bear population collapse simply do not reflect reality or account for the adaptability of these animals.

"Polar bears have survived several episodes of much warmer climate over the last 10,000 years than exists today," evolutionary biologist and paleozoologist Susan Crockford of the University of Victoria explains. "There is no evidence to suggest that the polar bear or its food supply is in danger of disappearing entirely with increased Arctic warming, regardless of the dire fairy-tale scenarios predicted by computer models." As her research shows, "Polar bears have not been harmed by sea ice declines in summer." And so she rejects predictions of doom: "While the decline in ice extent is greatest in September, all evidence suggests this is the least important month of the year for polar bears—the yearly ice minimum in September occurs after the critical spring/summer feeding period, after the spring/summer mating period and well before the winter birth of cubs," she added.[75]

The U.S. Fish and Wildlife Service has estimated that the polar bear population was as low as 5,000 to 10,000 bears in the 1950s and 1960s. A 2002 U.S. Geological Survey of wildlife in the Arctic Refuge Coastal Plain noted that the polar bear populations "may now be near historic highs."[76] And in 2016, the International Union for the Conservation of Nature estimated the current polar bear population at between 22,000 and 31,000, which according to Crockford is "the highest estimate in 50 years."[77]

As Crockford wrote in 2016, "So far there is no convincing evidence that any unnatural harm has come to them. Indeed, global population size appears to have grown slightly since 1993, as the maximum estimated number was 28,370 in 1993 but rose to 31,000."[78]

Climatologist Judith Curry has said, "It seems like the polar bears are doing well and have managed to evolve and adapt over a very long time. It's not clear what we're doing up in the Arctic that's particularly jeopardizing them."[79]

★ ★ ★

Porky the Polar Bear

A 2015 population survey of polar bears found that key populations had increased 42 percent over the past eleven years and noted that some of the bears are "as fat as pigs."[81]

According to geologist Don Easterbrook, "There are five times as many polar bears now as they were in the 1970s so doesn't look like they are hurting too much. And I can also tell you on a factual basis that the past 10,000 years we've had temperatures that were…a half to 5° warmer and Greenland and the polar bears survive[d] that so there's not any problem now."[80]

In 2008, scientists spoke out publicly against the polar bear climate fears and I wrote a report for the U.S. Senate Environment and Public Works Committee.[82] What follows is based up on that report.

Award-winning quaternary geologist Ólafur Ingólfsson, a professor at the University of Iceland, has also rejected bear fears. "We have this specimen that confirms the polar bear was a morphologically distinct species at least 100,000 years ago, and this basically means that the polar bear has already survived one interglacial period," said Ingólfsson, who has conducted extensive expeditions and field research in both the Arctic and Antarctic. "This is telling us that despite the on-going warming in the Arctic today, maybe we don't have to be quite so worried about the polar bear," he added.[83]

Biologist Matthew Cronin, a research professor at the School of Natural Resources and Agricultural Sciences at the University of Alaska

★ ★ ★

Getting Ahead of the Data

Internationally known forecasting pioneer J. Scott Armstrong of the Wharton School at the Ivy League University of Pennsylvania and his colleague, forecasting expert Kesten Green of Monash University in Australia, co-authored a January 27, 2008, paper with Harvard astrophysicist Dr. Willie Soon, which found that polar bear extinction predictions violate "scientific forecasting procedures." As they explained, their "study analyzed the methodology behind key polar bear population prediction and found that one of the two key reports in support of listing the bears had 'extrapolated nearly 100 years into the future on the basis of only five years data—and data for these years were of doubtful validity."[84]

Fairbanks, rejected climate fears as well. "Polar bear populations are generally healthy and have increased worldwide over the last few decades," Cronin said.[85]

Biologist Josef Reichholf, who heads the Vertebrates Department at the National Zoological Collection in Munich is also skeptical of bear fears. "In warmer regions it takes far less effort to ensure survival," Reichholf said. "How did the polar bear survive the last warm period? Look at the polar bear's close relative, the brown bear. It is found across a broad geographic region, ranging from Europe across the Near East and North Asia, to Canada and the United States. Whether bears survive will depend on human beings, not the climate."[86]

The Nunavut government in Canada is not concerned about the fate of polar bear populations. Territorial Environment Minister Daniel Shewchuk said, "Through direct consultation, [Inuit communities] are unanimous in their belief that polar bears have not declined.... Based on hunter observations, polar bears are presently still healthy and abundant across Nunavut—and for that reason, not a species of special concern."[88]

The *Los Angeles Times* reported in 2012, "Doomsday predictions of the polar bear's demise tend to draw an Inuit guffaw here in Nunavut, the remote Arctic territory where polar bears in some places outnumber people.... Heart-rending pictures of polar bears clinging to tiny islands of ice elicit nothing but derision."[89]

Polar bear expert Dennis Compayre, formerly of the conservation group Polar Bears International, who has studied the bears in their natural habitat for almost thirty years, weighs in. "I tell you there are as many bears here now as there were when I was a kid," Compayre, author of the 2015 book on polar bears *Waiting for Dancer*, said.[90] "Churchill [in Northern Canada] is full of these scientists going on about vanishing bears and thinner bears. They come here preaching doom, but I question whether some of them really have the bears' best interests at heart."[91]

Famed environmental campaigner David Bellamy—botanist, former lecturer at Durham University, and host of a popular UK TV series on wildlife—protested, "Why scare the families of the world with tales that polar bears are heading for extinction when there is good evidence that there are now twice as many of these iconic animals, most doing well in the Arctic, than there were 20 years ago?"[92]

In 2017, the case for polar bear alarm has grown so weak that it appeared even climate activists were finally abandoning the animal as an icon for their cause. "There have been no new reports of falling polar bear numbers, and images of fat, healthy polar bears abound," paleozoologist Susan Crockford noted. "A number of recent climate change reports even failed to mention polar bears in their discussion

★ ★ ★
"Extremely Unhelpful"

"Polar bear expert Mitch Taylor barred from conference over 'extremely unhelpful' skeptical global warming view," reported the UK Telegraph in 2009. "Canadian biologist Mitchell Taylor, the former director of wildlife research with the government of Nunavut who teaches at Lakehead University in Canada, has also debunked the warmists' polar bear claims." According to Taylor, the bears "appear to be as abundant and as productive as ever, in most populations." But a meeting of the Polar Bear Specialist Group refused to allow Taylor to attend. He "was voted down by its members because of his views on global warming. The chairman, Dr Andy Derocher, a former university pupil of Dr Taylor's, frankly explained in an email…that his rejection had nothing to do with his undoubted expertise on polar bears: 'it was the position you've taken on global warming that brought opposition.'"[87]

of Arctic sea ice decline. The polar bear does not get mentioned once in the draft of the US Climate Science Special Report, even in the fifty page discussion on changes in the Arctic. And NOAA's annual Arctic Report Card has not mentioned the polar bear since 2014, in spite of highlighting the dangers faced by bear populations in every issue since 2008."[93]

According to Crockford, "Even Al Gore seems to have forgotten to include the plight of polar bears in his newest climate change movie. The polar bear played a prominent role in his 2007 documentary, *An Inconvenient Truth*, the polar bear example was left out of [2017's] *An Inconvenient Sequel: Truth to Power*. It doesn't even get a mention. After years of campaigners' and researchers' claims that populations were in terminal decline, the 'canary in the coal mine' has been retired."

CHAPTER 6

The Hottest Temperatures in a Thousand Years! Michael Mann's Hockey Schtick

Penn State professor Michael Mann rose to climate fame with his 1998 study including the now infamous "hockey stick" graph, which purported to show that the Northern Hemisphere today is experiencing high temperatures unprecedented in almost a thousand years. The graph was called "the hockey stick" because it resembled a hockey stick lying on its side, with the blade being the supposed spike in temperatures in the twentieth century. Mann's ominous temperature graph was featured in the UN IPCC's Third Climate Assessment report in 2001, and he became a media celebrity.

Mann's hockey stick became the perfect poster illustrating the narrative of man-made global warming. As the BBC explained, "It is hard to overestimate how influential this study has been."[1]

Influential? Yes. Accurate? No.

Peer-reviewed research both before and after Mann's hockey stick show that both the Medieval Warm Period and the Roman Warm Period were as warm as or warmer than modern temperatures. In fact, the 1990 UN IPCC report originally featured a temperature chart showing a Medieval Warm Period that was *much warmer* than twentieth-century temperatures.

Did you know?

★ The 1990 UN climate report showed a Medieval Warm Period warmer than the twentieth century

★ Studies reveal the world was in a cooling trend from the Roman and Middle Ages to the twentieth century

★ Reputable scientists have called the hockey stick graph "exaggerated" and "an embarrassment"

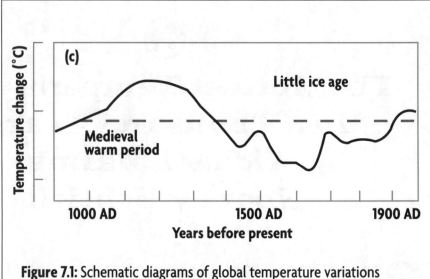

Figure 7.1: Schematic diagrams of global temperature variations since the Pleistocene on three time scales (a) the last million years (b) the last ten thousand years and (c) the last thousand years. The dotted line nominally represents conditions near the beginning of the twentieth century.

The 1990 UN climate report clearly showed a warmer Medieval Warm Period.[2]

While there were claims that the IPCC chart only represented temperatures in England, NASA climate researcher James Hansen also showed a very similar chart of medieval temperatures in his 1984 study, "Climate Sensitivity to Increasing Greenhouse Gases." Hansen and his co-authors write that the chart, which was from 1981, represents the "global temperature trend" for the past "millennium," and "is based on temperatures in central England, the tree limit in the White Mountains of California, and oxygen isotope measurements in the Greenland ice."[3]

Tony Heller of RealClimateScience.com wrote that Hansen's 1981 chart for his study shows "that claims the 1990 IPCC MWP graph was derived from CET (Central England Temperatures) only—are bogus. It came from Hansen 1981, and used multiple global proxies."[4]

Climate analysts "frequently claim that the 1990 IPCC temperature graph...showing the Medieval Warm Period (MWP) was only a representation of Central England Temperatures (CET) and was not global," Heller said. But Hansen's graph "was taken from temperatures in England, California and Greenland," he explained.

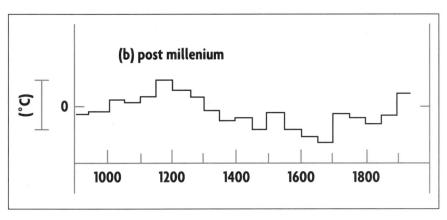

Hansen's 1981 global temperature graph showed a very warm medieval period.

On December 1, 1992, the *New York Times* reported that both the Medieval Warm Period and Little Ice Age were "global climate phenomena, not regional temperature variations." The paper quoted researchers who had compiled "a detailed record of the year-to-year variation in temperature and precipitation over the last thousand years" and found the "unmistakable signatures of the Medieval Warm Period, an era from 1100 to 1375 A.D. when, according to European writers of the time and other sources, the climate was so balmy that wine grapes flourished in Britain and the Vikings farmed the now-frozen expanse of Greenland; and the Little Ice Age, a stretch of abnormally frigid weather lasting roughly from 1450 to 1850."

The *Times* article quoted researcher Lisa J. Graumlich asking "how did we get those warmer temperatures during pre-industrial times, and what can we learn from those conditions about what is going on today?"[5]

Great question. But the climate campaigners are not interested in answering it. Instead they have waged a decades long battle to rewrite climate history and eliminate the inconvenient Medieval Warm Period.

Hiding the Medieval Warm Period

In 2006, I was the communications director for the U.S. Senate Environment and Public Works Committee. We held a hearing featuring geologist David Deming, who testified to our committee about attempts to get rid of the Medieval Warm Period. Deming had written about his surprising encounter, "With the publication of the article in *Science*, I gained significant credibility in the community of scientists working on climate change. They thought I was one of them, someone who would pervert science in the service of social and political causes. So one of them let his guard down. A major person working in the area of climate change and global warming sent me an astonishing email that said 'We have to get rid of the Medieval Warm Period.'"[6]

Who was the "major person" who wanted to "get rid of" the Medieval Warm Period, according to Deming? Deming "to the best of my recollection" fingered Jonathan Overpeck as the email culprit.[7] Overpeck, a researcher at the University of Arizona before moving to the University of Michigan, was a lead author on the 2007 and 2013 UN Intergovernmental Panel on Climate Change reports.

The emails leaked in the Climategate scandal, which we'll discuss in greater detail in chapter ten, revealed that Overpeck was concerned that Deming might be "taking the quote out of context." [8]

But Overpeck is on record denying the reality of the Medieval Warm Period. In 1999, he declared, "Now, high-resolution paleoclimate records stretching back 1,200 years confirm that the so-called Medieval Warm Period did not exist in the form of a globally synchronous period as warm, or warmer, than today." He was frustrated by the Medieval Warm Period's impact on the climate change debate, as a 2005 email from him, exposed in Climategate, revealed: "I'm not the only one who would like to deal a mortal blow to the misuse of supposed warm period terms and myths in the literature."[9]

But despite Overpeck's claims and what was implied by Mann's infamous hockey stick, the Medieval Warm Period—also called the Medieval Climate Optimum—is not a myth. The most recent research clearly shows that it was real—and was in fact global, not just confined to the Northern Hemisphere. The Center for the Study of Carbon Dioxide and Global Change reported in 2017 that "the Medieval Warm Period was: (1) global in extent, (2) at least as warm as, but likely even warmer than, the Current Warm Period, and (3) of a duration significantly longer than that of the Current Warm Period to date."[10]

And as the Science and Public Policy Institute reported in 2009, "More than 700 scientists from 400 institutions in 40 countries have contributed peer-reviewed papers providing evidence that the Medieval Warm Period (MWP) was real, global, and warmer than the present. And the numbers grow larger daily."[11]

A 2003 study by researchers from the Harvard-Smithsonian Center for Astrophysics, using more than two hundred paleoclimate studies from all over the planet, confirmed the existence of the MWP and LIA and found that twentieth-century warming was not unprecedented.[12]

A 2012 peer-reviewed study established medieval warming on the tropical island of New Caledonia in the Southern Hemisphere.[13] And a 2012 study published in the *Geophysical Research Letters* found that the subsequent

Little Ice Age (LIA) was probably caused by low solar and volcanic activity. The study's findings were "consistent with the idea that the LIA was a global event."[14] The LIA roughly occurred between 1300 to 1850.

Another 2012 study also added to the evidence that the MWP and LIA were both global in extent. Published by Elsevier, it examined the Antarctic Peninsula and "builds the case that the oscillations of the MWP and LIA are global in their extent and their impact reaches as far South as the Antarctic Peninsula.... "[15]

The Hockey Stick Breaks under Scrutiny

Geologist Robert Giegengack of the University of Pennsylvania detailed his initial reaction to the Mann's hockey stick graph: "I didn't like it when I first saw it. And when I saw that occur, two things occurred to me. One—I missed the medieval warm phase, which was very, very well documented. And most people who look at the medieval warm phase think that the temperature was higher than it is now. And the second thing I saw was a kink in his curve, and the kink exactly coincided with the change in the way the measurements were made." Giegengack's criticism of Mann's methodology was scathing. "He's not combining apples and oranges, he is combining apples and elephants and joining them on the same plot," Giegengack explained. "Where is the medieval warm phase? It has been detected in glaciers now in New Zealand. The medieval warm phase was real."[16]

Professor Ross Mckitrick of the University of Guelph in Canada and statistician Steve McIntyre were the duo instrumental in taking down the hockey stick.

"I had a problem when I looked at the [UN IPCC] report, first of all how promotional it was. This same graph reappears over and over throughout the report. It appears in different forms. Anywhere it appears it's full color. It's obvious that the people who put the report together wanted to promote

Protecting Their "Pet Findings" from Contradictory Evidence

Former UN IPCC climatologist John Christy ripped the UN's role in "misrepresenting the temperature record of the past 1000 years." As Christy testified to Congress in 2011, "Regarding the Hockey Stick of IPCC 2001 evidence now indicates, in my view, that an IPCC Lead Author working with a small cohort of scientists, misrepresented the temperature record of the past 1000 years by (a) promoting his own result as the best estimate, (b) neglecting studies that contradicted his, and (c) amputating another's result so as to eliminate conflicting data and limit any serious attempt to expose the real uncertainties of these data." "UN IPCC Lead Authors have virtually total control over the material and behave in ways that can prevent full disclosure of the information that contradicts their own pet findings and which has serious implications for policy in the sections they author." As Christy complained, "Lead Authors were transformed from serving as Brokers of science to Gatekeepers of a preferred point of view."[18]

it heavily," McKitrick, a senior fellow of the Fraser Institute and an adjunct scholar of the Cato Institute, explained.[17]

"I also had a problem with the fact that the author of that study (Mann) was the lead author of the [IPCC] chapter that promoted it and I didn't like the idea that a panel that was supposed to assess the whole literature operates in that way—where somebody assesses basically their own work and decides that it's the best thing out there and promote it heavily. It turns out the IPCC does that on lots of topics. They pick an author whose work they like, put them in charge of the chapter and then, what do you know? The chapter turns out to promote heavily that person's work and dismisses everything that says something else."

McKitrick teamed with Steve McIntyre to dig deeper into the hockey stick. The first problem they encountered was in getting access to the data that Mann had used to create it.

"It struck Steve as extremely odd that here is this famous study—governments around the world are relying on it, scientists were citing it intensively—and obviously nobody had ever asked to see the guy's data because they didn't have it in any one place," McKitrick said. "I got contacted by McIntyre who worked in the mineral finance field and he had actually unpacked some of the math, and got the data set from the authors and was finding all kinds of problems in the calculations."

McKitrick pointed out, "If you do your calculations correctly they'll show you that the uncertainties are so large you can't really say anything about what the results are. And in this case the method they used understated their uncertainties. So that they claim to have a lot more precision in their results [than] they really did. Also at a certain key point in the calculations, they—they just used the wrong formula," he added.[19]

Atmospheric physicist Fred Singer applauded the investigations of McIntyre and Mckitrick, noting that their research "showed that even random data fed into the faulty Mann theory would always yield a record-warmest 20th century."[20] Singer added, "Medieval temperatures were substantially greater—and so were temperatures during the earlier Roman Warm Period."

German professor Richard Dronskowski of Aachen University said, "No chart has been so falsified as the hockey stick chart. It's an embarrassment for the IPCC." Dornskowski called Mann's graph "a very, very nasty fabrication."[21]

A 2016 analysis at the website "No Tricks Zone" presented fifty "non-Hockey Stick Graphs" that had been published in peer-reviewed scientific papers and that refute the impression that modern temperatures are unusually warm. "All the graphs show that modern (post-1940s) temperatures aren't any warmer than the decades and centuries and millennia prior to the steep increase in anthropogenic $CO2$ emissions," the analysis concluded.[22]

Replicating Error

Despite the damning criticisms of Mann's work, many studies that followed, complete with hockey stick–style graphs of their own, appeared to support his conclusions.

But the 2006 Wegman Report commissioned by Congress exposed that Mann and the scientists whose work appeared to replicate his were operating in an echo chamber. "We found that at least 43 authors have direct ties to Dr. Mann by virtue of coauthored papers with him. Our findings from this analysis suggest that authors in the area of paleoclimate studies are closely connected and thus 'independent studies' may not be as independent as they might appear on the surface," the report explained.[23]

And in 2011, Steve McIntyre reported at Climate Audit that the leaked "Climategate documents confirm Wegman's hypothesis." As McIntyre wrote, "The Wegman Report was vindicated on its hypothesis about peer-review within the Mann 'clique.'"

"The Wegman Report hypothesized, but were [sic] unable to prove, that reviewers in the Mann 'clique' had been 'reviewing other members of the same clique'. Climategate provided the missing evidence, Climategate documents showed that clique member Phil Jones had reviewed papers by other members of the clique, including some of the articles most in controversy—confirming what the Wegman Report had only hypothesized," McIntyre wrote.

So the studies "weren't really independent," added Ross McKitrick. Instead, "A whole bunch of other researchers in the field started using their principal components series in their own reconstruction. So they're all sticking this biased hockey stick shape series in their data sets and getting a hockey stick result and then Mann turns around and says, 'Oh! Well, look at all these other studies. They get a hockey stick too.' Well, basically they're using the same data and some of the same methods," McKitrick explained.[24]

A Book You're Not Supposed to Read

The Hockey Stick Illusion by A. W. Montford (Anglosphere Books, 2015) exposes "in delicious detail, datum by datum, how a great scientific mistake of immense political weight was perpetrated, defended and camouflaged by a scientific establishment that should now be red with shame."

—British science journalist Matt Ridley[26]

Bo Christiansen, a Danish Meteorological Institute scientist came to a similar conclusion. "We cannot from these reconstructions conclude that the previous 50-year period has been unique in the context of the last 500-1000 years," Christiansen wrote. He dismissed claims that other reconstructions also show a "hockey stick," noting that "the different reconstructions all shared the same problems…all proxies are infected with noise."[25]

Mann has claimed that his study's temperature claims were vindicated by a 2006 National Academy of Sciences report, but as the Hockey Schtick website pointed out,

The NAS report did nothing of the sort, and in fact validated all of the significant criticisms of McIntyre & McKitrick (M&M) and the Wegman Report….

The NAS found that **Mann's methods had no validation (CE) skill significantly different from zero**….

Mann never mentions that a subsequent House Energy and Commerce Committee report chaired by Edward Wegman totally destroyed the credibility of the "hockey stick" and devastatingly ripped apart Mann's methodology as "bad mathematics." Furthermore, when Gerald North, the chairman of the NAS panel—which Mann claims "vindicated him"—was asked at the House Committee hearings whether or not they agreed with Wegman's harsh criticisms, **he said they did**:

CHAIRMAN BARTON: Dr. North, do you dispute the conclusions or the methodology of Dr. Wegman's report?

DR. NORTH [Head of the NAS panel]: No, we don't. We don't disagree with their criticism. In fact, pretty much the same thing is said in our report.

DR. BLOOMFIELD [Head of the Royal Statistical Society]: Our committee reviewed the methodology used by Dr. Mann and his co-workers and we felt that some of the choices they made were inappropriate. We had much the same misgivings about his work that was documented at much greater length by Dr. Wegman.

WALLACE [of the American Statistical Association]: "the two reports [Wegman's and NAS] were complementary, and to the extent that they overlapped, the conclusions were quite consistent." [Emphasis in the original][27]

Dissension in the Ranks

Mann's hockey stick and his research team have been subjected to intense scrutiny even by other warmist climate researchers and institutions.

A 2010 analysis by UK Royal Statistical Society head David Hand found that Mann's hockey stick graph was "exaggerated." According to Hand, "The particular technique they used exaggerated the size of the blade at the end of the hockey stick. Had they used an appropriate technique the size of the blade of the hockey stick would have been smaller." As the professor explained, "The change in temperature is not as great over the 20th century compared to the past as suggested by the Mann paper."[28]

One of Mann's colleagues had emailed about "Mike's Nature trick" to "hide the decline" in his temperature chart.

"I've just completed Mike's Nature trick of adding in the real temps to each series for the last 20 years (ie from 1981 onwards) and from 1961 for Keith's to hide the decline," Phil Jones, then director of Climatic Research Unit at University of East Anglia wrote in a 1999 email.[29]

"Mike" was Michael Mann, who had produced his hockey stick by grafting historical data from tree rings as proxies for northern hemisphere temperatures onto modern thermometer observations to produce the hockey stick temperature chart.

After about 1961 the tree ring data showed a decline in temperatures, contrary to the actual thermometer data, which showed an increase in temperatures. Switching from the proxy tree-ring data to the actual temperatures was Mann's "trick" to "hide the decline."

"Any scientist ought to know that you just can't mix and match proxy and actual data," said Philip Stott, professor emeritus of biogeography at London's School of Oriental and African Studies. "They're apples and oranges. Yet that's exactly what he did."[30]

Worse, the discrepancy between the tree-ring dataset, which showed a decline, and the actual temperature measurements, which rose, raised questions about the validity of the purported proxy data from the tree rings. Mann's graph and his use of his data came under harsh criticism from researchers.

The Climategate emails revealed that even Mann's UN colleagues did not all buy his hockey stick. And the Climategate 2.0 emails—a second set of emails featuring the leading scientists of the UN IPCC released in 2012—exposed more skepticism about Mann's work.

In a private 1999 email, Mann colleague Keith Briffa wrote, "I believe that the recent warmth was probably matched about 1000 years ago." Briffa explained, "I do not believe that global mean annual temperatures have simply cooled progressively over thousands of years as Mike [Mann] appears to and I contend that that there is strong evidence for major changes in climate over the Holocene (not Milankovich) that require explanation and that could represent part of the current or future background variability of our climate."[31]

UN IPCC scientist Tom Wigley had criticized the hockey stick in 2004. "I have just read the M&M [McIntyre and McKitrick] stuff criticizing MBH [Mann, Bradley, Hughes Hockey Stick]. A lot of it seems valid to me. At the very least MBH is a very sloppy piece of work—an opinion I have held for some time," Wigley, the former director of the Climatic Research Unit at the University of East Anglia, wrote in 2004, in one of the Climategate emails.[32] "Mike [Mann] is too deep into this to be helpful," he added.

In subsequent years, Mann has tried to use the courts to silence his critics, filing defamation lawsuits against Canadian Climatologist Timothy Ball, the Competitive Enterprise Institute, *National Review*, and author Mark Steyn.[33]

Meanwhile the scientific studies refuting the hockey stick claims of unprecedented warmth in the twentieth century just keep piling up.

A 2012 peer-reviewed study in the journal *Global and Planetary Change* showed that the temperatures were warmer than twentieth-century temperatures one to two thousand years ago, "revealing warmth during Roman and Medieval times were larger in extent and longer in duration than 20th century conditions."[34] A paper published in *Polar Research* in 2011 found the that two locations in the Arctic were much warmer in the Medieval Warm Period than at the end of twentieth century.[35] Yet another 2012 study, published in *Palaeogeography, Palaeoclimatology, Palaeoecology*, found, on the basis of isotopic records in Viking Age shells from Scotland, that the Medieval Warm Period "was warmer than the late 20th century by ~1°C."[36] A 2012, study published in *Nature Climate Change* added more evidence that medieval temperature were higher than today. "The international research team used these density measurements from sub-fossil pine trees in northern Scandinavia to create a sequence reaching back to 138 BC," Phys.org reported.[37] Their results showed a cooling trend since the Roman era. "For the first time, researchers have now been able to use the data derived from tree-rings to precisely calculate a much longer-term cooling trend that has been playing out over the past 2,000 years. Their findings demonstrate that this trend involves a cooling of -0.3°C per millennium due to gradual changes to the position of the sun and an increase in the distance between the Earth and the sun." Jan Esper of Johannes Gutenberg University, the study's lead author, explained, "We found that previous estimates of historical temperatures during the Roman era and the Middle Ages were too low."[38]

A 2013 study in *Science* magazine examined "10,000 years of layered fossil plankton in the western Pacific Ocean," and found that the Medieval Warm Period was 0.65 °C warmer than present temperatures. According to then–*New York Times* climate reporter Andrew Revkin, the paper reported "that several significant past climate ups and downs—including the medieval warm period and little ice age—were global in scope, challenging some previous conclusions that these were fairly limited Northern Hemisphere phenomena."[39] Revkin wrote of the study, "Michael Mann can't be happy about this work."[40]

CHAPTER 7

A Long Cool Pause

According to several analyses, global temperatures have been essentially holding nearly steady for almost two decades.

Not that you'd know it from the media and climate activists.

USA Today blared that "the planet sizzled to its third straight record warm year in 2016."[1] Former vice president Al Gore also touted 2016 as "the hottest year on record—confirmed by NASA and NOAA [the National Oceanographic and Atmospheric Administration]."[2] The *New York Times* claimed that the 2016 heat "record" was "trouncing" previous years' temperatures.[3] The *Times* used phrases like "blown past" to describe the alleged heat record.

Curiously, though, the actual numbers for the for the "hottest year" claim did not appear in the *Times* piece. "When you read a science report claiming that 2016 was the hottest year on record, you might expect that you will get numbers. And you would be wrong," wrote Robert Tracinski of The Federalist.[4] "We are not told what the average global temperature was, how much higher this is than last year's record or any previous records, or what the margin of error is supposed to be on those measurements."

Why did the *New York Times* omit the actual temperature data?

Did you know?

★ "Hottest year" claims in 2005, 2010, 2015, and 2016 were based on statistically meaningless year-to-year differences essentially within the margin of error

★ Global warming has seen a "slowdown" or "pause" since 1998

★ Tampering with the temperature record has been so widespread that the current climate era has been jokingly called the "Adjustocene" Era

While such years as 2005, 2010, 2014, and 2015 were declared the hottest years by global warming proponents, a closer examination revealed that the claims were based on year-to-year temperature data that differs by only a few hundredths of a degree to up to a few tenths of a degree—differences that were within the margin of error in the surface data. And the satellite data disagreed with the surface datasets. When an El Niño–fueled 2016 was declared "hottest year,"[5] the temperature rise from 2015 also failed to exceed the margin of error, or exceeded it just barely—depending on which of the multiple temperature datasets were reporting.

A 2017 analysis by astrophysicist David Whitehouse of the Global Warming Policy Foundation, said, "According to NOAA, 2016 was 0.07°F warmer than 2015, which is 0.04°C. Considering the error in the annual temperature is +/- 0.1°C this makes 2016 statistically indistinguishable from 2015, making any claim of a record using NOAA data specious."[6]

James Varney, writing at Real Clear Investigations, noted, "NOAA fixed the 2016 increase at 0.04 degrees Celsius. The British Met Office reported an even lower rise, of 0.01C. Both increases are well within the margin of error for such calculations, approximately 0.1 degrees, and therefore are dismissed by many scientists as meaningless."[7]

Physicist Steven E. Koonin, a former Obama administration official, mocked the "hottest year" claims, offering the media what he considered a more accurate way to present the temperature data headlines. "Global Temperatures Up 0.0X for 2016; Within Margin of Error for Last N Years," Koonin, who served as undersecretary for science in Obama's Department of Energy, wrote in 2017.[8]

In other words, global temperatures are holding basically steady. The media and climate activists are hyping supposed "record" temperatures that are not even outside the margin of error of the dataset as somehow meaningful. It is a fancy way of saying the "pause" or "slowdown," "hiatus" or "standstill" in temperatures is continuing.

Even former NASA climatologist James Hansen has admitted that "hottest year" declarations are "not particularly important."[9]

When 2014 was declared the "hottest year" based on surface data, the claim, which was within the margin of error from previous "hottest years," did not impress climatologist Judith Curry. "With 2014 essentially tied with 2005 and 2010 for hottest year, this implies that there has been essentially no trend in warming over the past decade. This 'almost' record year does not help the growing discrepancy between the climate model projections and the surface temperature observations," Curry told the *Washington Post*.[10]

Retired MIT climate scientist Richard Lindzen has ridiculed "hottest year" claims. "The uncertainty here is tenths of a degree. When someone points to this and says this is the warmest temperature on record, what are they talking about? It's just nonsense. This is a very tiny change period," Lindzen said.[11] "If you can adjust temperatures to 2/10ths of a degree, it means it wasn't certain to 2/10ths of a degree." Lindzen pointed out, "We're talking about less than a tenth of degree with an uncertainty of about a quarter of a degree. Moreover, such small fluctuations—even if real—don't change the fact that the trend for the past 20 years has been much less than models have predicted."[12] The former MIT professor believes the "hottest year" claims are returning us to a bygone era. "To imply that a rise of temperature of a tenth of a degree is proof that the world is coming to an end—has to take one back to the dark ages," he explained in 2017.[13] "As long as you can get people excited as to whether it's a tenth of a degree warmer or cooler, then you don't have to think, you can assume everyone who is listening to you is an idiot. The whole point is so crazy because the temperature is always going up or down a little. What is astonishing is that in the last 20 years it hasn't done much of anything," he added. "What they don't mention is there has been a big El Nino in 2016 and in recent months the temperature has been dropping back into a zero trend level."

★ ★ ★
Not So Scary

Extreme weather expert Roger Pielke Jr. noted that the media "hottest year" scare stories are simply not working: "It doesn't scare people."[14]

Satellites over Surface

And even the insignificant heating hyped in the media is often based on dodgy numbers.

Tony Heller of Real Climate Science noted in 2016 that "NOAA claimed record heat in numerous locations in September, like these ones in Africa and the Middle East." He added, "This is a remarkable feat, given that they didn't have any actual thermometer readings in those regions in September."[15]

The satellite temperature data is considered more accurate than surface measurements, but the climate campaigners prefer surface data.

Climate data analyst Paul Homewood observed that "both UAH [University of Alabama at Huntsville satellite data] and RSS [Remote Sensing Systems satellite measurements] say that atmospheric temperatures for 2016 statistically tied with 1998, at just 0.02C higher.[16] Neither 2014 or 2015 were anywhere near being a record." Homewood, who publishes the website Not a Lot of People Know That, pointed out, "Satellite measurements of global temperatures are regarded as much more comprehensive, accurate and unaffected by UHI [Urban Heat Island], as climatologist Roy Spencer explained in 2014."

While the media ignores satellite data in favor of surface data potentially distorted by the heat from the urban environments where many of the temperatures are measured, NASA has acknowledged that satellites are more accurate. A 1990 NASA Report found that "satellite analysis of upper atmosphere is more accurate, and should be adopted as the standard way to monitor temperature change."[17]

"Unfortunately, the surface temperature analysis contains several uncertainties and systematic biases when used to diagnose global warming," as former University of Colorado climatologist Roger Pielke Sr. explained in the *Washington Post*. "One of them is with respect to land minimum temperatures over land. Rather than measuring changes in heat content through depth in the atmosphere, even slight changes in vertical mixing of heat (even with no net heating) can produce warmer minimum temperatures."[18]

A Book You're Not Supposed to Read

The Deliberate Corruption of Climate Science by Tim Ball (Stairway Press, 2014).

NASA's temperature trends are "almost 20 times larger than the satellites," noted physicist Lubos Motl in 2017. "Both satellite-based teams quantifying the global mean temperature [University of Alabama at Huntsville and Remote Sensing Systems] concluded that 2016 was 0.02 °C warmer than 1998. These were otherwise very similar 'end of a strong El Niño years' separated by 18 years. According to these numbers and nothing else, one could estimate that the warming per century is some 0.11 °C, a negligible amount."[19]

But satellite data is also subject to "adjustments." Tampering with the temperature record has been so widespread that the current climate era has been jokingly called the "Adjustocene" Era.

One of the key satellite data sets, RSS, had basically showed a zero trend in global temperatures for over eighteen years, but in 2016, not only did a big warm El Niño impact temperatures, the dataset was also "adjusted," and hey, presto, the "pause" in global warming disappeared. This satellite temperature data was revised with "improved adjustments" and somehow "found" the missing warming.[20] AP reporter Seth Borenstein, who frequently serves as a mouthpiece for the climate change narrative, asked giddily, "Are satellites now contradicting the climate doubter community?"[21]

Christopher Monckton, a former science advisor to UK Prime Minister Margaret Thatcher, was livid at the RSS "adjustments," writing an analysis titled, "How They Airbrushed Out the Inconvenient Pause." "All gone! Vanished into thick air! Just like that! Amazing! Zowee! Look! A quarter of a degree of global warming where there was none before!" Monckton wrote in 2017.

"RSS now shows a warming almost 50% greater than the UAH warming," he explained. "On most of the global-temperature datasets, much of the warming of recent decades was not evident in the raw data and has been created by ex-post-facto manipulation of the data—whether for good reasons or bad," he added.[22]

Other temperature datasets have also seen revisions. "Temperature records have been altered considerably particularly the US Historical Climate Network. The alterations in general result in a cooler past and a warmer present," climatologist Pat Michaels noted.[23] Analysis by meteorologist Anthony Watts has revealed, "The temperature record has essentially doubled in trend over the last 30 years due to adjustments and citing issues. So if we look at the best stations in the network we have about half of the warming."[24]

And a 2015 analysis by Tony Heller of Real Climate Science accused NASA of reversing a post-1940 U.S. cooling trend by data adjustments or what he termed "tampering."

Hansen, NASA lead global warming scientist, wrote in 1999, "The U.S. has warmed during the past century, but the warming hardly exceeds year-to-year variability. Indeed, in the U.S. the warmest decade was the 1930s and the warmest year was 1934."[25] Hansen explained that "in the U.S. there has been little temperature change in the past 50 years, the time of rapidly increasing greenhouse gases—in fact, there was a slight cooling throughout much of the country."

Hansen showcased the following temperature graph of the continental U.S. since 1880, showing a cooling trend from the 1930s.

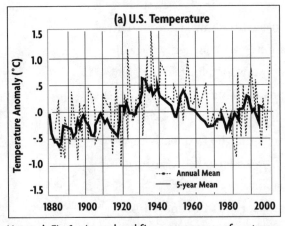

Hansen's Fig. 1a: Annual and five-year mean surface temperature for the contiguous forty-eight United States relative to 1951–1980, based on measurements at meteorological stations.[26]

Other temperature datasets agreed with NASA. Both NOAA and CRU (Climatic Research Unit of the University of East Anglia) also showed no warming trend over the twentieth century.

In 1989, the *New York Times* reported, "Last week, scientists from the United States Commerce Department's National Oceanic and Atmospheric Administration said that a study of temperature readings for the contiguous 48 states over the last century showed there had been no significant change in average temperature over that period. Dr. (Phil) Jones said in a telephone interview today that his own results for the 48 states agreed with those findings."[27]

But a change in this temperature data soon occurred, after Hansen wrote his 1999 analysis and presented his U.S. graph.

"Right after the year 2000, NASA and NOAA dramatically altered US climate history, making the past much colder and the present much warmer," Heller's analysis showed.[28] He explained that "NASA cooled 1934 and warmed 1998, to make 1998 the hottest year in US history instead of 1934. This alteration turned a long term cooling trend since 1930 into a warming trend."

The revised post-2000 NASA temperature chart for the U.S. looked very different from the 1999 version, with a noticeable cooling of the past.

Heller analogized the altering of the past temperature record to George Orwell's novel *1984*, quoting this passage from the book: "He who controls the past controls the future. He who controls the present controls the past."[30]

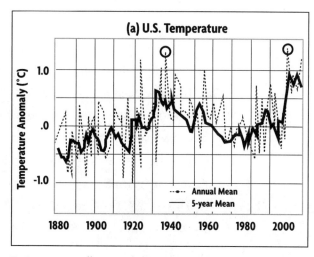

Fig.D — Tony Heller reveals how 1934 became cooler than 1998.[29]

★ ★ ★

A Meaningless Mean

A 2006 study published in the *Journal of Non-Equilibrium Thermodynamics* questioned the validity of a "global temperature." "It is impossible to talk about a single temperature for something as complicated as the climate of Earth," Bjarne Andresen, a professor at the Niels Bohr Institute at the University of Copenhagen, said. "The climate is not governed by a single temperature," he explained. Science Daily reported that the study had concluded "it is meaningless to talk about a global temperature for Earth. The Globe consists of a huge number of components which one cannot just add up and average. That would correspond to calculating the average phone number in the phone book. That is meaningless. Or talking about economics, it does make sense to compare the currency exchange rate of two countries, whereas there is no point in talking about an average 'global exchange rate.'"[31]

The Pause That Refreshes

Astrophysicist David Whitehouse of the UK Global Warming Policy Foundation has noted that, despite media claims, the "temperature pause never went away."[34]

The "pause," also known as the "hiatus" or "standstill" refers to the last two decades of essentially stable global temperatures, beginning in 1998.

In 2017, the "Pausebuster" scandal was exposed by NOAA whistleblower John Bates. Bates detailed how a federal study had "exaggerated global warming" and "was timed to influence" the UN Paris agreement. NOAA stands accused of manipulating temperature data to hype so-called "global warming."

The 2015 NOAA "Pausebuster" study—"Possible Artifacts of Data biases in the Recent Global Surface Warming Hiatus,"[35] published in *Science* magazine, had attempted to erase the "pause"—nearly twenty years of essentially stable temperatures—by rewriting the temperature history.[36]

Ironically, the global warming establishment was caught simultaneously trying to deny the temperature "pause" existed while at the same time making up endless excuses for it. They proposed at least sixty-six excuses to explain the "pause" or "halt" in warming. It was blamed on low solar activity, Chinese coal use, volcanic aerosols, faster trade winds, coincidence, and...the oceans ate global warming![37]

But as climatologist Roy Spencer pointed out, "Even if you assume all the extra energy that's not being used is being stored in the ocean, that's good news as far as I'm concerned because we really don't care if the ocean warms up by a tenth of a degree in 50 years."[38]

The Chinese coal excuse also took a run up the flagpole. A study published in the Proceedings of the National Academy of Science blamed Chinese burning of coal for the lack of global warming.[39] As the *New York Times* explained, "Coal releases greenhouse gases that will have a long-term warming effect, of course, but it also throws particles into the air that can reflect sunlight back to space over the short term."[40] The *Times* continued, "Can this account for the warming hiatus, or part of it? Experts simply do not know, and bad luck is one reason. A few years ago, NASA tried to send up a satellite that could have helped answer that question by carefully measuring particles in the air, but it blew up on launch."

Physicist Lubos Motl ridiculed the excuse of Chinese coal use, explaining that "the idea that warming predictions failed because of the Chinese coal is just a random guess, one among hundreds of possible explanations."

★ ★ ★

It's Not the Crime, It's the Cover-Up

NOAA's numerous temperature adjustments are now under investigation by the U.S. Congress.

"We're going to have a series of hearings on the extent and the degree to which data has been manipulated or falsified or even ignored," said Texas Congressman Lamar Smith, the chairman of the House Science Committee.[32]

But NOAA has not been cooperative, refusing to provide Congress with relevant e-mails.[33]

He concluded, "Why don't those people honestly admit that they simply have no clue what was happening since 1998 and what will be happening before 2020, 2030, or 2100?"[41]

The excuses piled up. Volcanic aerosols tamped down recent Earth warming, claimed a 2013 Colorado University study.[42]

But perhaps the most absurd excuse for the "pause" was given by the UK Energy Minister in 2014. Baroness Sandip Verma explained that government policies "may have slowed down global warming, but that is a good thing. It could well be that some of the measures we are taking today is helping that to occur. Warming may have decreased, which could support the effectiveness of green policies."[43]

Climate analyst Paul Homewood responded, "What has she been smoking to imagine that the UK's reduction of 16 million tonnes [in carbon dioxide emissions] can have had the slightest effect on climate, when the rest of the world has increased their emissions by 68 times as much?"[44]

A team of scientists including István Markó, Willie Soon, William Briggs, and David Legates noted that the temperature hiatus is real in a 2017 analysis. "In the last 20 years, we have released more than a third of all the CO_2 produced since the beginning of the industrial period. Yet global mean surface temperature has remained essentially constant for 20 years, a fact that has been acknowledged by the IPCC, whose models failed to predict it," they wrote. "NOAA's State of the Climate report for 2008 said that periods of 15 years or more without warming would indicate a discrepancy between prediction and observation—i.e., that the models were wrong. Just before the recent naturally occurring el Niño event raised global temperature, there had been 18 years and 9 months without any global warming at all."[45]

Climatologist Judith Curry points out that the temperature "pause" or "hiatus" since the late 1990s needs further examination: "So this is the big mystery and until we have a good answer for that I say we don't have any particular confidence in attributing the warming of the last quarter of the 20th century."[46]

Geologist Robert Giegengack has said, "I don't feel the need to explain the halt in warming because there's so many unknowns and there's so many variables in the climate system I could attribute it to almost anything. The only people who feel the need to explain that is the ones who have gone out on a limb and insisted CO_2 is the controlling factor."[47]

★ ★ ★
Quit Digging

"When will climate scientists say they were wrong?" climatologist Pat Michaels has asked. "Day after day, year after year, the hole that climate scientists have buried themselves in gets deeper and deeper. The longer that they wait to admit their overheated forecasts were wrong, the more they are going to harm all of science."[52]

The World Is Not Burning

Al Gore has likened the Earth to a sick child. "The Earth has a fever that is growing more, and more intense," the former vice president insisted.[48]

The Earth does not have a "fever."

Greenpeace co-founder Patrick Moore explains, "I do not believe the earth has a fever because it's colder now it has been through most of the history of life."[49]

As Nobel Prize–winning physicist Ivar Giaever points out, ".8 degrees is what we're discussing in global warming. .8 degrees. If you ask people in general what it is they think—it's 4 or 5 degrees. They don't know it is so little."[50]

And climatologist Pat Michaels explained that in any case the world's "temperature should be near the top of the record given the record only begins in the late 19th century when the surface temperature was still reverberating from the Little Ice Age."[51]

The late geologist Bob Carter dismissed warming claims: "I call this sort of stuff kindergarten science.... The fact that the temperature was warmer

Be Not Afraid

"We are creating great anxiety without it being justified...there are no indications that the warming is so severe that we need to panic.... The warming we have had the last 100 years is so small that if we didn't have meteorologists and climatologists to measure it we wouldn't have noticed it at all."

—Lennart Bengtsson, award-winning climate scientist[53]

at the end of the 20th century, than it was in the preceding hundred years, is such a piece of kindergarten science. It's true, and it is completely meaningless in telling you anything about climate change."[54]

Climatologist John Christy's research on the United States has found that "about 75% of the states recorded their hottest temperature prior to 1955, and over 50 percent of the states experienced their record cold temperatures after 1940."[55] Data from the Environmental Protection Agency agrees with Christy: the EPA website features a 2016 chart labelled "the U.S. Heat Wave Index from 1895 to 2015," and it reveals that the worst U.S. heatwaves by far happened in the 1930s.[56]

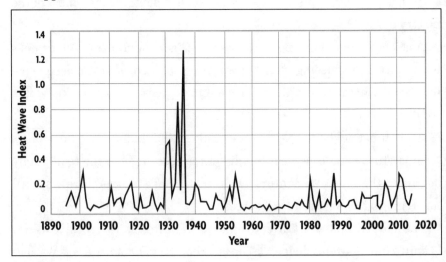

EPA: "This figure shows the annual values of the U.S. Heat Wave Index from 1895 to 2015."[57]

CHAPTER 8

Models Do Not Equal Evidence

The temperature of the Earth has held nearly steady for two decades. And yet the climate models continue to predict accelerated warming. In fact, they've already predicted much more warming for recent years than has actually occurred. Nevertheless, the warmists continue to tout their models' predictions for the future.

"Unfortunately the climate models, and this is very important for you to understand, you can take a look at the ensemble of the United Nations climate models they are failing at the 95% level—they're predicting too much warming," climatologist Pat Michaels has noted.[1]

"Predictions can never be 'falsifiable' in the present: we must ultimately wait to see whether they come true," climate activist professor Michael Mann of Penn State wrote in 2017.[2]

Scientists Willie Soon, István Markó, William Briggs, and David Legates pointed out in 2017, "The climate models relied upon by the [UN] IPCC and the politicians they advise have predicted warming at about twice the rate observed during the past 27 years, during which the Earth has warmed at 0.4 °C, about half of the 0.75 °C 27-year warming rate implicit

Did you know?

★ In recent decades, climate models have predicted much more warming than has actually occurred

★ Scientists have begun referring to climate model predictions as "evidence" and "data"

★ Energy secretary Steven Chu was so enamored with climate models that he boasted we know the future in IPCC's explicit 1990 prediction that there would be 1.0 °C warming from 1990–2025."[3]

To give just one example, a 2007 study found that actual Antarctic temperatures diverged from climate model predictions: "temperatures during the late 20th century did not climb as had been predicted by many climate models." The observed rates of precipitation were also out of line with the models' predictions.[4]

"The best we can say right now is that the climate models are somewhat inconsistent with the evidence that we have for the last 50 years from continental Antarctica," David Bromwich, a researcher with the Byrd Polar Research Center at Ohio State University explained. "We're looking for a small signal that represents the impact of human activity and it is hard to find it at the moment."

Predictions Are Suddenly "Evidence," Models are Now "Data"

And yet, such is the climate establishment's attachment to their computer models that they have begun to refer to their predictions as "evidence" and "data."

Scientists affiliated with the federal Oak Ridge National Laboratory in Tennessee claimed in 2011, "We find evidence from nine climate models that intensity and duration of cold extremes may occasionally, or in some cases quite often, persist at end-of-20th-century levels late into the 21st century in many regions."[5]

And Seth Wenger of the University of Georgia has said that "the most dire climate models show temperatures in Idaho rising an average of 9 degrees in 70 years.[6] That would make Boise pretty unpleasant. None of us want to believe that." But Wenger added, "I have to set aside my feelings and use the best data."

The assertion that models are now "evidence" raised the ire of former Colorado State Climatologist Roger Pielke Sr. "The use of the term 'evidence' with respect to climate models illustrates that this study is incorrectly assuming that models can be used to test how the real world behaves," Pielke explained.[7]

You Don't Scare Me

"I am a scientist. I deal with evidence. Not with frightening computer models."

—The late Australian geologist Bob Carter[8]

The climate change debate has morphed from focusing on actual data and evidence to misdirection based on climate model predictions. Here's how it works. If current reality fails to alarm, there is one simple way to make climate change sound scary. When anyone points out the current lack of warming, climate scientists and activists essentially make a bunch of frightening predictions about fifty to one hundred years from now: *Hey it's even worse than we thought—our predictions of the future are now much worse than they were a few years ago.*

President Obama's energy secretary, Steven Chu, was so confident of the climate models' predictions that he once boasted that we now know "what the future will be 100 years from now."

"For the first time in human history, science has shown that we are altering the destiny of our planet. At no other time in the history of science have we been able to say what the future will be 100 years from now," Chu said in a public speech. "It's quite alarming. Every year looks more alarming.... An irony of climate change is that the ones who will be hurt the most are the innocent—those yet to be born."[9]

Climate model predictions seem to have given many climate activists the illusion that they are prophets. Shouldn't Chu be touting these scary predictions of the year 2100 on a boardwalk somewhere with a full deck of tarot cards?

A 2009 Reuters article noted that scientists say the new Arctic sea ice predictions "should send a warning to world leaders meeting in Copenhagen in December for U.N. talks on a new climate treaty." Reuters cited Britain's Energy and Climate Change secretary Ed Miliband, who boldly announced that the new Arctic ice prediction "further strengthens the case for an ambitious global deal in Copenhagen."[10]

Sub-Standard Software

But an in-depth examination into climate models reveals them to be a particularly shadowy aspect of the climate change business.

In 2009, two prominent U.S. government scientists made two separate admissions calling into question the reliability of climate models used to predict warming decades and hundreds of years into the future. Gary Strand, a software engineer at the federally funded National Center for Atmospheric Research (NCAR), admitted that climate model software "doesn't meet the best standards available" in a comment he posted on the website Climate Audit.[11] "As a software engineer, I know that climate model software doesn't meet the best standards available. We've made quite a lot of progress, but we've still quite a ways to go."

Meteorologist Anthony Watts was prompted to ask, "Do we really want Congress to make trillion dollar tax decisions today based on 'software [that] doesn't meet the best standards available?'"[12] Watts noted, "NASA GISS model E written on some of the worst FORTRAN coding ever seen is a challenge to even get running. NASA GISTEMP is even worse. Yet our government has legislation under consideration significantly based on model output that Jim Hansen started. His 1988 speech to Congress was entirely based on model scenarios."

NASA climate modeler Gavin Schmidt also questioned the reliability of climate models. He admitted that the "chaotic component of the climate system—that is not predictable beyond two weeks, even theoretically." Schmidt noted that some climate models "suggest very strongly" that the

★ ★ ★

If It's Not a Prediction, It Can't Turn Out to Be Wrong

Kevin Trenberth, another high-profile UN IPCC lead author, referred to climate models' projections as "story lines." As he wrote on the blog of the journal *Nature* on June 4, 20017, "In fact there are no predictions by IPCC at all. And there never have been. The IPCC instead proffers 'what if' projections of future climate that correspond to certain emissions scenarios. There are a number of assumptions that go into these emissions scenarios. They are intended to cover a range of possible self consistent 'story lines' that then provide decision makers with information about which paths might be more desirable." Trenberth also admitted that the climate models have major shortcomings: "they do not consider many things like the recovery of the ozone layer, for instance, or observed trends in forcing agents. There is no estimate, even probabilistically, as to the likelihood of any emissions scenario and no best guess."[13]

American Southwest will dry in a warming world. But he also noted that "other models suggest the exact opposite." As Schmidt explained, "With these two models, you have two estimates—one says it's going to get wetter and one says it's going to get drier. What do you do? Is there anything that you can say at all? That is a really difficult question." The NASA climate modeler admitted that "there is so much unforced variability in the system" that it "takes about 20 years to evaluate" the "climate prediction and projections going out to 2030 and 2050."[14]

The credibility of these computer model predictions—which are still being used by governments to determine global warming policy based on future climate risks—has been under intense scrutiny for years. In 2007, top UN IPCC scientist Jim Renwick admitted that climate models do not account for half the variability in nature and thus are not reliable. "Half of the variability in the climate system is not predictable, so we don't expect to do terrifically well," Renwick conceded.[15]

A Book You're Not Supposed to Read

The Greenhouse Delusion: A Critique of "Climate Change 2001" by Vincent Gray (Multi-Science Publishing Co. Ltd., 2014).

IPCC reviewer and climate researcher Vincent Gray of New Zealand, the author of more than one hundred scientific publications and an expert reviewer on every single draft of the IPCC reports going back to 1990, declared in 2007 that IPCC claims were "dangerous unscientific nonsense"[16] because, "All the [UN IPCC does] is make 'projections' and 'estimates'. No climate model has ever been properly tested, which is what 'validation' means, and their 'projections' are nothing more than the opinions of 'experts' with a conflict of interest, because they are paid to produce the models. There is no actual scientific evidence for all these 'projections' and 'estimates,'" Gray noted.[17]

Atmospheric scientist Hendrik Tennekes, a scientific pioneer in the development of numerical weather prediction and former director of research at the Netherlands' Royal National Meteorological Institute, compared scientists who promote computer models predicting future climate doom to unlicensed software engineers. "I am of the opinion that most scientists engaged in the design, development, and tuning of climate models are in fact software engineers. They are unlicensed, hence unqualified to sell their products to society," Tennekes wrote in 2007.[18]

The late atmospheric scientist Augie Auer ridiculed climate model predictions, comparing them to video games: "Most of these climate predictions or models, they are about a half a step ahead of PlayStation 3. They're really not justified in what they are saying. Many of the assumptions going into [the models] are simply not right."[19] And atmospheric physicist James Peden compared the climate models to children's toys, calling them "computerized tinker toys with which one can construct any outcome he chooses."[20]

In Violation of Basic Principles

Top forecasting experts report that climate models violate the basic principles of forecasting. Forecasting pioneer Scott Armstrong of the Wharton School at University of Pennsylvania reported that the IPCC's forecasting procedures "clearly violated 72 [of the 89] scientific principles of forecasting." Armstrong and his colleague Kesten Green of the University of South

Australia reported, "As shown in our analysis experts' forecasts have no validity in situations characterized by high complexity, high uncertainty, and poor feedback. Without scientific support for their forecasting methods, the concerns of scientists should not be used as a basis for public policy."[21]

Prominent physicist Freeman Dyson, a professor emeritus of physics at the Institute for Advanced Study at Princeton University, a fellow of the American Physical Society, a member of the U.S. National Academy of Sciences, and a fellow of the Royal Society of London, has referred to climate models as "rubbish." Dyson is blunt in his criticism, mocking "the holy brotherhood of climate model experts and the crowd of deluded citizens who believe the numbers predicted by the computer models.... I have studied the climate models and I know what they can do. The models solve the equations of fluid dynamics, and they do a very good job of describing the fluid motions of the atmosphere and the oceans. They do a very poor job of describing the clouds, the dust, the chemistry, and the biology of fields and farms and forests."[22]

"I am not impressed by the ability of their models to either model the past or to model the future," Geologist Robert Giegengack of the University of Pennsylvania has said.[23]

A February 20, 2007, book review in the *New York Times* featured *Useless Arithmetic: Why Environmental Scientists Can't Predict the Future* by coast

If It's Not a Prediction, It Can't Turn Out to Be Wrong

In 2014, I asked the UN IPCC's Michael Oppenheimer about some of his predictions from the 1990s.

Morano: "In 1990 you predicted by '95 rising CO_2 impacts would quote "desolate the heartlands of North America."

Oppenheimer: "It is a very important distinction between a prediction and a scenario. A scenario is one of many possible futures. A prediction is a rather specific forecast of what's going to happen. Back in 1988, 1990 the science was rather uncertain, I picked from among a large number of possible futures to talk about some that were particularly threatening. There's a lot of things that were thought to be the case 30 years ago—that are no longer the case. Science marches on."[25]

Oppenheimer seems to think failed "scenarios" don't deserve scrutiny since "science marches on." Sadly, public policy is being made on these failed scenarios.

geologist and Duke professor emeritus Orrin H. Pilkey and Linda Pilkey-Jarvis, a geologist in the Washington State Department of Geology. According to the *Times*, the book presented "an overall attack on the use of computer programs to model nature. Nature is too complex, [the authors] say, and depends on too many processes that are poorly understood or little monitored—whether the process is the feedback effects of cloud cover on global warming or the movement of grains of sand on a beach." The review noted how models "may include coefficients (the authors call them 'fudge factors') to ensure that they come out right. And the modelers may not check to see whether projects performed as predicted."

So, "instead of demanding to know exactly how high seas will rise or how many fish will be left in them or what the average global temperature will be in 20 years, they argue, we should seek to discern simply whether seas are rising, fish stocks are falling and average temperatures are increasing. And we should couple these models with observations from the field. Models should be regarded as producing 'ballpark figures,' they write, not accurate impact forecasts." The coastal models are so flawed that Pilkey-Jarvis recommended dredging up a lot of sand and dumping it on the beach "willy-nilly"—a process that they suggested would yield the same results, minus the "false mathematical certitude."[24]

CHAPTER 9

The Eroding "Consensus"

The global warming establishment has attempted to silence skeptics, but its attacks on the "deniers"—an attempt to conflate global warming skepticism with Holocaust denial—have only backfired. Recent years have seen more scientists speaking out, in response to the smears.

Climatologist Roger Pielke Sr. is one of them. "Either they ignore you or they ridicule you, and that's not the scientific method," he has pointed out.[1]

Climate statistician William M. Briggs, who served on the American Meteorological Society's Probability and Statistics Committee, is another. "After reading UN IPCC chairman Pachauri's asinine comment [comparing skeptics to] Flat Earthers, it's hard to remain quiet," Briggs declared. "Anger. I was pretty angry," he said in response to insulting remarks from Al Gore about climate skeptics, which Briggs also called "asinine." The statistician complained, "I was being insulted by people who knew far less about these things than I did; and I found that preposterous. And that's why I started speaking out." There were consequences. "It did not help me in the least, it only hurt me professionally, for my career." Nevertheless, he explained, "You can't sit back and take that kind of thing."[2]

Did you know?

★ Smears against climate skeptics have inspired numerous scientists to re-evaluate the evidence and *become* skeptics

★ *Scientific American* branded a climatologist who changed her mind about global warming a "heretic"

★ Leftist scientists point out that the false global warming narrative hurts the world's poor

Goaded into Taking a Stand

Lord Christopher Monckton has noted that the intimidation has motivated many scientists to finally publicly speak out and dissent. "This hate speech, which is the one thing that the climate Left knows how to do is backfiring badly on them now," Monckton said.[3]

Climatologist Robert Durrenberger, past president of the American Association of State Climatologists has said that Gore's smears motivated him to look more deeply into the issue, and speak out. "Al Gore brought me back to the battle and prompted me to do renewed research in the field of climatology. And because of all the misinformation that Gore and his army have been spreading about climate change I have decided that 'real' climatologists should try to help the public understand the nature of the problem," Durrenberger wrote.[4]

So not only are already skeptical scientists speaking out because of the smears and attempted intimidation. More and more scientists—including many on the political Left—have reexamined the evidence for man-made global warming claims and reversed their views.

Re-Evaluating the Evidence

Award-winning geophysicist and French politician Claude Allègre was one of the first scientists to warn about the dangers of global warming. Allègre, a member of both the French and U.S. Academies of Science, has authored more than one hundred scientific articles, written eleven books, and received numerous scientific awards including the Goldschmidt Medal from the Geochemical Society of the United States. Allègre was one of fifteen hundred scientists who signed a November 18, 1992, letter titled "World Scientists' Warning to Humanity," arguing that the "potential risks" were "very great."[5]

But in recent years Allègre has taken another look at the evidence and reversed himself. He is now France's most outspoken global warming skeptic. Allègre explains that global warming hysteria is motivated by money. "The ecology of helpless protesting has become a very lucrative business for some people!" he has pointed out. Allègre mocked former vice president Al Gore's 2007 Nobel Prize, calling it "a political gimmick" and saying, "The amount of nonsense in Al Gore's film! It's all politics; it's designed to intervene in American politics."[6]

"The CO_2 is in a very short proportion in the atmosphere. But my point is nothing has proved this is man-made climate change," Allègre said in a French TV interview.[8]

Renowned geophysicist and green guru James Lovelock was one of the leading voices of man-made climate alarm. In 2006, he was predicting global warming doom: "Before this century is over, billions of us will die and the few breeding pairs of people that survive will be in the Arctic, where the climate remains tolerable."[9] He added, "We have given Gaia a fever and soon her condition will worsen to a state like a coma." Lovelock also warned, "Global warming is much more than just a real effect, it is something deadly that will threaten nearly all of us.... Anyone with an imagination can see the awful human consequences of that. And we're talking about something which is only about 30 years ahead." He prophesied "As many as 7 out of 8 [humans] are likely to be wiped out."[10]

But Lovelock grew steadily more skeptical and has now seemingly made a full U-turn on man-made global warming fears. The geophysicist, considered one of the pioneering scientists of the twentieth century for his Gaia hypothesis, declared in 2012, "I was WRONG and alarmist on climate. I swore Earth should be frying by now."[11]

★ ★ ★

Come On, Tell Us What You Really Think

Geophysicist Claude Allègre has ridiculed what he terms the "prophets of doom of global warming," calling them "the greenhouse-gas fanatics whose proclamations consist in denouncing man's role on the climate without doing anything about it except organizing conferences and preparing protocols that become dead letters."[7]

In an April 3, 2014, BBC TV interview, Lovelock came out swinging at his fellow environmentalists, accusing the 2014 UN IPCC report of plagiarizing his now retracted climate claims from his 2006 book *The Revenge of Gaia*. "The last IPCC report is very similar to the [now retracted] statements I made in my book about 8 years ago, called *The Revenge of Gaia*. It's almost as if they've copied it," Lovelock explained.[12]

BBC video interviewer Jeremy Paxman said, "Sure. But you then, after publishing these apocalyptic predictions, you then retracted them."

Lovelock answered, "Well, that's my privilege. You see, I'm an independent scientist. I'm not funded by some government department or commercial body or anything like that. If I make a mistake, then I can go public with it. And you have to, because it is only by making mistakes that you can move ahead." Lovelock dismissed any notion of "settled" science: "They all talk, they pass laws, they do things, as if they knew what was happening. I don't think anybody really knows what's happening. They just guess. And a whole group of them meet together and encourage each other's guesses."

Lovelock contended that the science supposedly justifying climate change was "overblown": "We haven't got the physics worked out yet," he said.[13] "It's very interesting because most of us in the, in what you might call the earth science game, climate science as well and I was one of them—have made quite a big mistake.... We all thought we knew how carbon dioxide in the air and climate were related."[14]

"CO_2 is going up, but nowhere near as fast as they thought it would. The computer models just weren't reliable. In fact, I'm not sure the whole thing isn't crazy, this climate change," Lovelock said in 2016.[15] When asked why he shifted his views on climate change in recent years, Lovelock responded, "I've grown up a bit since then."

Lovelock, who calls himself an "old-fashioned green,"[16] noted that climate activists were "scared stiff" of being exposed and warned that "the great climate science centers around the world are more than well aware how weak their science is."[17]

Others have joined Lovelock in re-examining the evidence and declaring their newfound dissent from climate change orthodoxy.

"The Evidence has switched from the mid-1980s to now. It's a much weaker case that there's human induced significant global warming," Daniel Botkin, professor emeritus at

★ ★ ★

You Gotta Have Faith

Renowned geophysicist and newly minted global warming skeptic James Lovelock described the modern green movement as "a religion, and religions don't worry too much about facts."[18]

the Department of Ecology, Evolution, and Marine Biology of the University of California, testified to Congress in 2014.[19]

"I have been concerned about global warming since 1968 and in the 1980s, it looked like the weight of evidence lent towards human induced climate change, to a significant extent, and since then it's moved against it," Botkin explained. "For example the temperature change is not tracking carbon dioxide very well. Then there is the information from the long term Antarctic ice core and some from a recent paper in the arctic, that suggest that carbon dioxide does not lead temperature change, it may actually lag it significantly or may not lead it at all, and if that is the case that is still an open but important scientific evidence. So there are several lines of evidence that are suggesting that it [anthropogenic global warming] is a weaker case today, not a stronger case."[20]

Meteorologist Hajo Smit of Holland, a former member of the Dutch UN IPCC committee, has reversed his belief in man-made warming and become a skeptic: "Gore prompted me to start delving into the science again and I quickly found myself solidly in the skeptic camp.... Climate models can at best be useful for explaining climate changes after the fact."[21]

Former UN IPCC lead author Richard Tol, a professor at the Vrije University in Amsterdam and "among the 25 most cited climate researchers

Telling It Like It Is

"Since 1968 I have published research on theoretical global warming, its potential ecological effects, and the implications for people and biodiversity. I have spent my career trying to help conserve our environment and its great diversity of species. In doing so I have always attempted to maintain an objective, intellectually honest, scientific approach in the best tradition of scientific endeavor. I have, accordingly, been dismayed and disappointed in recent years that this subject has been converted into a political and ideological debate.

"I want to state up front that we have been living through a warming trend driven by a variety of influences. However, it is my view that this is not unusual, and contrary to the characterizations by the IPCC and the National Climate Assessment, these environmental changes are not apocalyptic nor irreversible.

"The extreme overemphasis on human-induced global warming has taken our attention away from many environmental issues that used to be front and center but have been pretty much ignored in the 21st century."

—Univesity of California professor emeritus Daniel Botkin in testimony before Congress[22]

according to Google Scholar,"[23] testified at a 2014 Congressional hearing, "The climate has become a new religion, and that people who disagree would be treated as heretics."[24]

"I have been involved in the Intergovernmental Panel on Climate Change since 1994, serving in various roles in all three working groups, most recently as a Convening Lead Author in the economics chapter of Working Group II," Tol explained. But Tol had his name removed from the UN IPCC Fifth Assessment Summary for Policymakers report because of what he considered distortions of science for political reasons.[25] As he noted, the IPCC excludes scientists "because their views do not match those of their government." Plus, "There's a lot of people who didn't volunteer knowing full well what the IPCC is like. So it's not an open process at all." Tol cautions against UN IPCC media hype about warming: "Even if all the climate models were true, climate change is not an existential threat. We've been through much, much worse than the climate can throw at us."[26]

Tol has also said, "One of the startling facts about climate change is that there are very few facts about climate change. Climate change is mainly something of the future so we are really talking about model projections."[27]

More Skeptics on the Left

Ivar Giaever is a Nobel Prize–winning physicist who was one of President Obama's key scientific supporters in 2008. A former professor at the School of Engineering and School of Science Rensselaer Polytechnic Institute, he received the 1973 physics Nobel for his work on quantum tunneling. Giaever joined over seventy Nobel laureates in endorsing Barack Obama in an October 29, 2008 open letter, which read, in part, "The country urgently needs a visionary leader.... We are convinced that Senator Barack Obama is such a leader, and we urge you to join us in supporting him."[28]

But seven years after signing the letter, Giaever directly addressed the man he had campaigned for: "I say this to Obama: Excuse me, Mr. President, but you're wrong. Dead wrong." At the sixty-fifth Nobel Laureate Conference in Lindau, Germany, in 2015, which drew sixty-five recipients of the prize, Giaever called global warming "a non-problem." In his talk to the other laureates, titled "Global Warming Revisited," Giaever called the president's claim that "no challenge poses a greater threat to future generations than climate change" a "ridiculous statement.... How can he say that? I think Obama is a clever person, but he gets bad advice. Global warming is all wet." Echoing Lovelock and Tol, he said, "Global warming has become a new religion."

Giaever, who is originally from Norway, declared "I am a skeptic.... I don't see that CO_2 is the cause of all this problem."

He was embarrassed that Gore and UN IPCC chief Rajendra Pachauri had shared a Nobel prize in 2007: "These two people got the Nobel prize in peace, and I am ashamed of the Norwegian government who did that."

The Nobel physicist questioned the basis for rising carbon dioxide fears. "When you have a theory and the theory does not agree with the experiment then you have to cut out the theory. You were wrong with the theory," Giaever explained.

"The facts are that in the last 100 years we have measured the temperatures it has gone up .8 degrees and everything in the world has gotten better. So how can they say it's going to get worse when we have the evidence? We live longer, better health, and better everything. But if it goes up another .8 degrees we are going to die I guess," he noted.

"I would say that basically global warming is a non-problem. Just leave it alone and it will take care of itself. It is almost very hard for me to understand why almost every government in Europe—except for the Polish government—is worried about global warming. It must be politics."

Giaever pointed out that science is not by majority rule: "We frequently hear about the number of scientists who support it. But the number is not important: only whether they are correct is important."

Giaever resigned from the American Physical Society to protest the group's promotion of man-made climate change. "I resigned from the American Physical Society because of this statement. 'The evidence is incontrovertible.' That's religion, That's a religious statement.... The temperature has been amazingly stable, and both human health and happiness have definitely improved in this 'warming' period," Giaever said.[29]

"We have to stop wasting huge, I mean huge amounts of money on global warming," he added.

Robert Giegengack, former chair of the Department of Earth and Environmental Science at the University of Pennsylvania, publicly announced that he had voted for Gore in 2000 and said he could do so again: "I voted for Gore in 2000, yeah. I think that if he ran again, depending on who he ran against, I might vote for him. He's a smart man."

But he is not a fan of Gore's film. After viewing *An Inconvenient Truth*, Giegengack said, "I was appalled. I was appalled because he either deliberately misrepresented the point he was making or didn't understand it. So it was irresponsible of Al Gore." As we have already seen from the interview of Giegengack I did for my film *Climate Hustle*, the Ivy League scientist sees nothing in the geological record to justify the belief that CO_2 controls the climate.

"It was too bad, because I thought Al Gore had a brilliant opportunity that he blew it," Giegengack said. "He could have done us all a real service, and what he chose to do instead was to polarize the argument. And I think the polarity in the argument really began with the Al Gore film," he added. "I think he was a true-believer, he was a zealot. And he disappointed me because he did not give his audience credit for enough intelligence."[30]

Renowned Princeton University physicist Freeman Dyson, who has been hailed as "Einstein's successor,"[31] is another scientific dissenter on the political Left. "'I'm 100% Democrat and I like Obama. But he took the wrong side on climate issue, and the Republicans took the right side," Dyson explained in 2015. He called the UN climate pact "pointless" and explained, "pollution is quite separate to the climate problem: one can be solved, and the other cannot, and the public doesn't understand that."[32]

According to Dyson, "The effects of CO_2 on climate are really very poorly understood. The experts all seem to think they understand it, I don't think they do.... I like carbon dioxide, it's very good for plants. It's good for the vegetation, the farms, essentially carbon dioxide is vital for food production, vital for wildlife. It would be crazy to try to reduce CO_2. Earth is growing greener as a result of carbon dioxide."[33]

"I just think they don't understand the climate," he said, referring to climatologists. "Their computer models are full of fudge factors."[34]

Who Will Guard the Guardians?

"Climate is a very complicated story. And we may or may not understand it better (in the future). The main thing that is lacking at the moment is humility. The climate experts have set themselves up as being the guardians of the truth and they think they have the truth and that is a dangerous situation."

—**Freeman Dyson, retired professor of physics at the Institute for Advanced Study in Princeton**[37]

"I have strong views about climate because I think the majority is badly wrong," he said.

"I think the notion that I always like to oppose the consensus in science is totally wrong. The fact is there's only one subject that I've been controversial, which is climate," he added. Dyson explained that climate change is "the only field in which I'm opposed to the majority. Generally speaking, I'm much more of a conformist, but it happens I have strong views about climate because I think the majority is badly wrong, and you have to make sure if the majority is saying something that they're not talking nonsense."[35]

Like many of his fellow skeptics on the Left, Dyson sees the faith in climate change as a religion: "There certainly is an enormous religion in which there are lots of true believers who think that climate change is evil and that we're going to run into big catastrophes if we don't do something drastic. That's a sort of belief system which exists."[36]

From True Believers to Heretics

Other prominent scientists have also reviewed the evidence and now dissent from the climate change narrative.

"My position on this has evolved over time," explained climatologist Judith Curry, the former chair of the School of Earth and Atmospheric Sciences at the Georgia Institute of Technology.[38]

Judith Curry was branded a "heretic" by *Scientific American* for reversing her position on global warming."[39] She reconsidered man-made climate change in late 2009 when the emails of top UN scientists were leaked during the Climategate scandal. "Climategate and the weak response of the IPCC and other scientists triggered a massive re-examination of my support of the IPCC, and made me look at the science much more skeptically," Curry explained, adding she felt she had been "duped" by the IPCC.

The 2009 Climategate scandal, in which emails from the Climatic Research Institute at the University of East Anglia were hacked or leaked, revealed that the upper echelon of UN scientists were colluding with each other to suppress studies and data that did not support the carefully crafted narrative on global warming that the UN was pushing on the public—as we shall see in more detail in the next chapter. Lead climate scientists had threatened journal editors if they published studies that did not fit the climate change narrative.

"I started saying that scientists should be more accountable, and I began to engage with skeptic bloggers. I thought that would calm the waters. Instead I was tossed out of the tribe," Curry explained. "There's no way I would have done this if I hadn't been a tenured professor, fairly near the end of my career. If I were seeking a new job in the US academy, I'd be pretty much unemployable."[40]

Curry explained, "By the time you get to late spring of 2010 I had been ostracized by the mainstream of 'consensus' and had really been pushed over to the other side if you will by attacks."[41]

In 2013, Curry called for disbanding the UN IPCC. "'Given the widespread nature of the infection and intrinsically motivated reasoning. We

★ ★ ★
Science or Religion?

The November 2010 issue of *Scientific American* featured an article revealing how Curry was shifting her views on climate change. The article was titled "Climate Heretic: Judith Curry Turns on Her Colleagues."[42]

"So I responded," said Curry, "with a rather blistering blog post. And the punchline of that is that, 'If the IPCC is dogma, then count me in as a heretic.... I was sort of booted out of the tribe, if you will.'"[43]

need to put down the IPCC as soon as possible—not to protect the patient who seems to be thriving in its own little cocoon, but for the sake of the rest of us whom it is trying to infect with its disease," she wrote. "There is a growing realization that you can't control climate by emissions reductions," she added.[44]

According to Curry, "Not only is more research needed to clarify the sensitivity of climate to carbon dioxide and understand the limitations of climate models, but more research is needed on solar variability, sun-climate connections, natural internal climate variability and the climate dynamics of extreme weather events."[45]

We have already met one of the highest-profile environmentalists to speak out skeptically on global warming, Greenpeace co-founder Patrick Moore. Moore, an ecologist, explained in testimony in the U.S. Senate in 2014 why he was originally involved in Greenpeace and why he ultimately left.

"For Greenpeace, which I was involved in the beginning of, it was the threat of all out nuclear war. We cared about humans because our focus was to stop nuclear war and the destruction of human civilization. That's the peace in 'Greenpeace.' The green part, of course, is nature, and over the years, gradually, Greenpeace lost the peace part and drifted into a position of depicting humans as the enemies of the earth," Moore said. "And in 1986, I left to become an independent environmentalist basing my positions on science and logic rather than sensationalism, misinformation, and fear."[46]

Moore explained what he thinks is behind the push for global warming: "It is a powerful convergence of interests among a very large number of elites, including politicians who want to make it seem as though they're saving the world, environmentalists who want to raise money and get control over very large issues like our entire energy policy, media, for sensationalism, Universities and professors for grants. You can hardly get a science grant these days without saying it has something to do with climate change."[47]

"People need to study the long history of the Earth. These true believers in climate change are only looking at the last 100 years. That is a blink in nature's eye. We've had billions of years of climate history in this world, and if you look at just the half billion of them, the most recent half billion, you will see that CO_2 is lower now than it has been through most of the history of life on Earth, and so is the temperature," he added.

Another former Greenpeace member, Finnish scientist Jarl R. Ahlbeck of Åbo Akademi University, who has authored two hundred scientific publications, is also a climate dissenter. "So far, real measurements give no ground for concern about a catastrophic future warming," he explained.[48]

Yet another left-of-center scientist to bail out of the global warming movement is physicist Denis Rancourt, a former professor and environmental science researcher at the University of Ottawa. Rancourt has declared global warming a "corrupt social phenomenon. Strictly an imaginary problem of the 1st World middle-class. It is as much psychological and social phenomenon as anything else."[49]

Rancourt, who has authored over one hundred articles in scientific journals, argues, "that by far the most destructive force on the planet is power-driven financiers and profit-driven corporations and their cartels backed by military might; and that the global warming myth is a red herring that contributes to hiding this truth. In my opinion, activists who, using any justification, feed the global warming myth have effectively been co-opted, or at best neutralized."

Rancourt's dissent on man-made climate fears does not sit well with many of his green friends: "When I tell environmental activists that global warming is not something to be concerned about, they attack me—they shun me, they do not allow me to have my materials published in their magazines."[50]

Rancourt's explanation of why his fellow environmentalists are wrapped up in promoting climate alarm is blunt. "They look for comfortable lies that they can settle into and alleviate the guilt they feel about being on the privileged end of the planet—a kind of survivor's guilt. A lot of these environmentalists are guilt laden individuals who need to alleviate the guilt without taking risks," he said. "The modern environmental movement has hijacked itself by looking for an excuse to stay comfortable and stay away from actual battle. Ward Churchill has called this pacifism as pathology. If you are really concerned about saving world's forests or habitat destruction, then fight against habitat destruction, don't go off in tenuous thing about CO_2 concentration in the atmosphere. Actually address the question; otherwise you are weakening your effect as an activist."[51]

"Climate change 'science' is part of just another screw-the-brown-people scam," he explained. "Carbon trading will be the largest financial extortion enterprise.... The whole climate change scam is now driven by the top-level financiers newly eyeing a multi-trillion-dollar paper economy of carbon trading and that this is the reason it's now a dominant mainstream media and corporate messaging presence."[52]

Climate statistics professor Caleb Rossiter of American University is an outspoken anti-war activist with a flawless progressive record on a range of political issues. He is a former Democratic congressional candidate who campaigned against U.S.-backed wars in Central America and Southern Africa.

"I've spent my life on the foreign-policy left. I opposed the Vietnam War, U.S. intervention in Central America in the 1980s and our invasion of Iraq. I have headed a group trying to block U.S. arms and training for 'friendly' dictators, and I have written books about how U.S. policy in the developing

★ ★ ★

Rotten Tomatoes

Al Gore's film made physicist Denis Rancourt sick: "I felt ill walking out of the theatre. It's terrible. It does not respect the intelligence of the viewer. The film does not acknowledge people can think for themselves at all." Gore, he said, "strikes me as someone working for someone—as someone who will financially benefit from this. He does not give me [the] impression of someone who genuinely cares about environmental or social justice."[53]

world is neocolonial," he said. Rossiter is also outspoken about being a global warming skeptic.

As he told me in 2014, "I would say since 2004 I've been very lonely, Marc. I've been lonely working on the Hill for the Democratic Party."[54]

"My blood simply boils too hot when I read the blather, daily, about climate catastrophe. It is so well-meaning, and so misguided," Rossiter, an adjunct professor in American University's Department of Mathematics and Statistics, explained.[55] He is particularly frustrated by the impact on the world's poor. "Climate Justice in limiting carbon dioxide emissions is a crime against Africa, and it's what motivated me to get involved again in this debate," he said.

"I have assigned hundreds of climate articles as I taught and learned about the physics of climate, the construction of climate models, and the statistical evidence of extreme weather. I started to suspect that the climate-change data were dubious a decade ago while teaching statistics. Computer models used by the UN Intergovernmental Panel on Climate Change to determine the cause of the six-tenths of one degree Fahrenheit rise in global temperature from 1980 to 2000 could not statistically separate fossil-fueled and natural trends."[56]

Rossiter dismisses CO_2 as the climate control knob. "We always, as humans, are looking for cause-and-effect, but it's extremely difficult to find it in a complex system like the Earth's climate over thousands of years," he explained. "For the IPCC to say nothing else can explain [global warming except mankind's CO_2] is the opposite of what we do in science. We are trying to test the known hypothesis that there is no effect to anthropogenic warming. And in order to do that, you have to have data that removes all the other causes—factor out all the other elements, and isolate yours. It is simply not true that you can only model how temperature has changed from 1850 to today using a doubling of carbon dioxide levels. I can model it for you with baseball statistics from that same period if you give me enough time to scrub the models."

Rossiter's failure to follow his colleagues on the Left on the claims of global warming has isolated him. "What we are supposed to do as professors is follow the data to our conclusion, and then put it out there to be debated," Rossiter explained. But his colleagues refuse to debate global warming. "I have invited the Union of concerned scientists, Greenpeace, Institute for policy studies, random members of Congress who I knew from when I worked up there on the Hill, to come to my classes at A.U. to debate—they simply refused," he said. "There was an agreement among the groups who believe strongly that there's catastrophic climate change not to debate because it gives credit to those of us who have questions about the certainty with which they operate."

Rossiter dismissed Gore's winning of the Nobel Peace Prize as the "worst Nobel Prize for peace since Henry Kissinger." Rossiter chastised his colleagues on the political Left for "hopping into bed" with Gore when it comes to climate change. "I know why the Left is supporting Al Gore on this when they didn't on anything else, it's because it gives them the lever to move away from an industrial society to what they call a postindustrial society," he said. Rossiter says the political Left in the United States is using climate fears to achieve a "welcome license to dismember the carbon-driven capitalism."[57]

Rossiter attempted to convince the liberal think tank Institute for Policy Studies (IPS) to allow an open debate on climate change.

"I wrote a very long memo," he said explaining to the director that "we need to stop and look at the data, I want to have a debate with the staff and the board. And he said, 'No, we know your views Caleb.'...So it was because I could not reach the board and them directly, that I wrote the piece for the *Wall Street Journal*."[58]

When Rossiter called global warming "unproved science" in a *Wall Street Journal* op-ed, he found that his credentials as a long-time progressive could not protect him from the consequences of his climate skepticism.[59]

"Two days later, I was handed my walking papers from a twenty-three-years association with that think-tank," he explained. "They felt it was best that I be terminated because my views on African Development, and Climate Change, and Climate Justice were divergent from theirs."

The IPS email to Rossiter explained, "We would like to inform you that we are terminating your position as an Associate Fellow of the Institute for Policy Studies.... Unfortunately, we now feel that your views on key issues, including climate science, climate justice, and many aspects of U.S. policy to Africa, diverge so significantly from ours."

Many other politically left-of-center scientists are converts to skepticism.

Philip Stott, a professor emeritus of biogeography at London's School of Oriental and African Studies has explained, "I come from the left wing politically. I am fed up with environmentalists putting regressive costs and taxes on the poor."

He points out that global warming is "actually not very much about the science. It's always been about economic and political choice." According to Stott, "Climate science and these costs are sub-prime science, subprime economics and above all subprime politics." He urges, "Let's give this global warming nonsense its Waterloo."[60]

According to Stott, "The fundamental point has ever been this: climate change is governed by hundreds of factors. The very idea that we can manage climate change predictably by understanding and manipulating at the margins just one politically-selected factor is about as bonkers as it gets. How on Earth have folk been conned into believing such hubris? It is so like The Prophecies by Nostradamus—the vagueness and lack of dating make it easy to quote 'evidence' selectively after every major dramatic event, and retrospectively claim them as a 'hit'!"[61]

Stott has described in vivid terms how futile attempting to control the climate would be. "I want you to think of the world. I want you to think of the world from inner Siberia, to Greenland, then to Singapore, and then come to the Arab states and to Sahara. Ladies and gentlemen, in the temperature range I have just covered, it is from minus 20 degrees C, to nearly

50 degrees C, a range of 70 degrees C, in which humanity has adapted and learnt to live. We are talking about…a prediction of 2 to 3 degrees C, what a funk!" Stott explained. "Humanity lives successfully from Greenland to Singapore to Saudi Arabia. 70 degrees C. And what is more, the carbon reductions will not produce an outcome that is predictable."

He contended "I would love to be able to think we can control climate, when of course it is indeed going to have to be adaptation, flexibility put to an outcome that we don't know 'cause I actually don't know what climate they're wanting to produce for us. And actually I don't think they know either," Stott said.[62]

As Stott pointed out, "The global warming 'crisis' is misguided. In hubristically seeking to 'control' climate, we foolishly abandon age-old adaptations to inexorable change. There is no way we can predictably manage this most complex of coupled, nonlinear chaotic systems. The inconvenient truth is that 'doing something' (emitting gases) at the margins and 'not doing something' (not emitting gases) are equally unpredictable." He concluded, "We can no longer afford to cling to the anti-human doctrines of outdated environmentalist thinking. The 'crisis' is the global warming political agenda, not climate change."[63]

Martin Hertzberg, a retired Navy meteorologist with a Ph.D. in physical chemistry, has also diverged from his progressive colleagues on man-made climate change. "As a scientist and life-long liberal Democrat, I find the constant regurgitation of the anecdotal, fear mongering clap-trap about human-caused global warming to be a disservice to science," Hertzberg wrote.[64]

The stream of scientists from around the world who publicly dissent on man-made global warming claims continues to swell.

EPA award–winning geologist Leighton Steward has reversed his views on climate change over the past decade. "I started to write a book on the humongous impact that Earth's old CO_2 levels used to have," he said. "After about 4 months, I said, 'Leighton, you are the dumbest researcher on earth.

You're not finding any evidence of the humongous impact that CO_2 had on Earth's whole climate because the correlations were not that good.'" He now regrets that the public is not getting the full story about climate change: "I think it is a damn catastrophe that people are being misled on this issue."[65]

Environmental physicist Jean-Louis Pinault has now publicly joined the dissenters. "This is a very uneven debate, skeptics cannot enforce their arguments in scientific journals that are subject to censorship," he said in 2014. Pinault pointed out that global warming concerns have produced an "economic and political media frenzy unprecedented in the history of science."[66]

Pinault lamented the current state of peer review. "I resigned myself to suspend submitting my work in peer-reviewed scientific journals, giving priority to dissemination of results."

German meteorologist and physicist Klaus-Eckart Puls reversed his belief in man-made global warming and now calls the idea that CO_2 can regulate climate "sheer absurdity." He explained, "Ten years ago I simply parroted what the IPCC told us. One day I started checking the facts and data—first I started with a sense of doubt but then I became outraged when I discovered that much of what the IPCC and the media were telling us was sheer nonsense and was not even supported by any scientific facts and measurements. To this day I still feel shame that as a scientist I made presentations of their science without first checking it," Puls explained in 2012.[67]

Geoscientst and former UN consultant David Kear has also spoken out. In 2014, Kear said global warming fears are "based on unfounded unscientific beliefs" and lamented the "astronomical cost" of proposed measures to "combat a non-existent threat."[68]

Former NASA Scientist Les Woodcock now "laughs" at man-made climate change claims. Woodcock, professor emeritus of Chemical Thermodynamics at the University of Manchester, is a former NASA researcher. Woodcock,

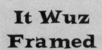

It Wuz Framed

"Perhaps the saddest part has been that the essential and innocent gas, carbon dioxide, has been demonized and criminalized."

—**former UN consultant David Kear**[69]

who has published more than seventy journal papers, declared there has been "professional misconduct by Government advisors around the world" when it comes to man-made climate change claims. "The theory of 'man-made climate change' is an unsubstantiated hypothesis—water is a much more powerful greenhouse gas and there is 20 time[s] more of it in our atmosphere (around one per cent of the atmosphere) whereas CO_2 is only 0.04%—Carbon dioxide has been made out to be some kind of toxic gas but the truth is it's the gas of life. We breath [sic] it out, plants breath [sic] it in. The green lobby has created a do-good industry and it becomes a way of life, like a religion," he explained in 2014. "The temperature of the earth has been going up and down for millions of years, if there are extremes, it's nothing to do with carbon dioxide in the atmosphere, it's not permanent and it's not caused by us. Global warming is nonsense. If you talk to real scientists who have no political interest, they will tell you there is nothing in global warming. It's an industry which creates vast amounts of money for some people."[70]

Woodcock believes the agenda of money and politics is driving climate change fears: "This is not the way science works. If you tell me that you have a theory there is a teapot in orbit between the earth and the moon, it's not up to me to prove it does not exist, it's up to you to provide the reproducible scientific evidence for your theory. Such evidence for the man-made climate change theory has not been forthcoming."

Noted theoretical physicist Steven E. Koonin, who served as undersecretary for science at the Department of Energy during Obama's first term, declared his climate dissent in 2014: "The Obama administration relentlessly politicized science and it aggressively pushed a campaign about that politicized science." According to Koonin, a "happily complicit" media never challenged the climate claims.[71]

Koonin accused the administration he had worked for of manipulating climate data. "What you saw coming out of the press releases about climate data, climate analysis, was, I'd say, misleading, sometimes just wrong," the physicist said.[72]

"The public is largely unaware of the intense debates within climate science. At a recent national laboratory meeting, I observed more than 100 active government and university researchers challenge one another as they strove to separate human impacts from the climate's natural variability," Koonin, currently the director of the Center for Urban Science and Progress at New York University, wrote. "At issue were not nuances but fundamental aspects of our understanding, such as the apparent—and unexpected—slowing of global sea-level rise over the past two decades. Summaries of scientific assessments meant to inform decision makers, such as the United Nations' Summary for Policymakers, largely fail to capture this vibrant and developing science," he added.[73]

★ ★ ★

The Ranks of the Dissenters

Below is a small sample of the dissenting opinions of scientists featured in the reports I authored for the U.S. Senate Environment and Public Works Committee and later for the Climate Depot website. The first report for the U.S. Senate, published in 2007, included over four hundred dissenting scientists. It was updated with hundreds more scientists in 2008 and 2009.[74] In 2010, the number of dissenting scientists exceeded one thousand.[75]

"Any reasonable scientific analysis must conclude the basic theory wrong!!"

—**Leonard Weinstein, who worked thirty-five years at the NASA Langley Research Center,** finished his career there as a **Senior Research Scientist, and is presently a Senior Research Fellow at the National Institute of Aerospace**

"Please remain calm: The Earth will heal itself—climate is beyond our power to control.... Earth doesn't care about governments or their legislation. You can't find much actual global warming in present-day weather observations. Climate change is a matter of geologic time, something that the earth routinely does on its own without asking anyone's permission or explaining itself."

—Stanford University physicist Robert B. Laughlin, who won the Nobel Prize for physics in 1998, and was formerly a research scientist at Lawrence Livermore National Laboratory

"Hundreds of billion dollars have been wasted with the attempt of imposing a Anthropogenic Global Warming (AGW) theory that is not supported by physical world evidences.... AGW has been forcefully imposed by means of a barrage of scare stories and indoctrination that begins in the elementary school textbooks."

—Brazilian geologist Geraldo Luís Lino, who authored the 2009 book *The Global Warming Fraud: How a Natural Phenomenon Was Converted into a False World Emergency*

"I am an environmentalist," but "I must disagree with Mr. Gore."

—Chemistry professor Dr. Mary Mumper, chair of the Chemistry Department at Frostburg State University in Maryland, during a presentation titled "Anthropogenic Carbon Dioxide and Global Warming, the Skeptic's View"

"I am ashamed of what climate science has become today." The science "community is relying on an inadequate model to blame CO_2 and innocent citizens for global warming in order to generate funding and to gain attention. If this is what 'science' has become today, I, as a scientist, am ashamed."

—Research chemist William C. Gilbert, author of "The Thermodynamic Relationship between Surface Temperature and Water Vapor Concentration in the Troposphere" published in 2010 in the journal *Energy & Environment*

"The dysfunctional nature of the climate sciences is nothing short of a scandal. Science is too important for our society to be misused in the way it has been done within the Climate Science Community." The global warming establishment "has actively suppressed research results presented by researchers that do not comply with the dogma of the IPCC."

—Swedish climatologist Dr. Hans Jelbring of the Paleogeophysics & Geodynamics Unit at Stockholm University

"Those who call themselves 'Green planet advocates' should be arguing for a CO_2-fertilized atmosphere, not a CO_2-starved atmosphere.... Diversity increases when the planet was warm AND had high CO_2 atmospheric content.... Al Gore's personal behavior supports a green planet—his enormous energy use with his 4 homes and his bizjet, does indeed help make the planet greener. Kudos, Al for doing your part to save the planet."

> —Renowned engineer and aviation and space pioneer Burt Rutan, who was named one of the "100 most influential people in the world, 2004" by *Time* magazine and called "the man responsible for more innovations in modern aviation than any living engineer" by *Newsweek*

"Global warming is the central tenet of this new belief system in much the same way that the Resurrection is the central tenet of Christianity. Al Gore has taken a role corresponding to that of St Paul in proselytizing the new faith.... My skepticism about AGW arises from the fact that as a physicist who has worked in closely related areas, I know how poor the underlying science is. In effect the scientific method has been abandoned in this field."

> —Atmospheric physicist John Reid, who worked with Australia's CSIRO's (Commonwealth Scientific and Industrial Research Organization) Division of Oceanography

"The IPCC has actually become a closed circuit; it doesn't listen to others. It doesn't have open minds.... I am really amazed that the Nobel Peace Prize has been given on scientifically incorrect conclusions by people who are not geologists."

> —Indian geologist Dr. Arun D. Ahluwalia of Punjab University, a board member of the UN-supported International Year of the Planet

"Anyone who claims that the debate is over and the conclusions are firm has a fundamentally unscientific approach to one of the most momentous issues of our time."

> —Solar physicist Dr. Pål Brekke, senior advisor to the Norwegian Space Centre in Oslo, who has published more than forty peer-reviewed scientific articles on the sun and solar interaction with the earth

The models and forecasts of the UN IPCC "are incorrect because they only are based on mathematical models and presented results at scenarios that do not include, for example, solar activity."

> —Victor Manuel Velasco Herrera, a researcher at the Institute of Geophysics of the National Autonomous University of Mexico

"It is a blatant lie put forth in the media that makes it seem there is only a fringe of scientists who don't buy into anthropogenic global warming."

> —Atmospheric scientist Stanley B. Goldenberg of the Hurricane Research Division of NOAA

CHAPTER 10

Climategate:
The UN IPCC Exposed

"I view Climategate as science fraud, pure and simple." That's Princeton physicist Robert Austin's take on the scandal that exposed the very unscientific conduct of UN IPCC scientists.[1]

But what the hacked emails from the University of East Anglia's Climate Research Institute revealed was more than just a shocking case of dishonesty in science. It was the fraudulence of the entire man-made climate change narrative. The Climategate emails showed that UN IPCC scientists were holding together the global warming narrative and the supposed scientific "consensus" that supported it by subterfuge and intimidation. The Climategate scandal opened a lot of eyes to the fact that the UN's Intergovernmental Panel on Climate Change was more political than scientific.

The Climategate scandal pulled back the curtain on the upper echelon of UN IPCC scientists, who were caught artificially propping up the climate change narrative via a partisan campaign to boost only the science and scientists that support their cause and exclude science and scientists that don't fit. Data manipulation, manipulation of the peer-review process, blacklisting, data destruction, and willful violation of Freedom of Information Act requests were some of the key revelations in the Climategate emails.

Did you know?

★ Leading UN IPCC scientists were caught manipulating the peer review process to create an artificial "consensus"

★ Penn State's "investigation" into Michael Mann's Climategate emails asked Mann to supply the evidence against himself

★ One of the Royal Society's "investigators" was a Mann co-author

CBS News reported on the Climategate scandal in December of 2009: "Those files show that prominent scientists were so wedded to theories of man-made global warming that they ridiculed dissenters who asked for copies of their data, plotted how to keep researchers who reached different conclusions from publishing, and discussed how to conceal apparently buggy computer code from being disclosed under the Freedom of Information law."[2]

When NBC News reported on "A scandal called 'Climategate'" in 2009, it was introduced as "a scandal involving some stolen emails." NBC noted that "the language in the emails suggest these scientists manipulated their findings."[3]

The thousands of emails, either hacked or more likely leaked from the Climate Research Unit at the University of East Anglia, revealed the behind-the-scenes collusion of the climate change leadership. The leading UN IPCC scientists were caught red-handed artificially manufacturing the "scientific consensus" for the global warming narrative. Their own words betrayed that they were acting like political partisans, not scientists—crafting a predetermined message rather than following the evidence. Climategate exposed the work product of the IPCC as the best science that politics and activism could manufacture.

Emails between Climategate scientists showed a concerted effort to hide rather than disseminate underlying evidence and procedures.

As Forbes reported on the emails released in both the original 2009 Climategate scandal and a second release in 2011 dubbed "Climategate 2.0," "'I've been told that IPCC is above national FOI [Freedom of Information] Acts. One way to cover yourself and all those working in AR5 would be to delete all emails at the end of the process,' writes Phil Jones, a scientist working with the United Nations Intergovernmental Panel on Climate Change (IPCC), in a newly released email. 'Any work we have done in the past is done on the back of the research grants we get—and has to be well hidden,' Jones writes in another email. 'I've discussed this with the main funder (U.S. Dept of Energy) in the past and they are happy about not releasing the original station data.... ""[4]

Chris Horner, author of the 2007 *Politically Incorrect Guide® to Global Warming*, reported on the efforts to delete correspondence by Climategate scientists. "Phil Jones in the United Kingdom asked Mann, now at Penn State, by email to delete records being sought under the UK's Freedom of Information Act, and to get a colleague to do so as well," Horner explained in 2011.[5]

Jones had emailed,

> Mike:
>
> Can you delete any emails you may have had with Keith [Briffa] re AR4 [UN IPCC 4th Assessment]? Keith will do likewise. He's not in at the moment—minor family crisis. Can you also email Gene and get him to do the same? I don't have his new email address. We will be getting Caspar [Ammann] to do likewise.

"'Gene' is Eugene Wahl, who now works for the federal government," explained Horner. Mann's terse reply included in pertinent part: "I'll contact Gene about this ASAP."

According to Wahl, Mann did contact him. "For the record, while I received the email from CRU [Phil Jones Climatic Research Unit] as forwarded by Dr. Mann, the forwarded message came without any additional comment from Dr. Mann; there was no request from him to delete emails," Wahl explained in 2011.[6]

The *Telegraph* reported that CRU director Jones was "accused of making error of judgment by colleague" Mann for asking their colleagues to "delete sensitive emails to evade Freedom of Information requests." Mann tried to distance himself from Jones, "I can't justify the action, I can only speculate that he was feeling so under attack that he made some poor decisions frankly and I think that's clear."[7] Jones retired in 2016.

A *Washington Post* editorial on November 25, 2009, summed up the unfolding scandal:

> According to one of the stolen e-mails, CRU [Climate Research Unit] Director Phil Jones wrote that he would keep papers questioning the connection between warming and human activity out of the authoritative Intergovernmental Panel on Climate Change report "even if we have to redefine what the peer-review literature is!" In another, Mr. Jones and Pennsylvania State University's Michael E. Mann write about an academic journal and its editor, with Mr. Mann discussing organizing a boycott of the publication and Mr. Jones saying, "I will be emailing the journal to tell them I'm having nothing more to do with it until they rid themselves of this troublesome editor." Other e-mails speak of withholding data from climate-change skeptics.... Climate scientists should not let themselves be goaded by the irresponsibility of the deniers into overstating the certainties of complex science or, worse, censoring discussion of them.[8]

Mann joined Jones in planning to punish a scientific journal that he did not consider faithful to the climate narrative: "I think we have to stop considering Climate Research as a legitimate peer-reviewed journal. Perhaps we should encourage our colleagues in the climate research community to no longer submit to, or cite papers in, this journal."[9]

Climate blogger Tom Nelson dug through and collected a slew of the Climategate emails on his website:[10]

- Email 1819, Nov 2003, warmist Tom Wigley to Mann et al on possible responses to McIntyre and McKitrick's request for data: "The second is to tell them to go to hell"

- Email 4868, Sept '05: IPCC reviewer McIntyre asks to see the data underlying a paper; warmists complain this is a "major abuse of his position"
- Email 1897, Dec 2008: After Phil Jones admits deleting material, UEA's FOI officer David Palmer writes: "Phil, you must be very careful about deleting material, more particularly when you delete it"
- 2000: Warmist Phil Jones goes to "solar variability and climate" conference in Tenerife; finds that "Many in the solar terrestrial physics community seem totally convinced that solar output changes can explain most of the observed changes we are seeing"; laments that THEY are "so set in their ways"
- Email 4657, Oct 2000, It's a small world after all: Editor of Journal of Climate, Michael Mann, gets Phil Jones to review a paper by Tom Wigley and Ben Santer
- 2004 email: Phil Jones on why he thought the last 20 years was warmer than the Medieval Warm Period: "This is all gut feeling, no science"; warmist Tom Wigley also calls the hockey stick "a very sloppy piece of work"

Climategate exposed the manufactured consensus and gave the lie to the endlessly repeated mantra that all scientists agree on anthropogenic global warming.

Breaking Ranks

Climate skeptics hailed the release of the emails as a victory for science. But even more significant, Climategate ultimately prompted UN scientists to turn on UN scientists, and on the UN IPCC process.

How the Global Warming Narrative Undermines Genuine Scientific Research (an Insider Explains)

"In this atmosphere, Ph.D. students are often tempted to tweak their data so as to fit the 'politically correct picture'. Some, or many issues, about climate change are still not well known. Policy makers should be aware of the attempts to hide these uncertainties under a unified picture. I had the 'pleasure' to experience all this in my area of research."

—Eduardo Zorita, UN IPCC contributing author[12]

UN IPCC scientist Eduardo Zorita, for example, publicly declared that his colleagues Michael Mann and Phil Jones, who had both been implicated in Climategate, "should be barred from the IPCC process.... They are not credible anymore."[11]

Zorita also noted how petty and punitive the global warming science had become: "By writing these lines I will just probably achieve that a few of my future studies will, again, not see the light of publication." Zorita was making reference to Climategate emails in which IPCC scientists had discussed how to suppress data and scientific studies that did not agree with the UN IPCC line. He noted how scientists who deviated from the UN IPCC's position were "bullied and subtly blackmailed."

Zorita was a contributing author to the UN IPCC Fourth Assessment Report in 2007. He has published more than seventy peer-reviewed scientific studies.

We have already met UN lead author Richard Tol, now a dissenter. In the wake of Climategate he lamented that the IPCC had been "captured" and demanded that "the Chair of IPCC and the Chairs of the IPCC Working Groups should be removed."[13] Despite the fact that Tol publicly called to "suspend" the IPCC process in 2010, he once again served as lead author for the Fifth Assessment Report.[14] Over subsequent years, Tol grew even more disillusioned with the UN and appeared in my 2016 film *Climate Hustle*.

Another scientist suggested disbanding the United Nations climate panel altogether. Mike Hulme, Professor of Climate Change at the University of East Anglia, which was ground zero of the Climategate scandal, suggested that the UN IPCC had "run its course." He complained about its "tendency to politicize climate science" and suggested that it had "perhaps helped to foster a more authoritarian, exclusive form of knowledge production."[15]

Hulme warned, "It is possible that climate science has become too partisan, too centralized. The tribalism that some of the leaked emails display is something more usually associated with social organization within primitive cultures; it is not attractive when we find it at work inside science."

Pat Michaels, a climate scientist and IPCC reviewer, commented, "This is what everyone feared. Over the years, it has become increasingly difficult for anyone who does not view global warming as an end-of-the-world issue to publish papers. This isn't questionable practice, this is unethical."[16]

Yet another UN IPCC reviewer, Vincent Gray, declared in November 2009, "I long ago realized that they were faking the whole exercise."[17]

Other UN scientists were even more blunt. Will Alexander, professor emeritus at the Department of Civil and Biosystems Engineering at the University of Pretoria in South Africa and a former member of the UN Scientific and Technical Committee on Natural Disasters, called the UN IPCC a "worthless carcass" and then–IPCC chair Rajendra Pachauri a "disgrace." He complained of the IPCC's "deliberate manipulation to suit political objectives" and "fraudulent science" that "continue[d] to be exposed" and explained, "I was subjected to vilification tactics.... I persisted. Now, at long last, my persistence has been rewarded.... There is no believable evidence to support [the IPCC] claims. I rest my case!"[18]

Geologist Don Easterbrook, a professor at Western Washington University, summed up the scandal: "The corruption within the IPCC

★ ★ ★

Now We Can See the Force of Your Argument

Warmists' tendency to resort to insults in the climate debate suggests that they may not have scientific evidence and rational arguments to back up their position.

On December 4, 2009, at the height of the Climategate scandal, I appeared on BBC TV—which described me as "one of America's leading climate change skeptics"—to debate Andrew Watson, professor at the School of Environmental Sciences at the University of East Anglia, whose emails appear in the Climategate files. Watson, bent on defending his colleagues, was the climate activist who called me an asshole on live television. As CBS News reported, "Professor Andrew Watson of the University of East Anglia in eastern England. It didn't take long before the two got in each other's face and Watson became increasingly annoyed with Morano's loud interruptions. He finally lost it by the end when the anchor thanked the participants. 'What an asshole,' Watson said."[22]

His remark prompted an on-air apology to viewers from the BBC for the offensive language.

During the live debate, I charged Professor Watson with being in "denial" over the importance of Climategate and noted that "you have to feel sorry for Professor Watson." I explained that Professor Watson's "colleague, Mike Hulme at the University of East Anglia is saying this is authoritarian science, he is suggesting the IPCC should be disbanded based on what Climategate reveals."

A clearly agitated Watson, whose university was at the epicenter of the Climategate scandal, blurted out, "Will you shut up just a second!?" right before dropping the A bomb on me. He later apologized to me via email.

I myself was actually mentioned in one of the Climategate scandal emails. On July 23, 2009, AP reporter Seth Borenstein had emailed one of the Climategate scientists, Penn State professor Michael Mann of hockey stick fame, about a "a paper in JGR [*Journal of Geophysical Research*] today that Marc Morano is hyping wildly." Mann wrote back to Borenstein, "The aptly named Marc 'Morano' has fallen for it!"

As Breitbart News reported, "Borenstein's email is hardly a neutral 'standard step for journalists.' Borenstein criticizes Marc Morano, a critic of man-made global warming claims, of 'hyping wildly' the study that Borenstein was asking for comments on. The email looks as if Borenstein was working with others involved in Climategate to discredit critics of man-made global warming."[23]

Associated Press climate reporter Seth Borenstein's reputation as a foot soldier in the global warming cause was further cemented by the Climategate revelations.

revealed by the Climategate scandal, the doctoring of data and the refusal to admit mistakes have so severely tainted the IPCC that it is no longer a credible agency."[19]

Kiminori Itoh, an award-winning environmental physical chemist from Japan, is another UN IPCC scientist who has turned his back on the UN climate panel. Kiminori declared that global warming fears are the "worst scientific scandal in the history.... When people come to know what the truth is, they will feel deceived by science and scientists."[20]

Berkeley professor Richard A. Muller presented a video lecture in 2011 on his disgust with the "hide the decline" temperature alterations, which came to light in Climategate, and which we discussed in chapter six. "They are not allowed to do this in science. It isn't up to our standards," Muller declared. "As a scientist, I now have a list of scientists whose papers I will not read anymore."[21]

In November 2009, I was one of the first reporters to publicize the Climategate scandal, after first being alerted to it by a phone call from fellow skeptic Anthony Watts. I devoted my website to the unfolding revelations. The mainstream media's initial attempts to ignore or downplay Climategate allowed skeptics to report on the reality of the scandal without the filter of the warmist defenders in the media. "My fervent hope is the mainstream media continues to ignore Climategate, as this will ensure the public will continue to receive the most accurate and balanced information about the scandal," I told *Newsweek* magazine in December 2009.[24]

The *Newsweek* profile of me noted, "While on the Hill, Morano was more like a wire service than a spokesman, pumping out scads of e-mails each week, sometimes each day, to reporters covering climate change" adding, "Morano was influential if not [sic] just through sheer relentlessness. With 'Climategate'—the release last month of thousands of hacked e-mails showing debate about climate change may have been stifled—he is now getting more attention than ever before."

Circling the Wagons

When the scandal broke, the global warming establishment—led by the UN, academia, and the media—immediately went into move-along-nothing-to-see-here mode. There were several high-profile "investigations" of Climategate that were obviously designed simply to restore credibility to the UN and climate scientists

The global warming industry investigated itself and exonerated itself. The pre-determined goal was to declare that Climategate was much ado about nothing. The investigations were hopelessly compromised—lacking thoroughness and riddled with conflicts of interest.

The Hockey Stick Illusion author Andrew Montford analyzed four of the Climategate investigations and found that they were "rushed, cursory and largely unpersuasive."[25]

Clive Crook, writing for the *Atlantic*, also slammed the Penn State investigation: "The Penn State inquiry exonerating Michael Mann—the paleoclimatologist who came up with 'the hockey stick'—would be difficult to parody. Three of four allegations are dismissed out of hand at the outset: the inquiry announces that, for 'lack of credible evidence', it will not even investigate them…. You think I exaggerate?…In short, the case for the prosecution is never heard. Mann is asked if the allegations (well, one of them) are true, and says no."

As Crook explained, "The [Penn State] report…says, in effect, that Mann is a distinguished scholar, a successful raiser of research funding, a man admired by his peers—so any allegation of academic impropriety must be false."

But the coup de grâce was the report's conclusion that anyone as respected (and as lucrative for Penn State) as Mann couldn't possibly be guilty. Penn State was touting Mann's cash cow status for the university as some sort of guarantee that he could do no wrong. As the report explained,

> This level of success in proposing research, and obtaining funding to conduct it, clearly places Dr. Mann among the most respected

scientists in his field. Such success would not have been possible had he not met or exceeded the highest standards of his profession for proposing research.... Had Dr. Mann's conduct of his research been outside the range of accepted practices, it would have been impossible for him to receive so many awards and recognitions, which typically involve intense scrutiny from scientists who may or may not agree with his scientific conclusions.... Clearly, Dr. Mann's reporting of his research has been successful and judged to be outstanding by his peers. This would have been impossible had his activities in reporting his work been outside of accepted practices in his field.[26]

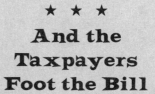

★ ★ ★

And the Taxpayers Foot the Bill

The *Wall Street Journal* reported in 2010 that Michael Mann has racked up "more than $2.4 million" in stimulus money from the U.S. government. "He received another grant worth nearly $1.9 million to investigate the role of 'environmental temperature on the transmission of vector-borne diseases,'" the paper noted.[27]

At the Watts Up with That blog, Willis Eschenbach pointed out the fact that the Penn State investigators had tasked Mann, the man under investigation, with gathering and presenting the evidence against himself. The university simply exonerated Mann by making sure that "none of the important questions are ever answered."[28]

Steve McIntyre at Climate Audit pointed out that the UK Royal Society's Climategate investigation was "tainted" by the fact that the investigators— including one of Mann's co-authors—had an obvious stake in declaring there was nothing to see here.[29]

Another one of the Climategate investigations was the Muir Russell investigation, which the UK *Register*'s Andrew Orlowski called "shameful"—its main goal was to urge a "campaign to win hearts and minds" to restore confidence in global warming science.[30] Climate Audit's Steve

McIntyre noted that the Muir Russell report "adopted a unique inquiry process in which they interviewed only one side—CRU [the Climatic Research Unit of East Anglia University]. As a result, the report is heavily weighted towards CRU apologia."[31]

An East Anglia University inquiry chaired by Lord Oxburgh was characterized by the *Register* as "Dracula's in charge of the blood bank" because of conflicts of interest. As Andrew Orlowski reported, "The peer leading the second Climategate enquiry at the University of East Anglia serves as a director of one of the most powerful environmental networks in the world, according to Companies House documents—and has failed to declare it. Lord Oxburgh, a geologist by training and the former scientific advisor to the Ministry of Defence, was appointed to lead the enquiry into the scientific aspects of the Climategate scandal on Monday. But Oxburgh is also a director of GLOBE, the Global Legislators Organisation for a Balanced Environment."[32]

Thoroughly Discredited

But despite these whitewashes coming from the global warming establishment, Climategate did have a major effect. We have already seen how it triggered the conversion of numerous scientists—including UN IPCC scientists—from true believers to more skeptical. Climatologist Judith Curry, for example, said in 2014, "Bottom line: Climategate was career changing for me." She explained, "Climategate shed a public light on the lack of transparency in climate science, which was deemed intolerable by pretty much everyone (except for some people who 'owned' climate data sets)."

Curry noted that "in the U.S., it seems that Climategate had a more palpable impact on climate legislation. Senator James Inhofe stated that Climategate was the death knell of carbon cap and trade legislation."[33]

Rex Murphy of the Canadian Broadcasting Corporation summed it up this way: Climategate "pulls back the curtain on a scene of pettiness, turf protection, manipulation, defiance of freedom of information, lost or destroyed data and attempts to blacklist critics or skeptics of the global warming cause." He continued, "You wouldn't accept that at a grade 9 science fair." Murphy added, "Science has gone to bed with advocacy and both have had a very good time."[34]

The Climategate scandal revealed that the UN IPCC was simply a lobbying organization portraying itself as a science panel. If the UN failed to find carbon dioxide was a problem, it would no longer have a reason to continue studying it—or to be in charge of offering "solutions."

Professor Roger Pielke Jr. of the University of Colorado has noted, "I think we can get past the lie—and it was a lie—that these activist scientists, in the words of RealClimate.org's Gavin Schmidt, are not taking a political stand."[35]

The UN IPCC reports are often used to claim the science is "settled." *New Scientist* magazine once dubbed the IPCC "the gold standard of consensus on climate change science." Well, if there was any doubt before, Climategate exposed the IPCC to be fool's gold.[36]

But even before Climategate, there was good reason to realize that the UN IPCC was more political than scientific.

On July 23, 2008, more than a year before the Climategate emails were leaked, John Brignell, an engineering professor emeritus at the University of Southampton who had held the chair in Industrial Instrumentation, accused the UN of censorship. "The creation of the UN IPCC was a cataclysmic event in the history of science. Here was a purely political body posing as a scientific institution. Through the power of patronage it rapidly attracted acolytes. 'Peer review' soon rapidly evolved from the old style refereeing to a much more sinister imposition of The Censorship," wrote Brignell. "As [the] Wegman [report] demonstrated, new circles of like-minded propagandists formed, acting as judge and jury for each other. Above all, they acted in concert to keep out alien and hostile opinion. 'Peer review' developed into a mantra that was picked up by political activists who clearly had no idea of the procedures of science or its learned societies. It became an imprimatur of political acceptability, whose absence was equivalent to placement on the proscribed list."[37]

In 2007, Australian climate data analyst John McLean did research into the IPCC's peer-review process. McLean's study found that very few scientists are actively involved in the UN's peer-review process, which he called "an illusion."[38]

"More than two-thirds of all authors of chapter 9 ('Understanding and Attributing Climate Change') of the IPCC's 2007 climate-science assessment are part of a clique whose members have co-authored papers with each other," McLean found. "Of the 44 contributing authors, more than half have co-authored papers with the lead authors or coordinating lead authors of chapter 9."

According to McLean, "Governments have naively and unwisely accepted the claims of a human influence on global temperatures made by a close-knit clique of a few dozen scientists, many of them climate modellers, as if such claims were representative of the opinion of the wider scientific community."

As McLean explained, "To sum up, the IPCC is a single-interest organisation, whose charter assumes a widespread human influence on climate, rather than consideration of whether such influence may be negligible or missing altogether.[39]

For example, the IPCC Summary had asserted that "it is very highly likely that greenhouse gas forcing has been the dominant cause of the observed global warming over the last 50 years." But as McLean discovered, "The IPCC leads us to believe that this statement is very much supported by the majority of reviewers. The reality is that there is surprisingly little explicit support for this key notion. Among the 23 independent reviewers just 4 explicitly endorsed the chapter with its hypothesis, and one other endorsed only a specific section. Moreover, only 62 of the IPCC's 308 reviewers commented on this chapter at all."[40]

Many UN scientists have publicly rejected the IPCC's methods. (The following material on UN scientists who have turned on the UN has been adapted and updated from a speech I wrote for Senator Jim Inhofe in 2007, while working at the Senate Environment and Public Works Committee.)

"I have found examples of a Summary saying precisely the opposite of what the scientists said," noted South African nuclear physicist and chemical engineer Philip Lloyd, a UN IPCC co-coordinating lead author who has authored over 150 refereed publications. "The quantity of CO_2 we produce is insignificant in terms of the natural circulation between air, water and soil.... I am doing a detailed assessment of the UN IPCC reports and the Summaries for Policy Makers, identifying the way in which the Summaries have distorted the science."[41]

Andrei Kapitsa, a Russian geographer and Antarctic ice core researcher, has claimed, "A large number of critical documents submitted at the 1995 U.N. conference in Madrid vanished without a trace. As a result, the discussion was one-sided and heavily biased, and the U.N. declared global warming to be a scientific fact."[42]

UN IPCC expert reviewer Madhav Khandekar, a retired Environment Canada scientist, lamented that many "seem to naively believe that the climate change science espoused in the [UN's] Intergovernmental Panel on Climate Change (IPCC) documents represents 'scientific consensus.'" In fact, "Nothing could be further than the truth! As one of the invited expert reviewers for the 2007 IPCC documents, I have pointed out the flawed review process used by the IPCC scientists in one of my letters. I have also pointed out in my letter that an increasing number of scientists are now questioning the hypothesis of Greenhouse gas induced warming of the earth's surface and suggesting a stronger impact of solar variability and large-scale atmospheric circulation

★ ★ ★
Withdrawing in Disgust

Paul Reiter, a malaria expert formerly of the Centers for Disease Control and Prevention, was part of the UN IPCC assessments. But Reiter resigned in disgust and declared the "consensus" claims a "sham." Reiter, a professor of entomology and tropical disease with the Pasteur Institute in Paris, threatened legal action to have his name removed from the IPCC. "That is how they make it seem that all the top scientists are agreed," he said on March 5, 2007. "It's not true," he added.[44]

patterns on the observed temperature increase than previously believed.... Unfortunately, the IPCC climate change documents do not provide an objective assessment of the earth's temperature trends and associated climate change."[43]

Hurricane scientist Christopher W. Landsea, formerly of NOAA's National Hurricane Center, was an author for the IPCC's Second Assessment Report in 1995 and the Third Assessment Report in 2001, but he resigned from the Fourth Assessment Report, accusing the IPCC of distorting hurricane science. "I am withdrawing because I have come to view the part of the IPCC to which my expertise is relevant as having become politicized. In addition, when I have raised my concerns to the IPCC leadership, their response was simply to dismiss my concerns," Landsea wrote in a January 17, 2005, public letter. "I personally cannot in good faith continue to contribute to a process that I view as both being motivated by pre-conceived agendas and being scientifically unsound." Landsea is currently with the Science and Operations Officer at the National Hurricane Center.

The process in which UN IPCC documents are produced is simply not compatible with good science. The UN IPCC's guidelines stipulate that the scientific reports have to be "change[d]" to "ensure consistency with" the media-hyped Summary for Policymakers.[45]

As Senator Inhofe, the former chair of the Senate Environment and Public Works Committee has noted, "The IPCC more closely resembles a political party's convention platform battle—not a scientific process." Inhofe explained, "During an IPCC Summary for Policymakers process, political delegates and international bureaucrats squabble over the specific wording of a phrase or assertion."[46]

The *Guardian* detailed the process in a 2014 article. "Government officials and scientists are gathered in Yokohama this week to wrangle over every line of a summary of the report before the final wording is released on Monday—the first update in seven years. Nearly 500 people must sign off on the exact wording of the summary, including the 66 expert authors, 271 officials from 115 countries, and 57 observers."[47]

Steve McIntyre of Climate Audit analyzed the process behind the IPCC Summary for Policymakers and discovered that "the purpose of the three-month delay between the publication of the (IPCC) Summary for Policy-Makers and the release of the actual WG1 (Working Group 1) is to enable them to make any 'necessary' adjustments to the technical report to match the policy summary. Unbelievable. Can you imagine what securities commissions would say if business promoters issued a big promotion and then the promoters made the 'necessary' adjustments to the qualifying reports and financial statements so that they matched the promotion. Words fail me."[48]

Former Colorado State Climatologist Roger Pielke Sr. revealed his personal experience dealing with the UN IPCC: "The same individuals who are doing primary research in the role of humans on the climate system are then permitted to lead the [IPCC] assessment! There should be an outcry on this obvious conflict of interest, but to date either few recognize this conflict, or see that since the recommendations of the IPCC fit their policy and political agenda, they chose to ignore this conflict. In either case, scientific rigor has been sacrificed and poor policy and political decisions will inevitably follow."[49]

Years before the Climategate scandal broke, Pielke was warning the public, "We need recognition among the scientific community, the media, and policymakers that the IPCC process is obviously a real conflict of interest, and this has resulted in a significantly flawed report."[50]

★ ★ ★

UN Chief's Climate Religion

In 2015, former UN IPCC Chief Rajendra Pachauri, whose organization shared the Nobel Peace Prize with Al Gore, literally called global warming his religion. Pachauri, who was forced out of his position at the UN by a sexual harassment scandal, said in his resignation letter, "For me the protection of Planet Earth, the survival of all species and sustainability of our ecosystems is more than a mission. It is my religion and my dharma."[52]

Journalist Donna Laframboise, who has written two books critical of the UN climate panel responded to Pachauri's admission: "Yes, the IPCC—which we're told to take seriously because it is a scientific body producing scientific reports—has, in fact, been led by an environmentalist on a mission. By someone for whom protecting the planet is a religious calling."[53]

Any remaining doubts that the IPCC is a political organization were eliminated when former UN IPCC chief Rajendra Pachauri admitted the IPCC is an arm of world governments and serves at their "beck and call." "We are an intergovernmental body and we do what the governments of the world want us to do," Pachauri told the *Guardian* in 2013. "If the governments decide we should do things differently and come up with a vastly different set of products we would be at their beck and call.

Pachauri freely told the world that the purpose of the UN IPCC reports is to make the case for "action" on global warming. As he explained, "There will be enough information provided so that rational people across the globe will see that action is needed on climate change."[51]

In 2017, climate policy researcher and author Donna Laframboise issued an analysis finding that U.S. government rules "in no uncertain terms, repudiate the process by which UN climate reports are produced. The US government says political tampering with scientific findings is a violation of scientific integrity. But political revision is central to how IPCC reports get produced."

Laframboise, who authored the 2011 book exposing the IPCC titled *The Delinquent Teenager Who Was Mistaken for the World's Top Climate Expert*," reported:

> IPCC reports therefore lack scientific integrity.
>
> People who rely on IPCC reports are basing their decisions on documents that have no scientific integrity.
>
> The IPCC goes back, after the fact, and changes the original scientific report so that it aligns with the politically negotiated summary.

She also noted, "After the summaries are haggled over, the IPCC alters what the scientists wrote. That's the reason the IPCC routinely releases its summaries before it releases the underlying scientific report. In this 2007 news clipping, the IPCC chairman explains: "we have to ensure that the underlying report conforms to the refinements."[54]

Greenpeace co-founder turned climate skeptic Dr. Patrick Moore commented on Laframboise's report, noting this is the "perfect reason for the US to abandon the UN Paris climate 'agreement.'"[55]

Insiders Speak Out

An impressive array of former UN IPCC scientists are completely disillusioned with the climate panel and its politically manufactured "scientific" conclusions. They've seen how the sausage is made, and they're willing to testify to the dishonesty of the process.

Indian geologist Arun D. Ahluwalia of Punjab University, a board member of the UN-supported International Year of the Planet, has charged, "The IPCC has actually become a closed circuit; it doesn't listen to others. It doesn't have open minds.... I am really amazed that the Nobel Peace Prize has been given on scientifically incorrect conclusions by people who are not geologists."[56]

Steven M. Japar, an atmospheric chemist who was part of the UN IPCC's Second (1995) and Third (2001) Assessment Reports and has authored eighty-three peer-reviewed publications in the areas of climate change, atmospheric chemistry, air pollutions, and vehicle emissions, explained, "Temperature measurements show that the [climate model–predicted mid-troposphere] hot zone is non-existent. This is more than sufficient to invalidate global climate models and projections made with them!"[57]

Kenneth P. Green, who was a Working Group 1 expert reviewer for the IPCC in 2001, has declared, "We can expect the climate crisis industry to grow increasingly shrill, and increasingly hostile toward anyone who questions their authority."[58]

Climatologist John Christy of the University of Alabama in Huntsville was a lead author on the 2001 UN IPCC report. Christy explained how his colleagues were telegraphing the science to support politics. "I was at the table with three Europeans, and we were having lunch. And they were talking about their role as lead authors. And they were talking about how they were trying to make the report so dramatic that the United States would just have to sign that Kyoto Protocol."[59]

Top United Nations officials apparently know years in advance that each UN climate report will be more alarming—an exercise in making the science fit their political agenda. In 2010, AFP reported that Robert Orr, UN undersecretary general for planning, had declared that the "next Intergovernmental Panel on Climate Change report on global warming will be much worse than the last one."[60]

In 2017, the IPCC, realizing how damaging these slips of the tongue from UN officials could be to public support, attempted to dismiss this Orr's comments, saying that Orr "was UN Under Secretary-General, not working with IPCC."[61]

So according to the IPCC, if Orr was not an official IPCC executive, then his comments had no bearing on its work. But how do they explain the then-head of the IPCC, Pachauri, making very similar comments in 2009, a full four years ahead of the next report? "When the IPCC's fifth assessment comes out in 2013 or 2014, there will be a major revival of interest in action that has to be taken. People are going to say, 'My God, we are going to have to take action much faster than we had planned.'"[62]

Pachauri told the BBC in 2013, "I hope that [the report] will reassure every-one that human influence is having a major impact on the Earth's climate."[63]

It does not stop there. In 2012, a year before the report came out, former UN climate chief Yvo de Boer announced that the next IPCC report "is going to scare the wits out of everyone." He added, "I'm confident those scientific findings will create new political momentum."[64] Australia's the *Age* newspaper reported that de Boer believes the scary IPCC report "should provide the impetus needed for the world to finally sign an agreement to tackle global warming."

In 2014, I became a bit bored with the whole IPCC scare the public and media hype routine. "After years of covering this debate for well over a decade as a reporter, researcher, and U.S. Senate staffer, I find myself com-pletely bored by the UN's same old ramp up the alarm approach," I responded to media inquiries. "I have covered this debate on a daily basis, hourly basis and sometimes minute by minute basis. I am trying to get excited, but alas, even the alarmism and apocalyptic claims fail to excite me. Can't the UN think of more effective ways to get attention? Can't the UN try something different?"[65]

Let's let IPCC reviewer and climate researcher Vincent Gray of New Zea-land have the last word. Gray, the author of more than one hundred scien-tific publications, was an expert reviewer on every single draft of the IPCC reports going back to 1990. And he says, "The claims of the IPCC are danger-ous unscientific nonsense."[66]

CHAPTER 11

What's in a Name? How "Global Warming" Became "Climate Change"

"**G**lobal warming"—that was the popular label that the activists substituted for the much less catchy "greenhouse effect," which was the original term used by the media in the 1980s. "Global warming" caught on in the 1990s.

Then the activists decided to change it again. Former Obama science czar John Holdren suggested "global climate disruption."[1] "I think one of the failures of the scientific community was in embracing the term 'global warming'. Global warming is in fact a dangerous misnomer," Holdren said in 2010. Others suggested a "new normal" or "global weirding."[2]

The climate movement has steadily and quietly been morphing the phrase "global warming" into "climate change."

Why the change?

As we have already seen, global temperatures have essentially been holding steady for the past two decades. The hysterical predictions about polar bears' habitat melting and rising sea levels inundating our coastal cities were all a bust. Given those facts, "global warming" is an increasingly hard sell. Besides, "global warming" sounds so gentle, almost cozy—especially to folks in the major population centers on the East Coast of the United

Did you know?

★ In 2000 a senior research scientist at the University of East Anglia said "children just aren't going to know what snow is"

★ Ten years later, environmentalist George Monbiot declared, "That snow outside is what global warming looks like"

★ Prostitution, barroom brawls, and airplane turbulence have all been blamed on climate change

A Name Change That Didn't Catch On

"Avoid the term 'global warming.' I prefer the term 'global weirding.'"

—**Thomas Friedman of the *New York Times***

Friedman is one of the biggest pushers of the "extreme weather" meme that has replaced the more traditional scare stories of polar bear extinction and rising sea levels.

"What actually happens in climate change is the weather gets weird. The hots get hotter, the dries get drier and longer and wider, the rains get heavier, etc. the snows get thicker," Friedman claims.[3]

States, who have experienced record-setting cold weather in recent winters. The climate activists simply had to move the goal posts.

Unfalsifiable

"Global warming" means something actually falsifiable: the climate establishment may hide and manipulate the evidence, as we have seen they do, but if the warming pauses, or—worse—if temperatures actually drop, that messes up the "global warming" narrative.

Any and all weather events, on the other hand—hotter weather, colder weather, wetter weather, drier weather, any kind of storm—can be attributed to "climate change." So now the media jumps on every heat wave, cold snap, hurricane, tornado, drought, flood, or other example of weather in the news to claim that the "unprecedented" weather is due to human activity. Blizzards and record cold temperatures are now caused by global warming—or what used to be "global warming," but is now "climate change." No matter what climate or weather event occurs, it seems that climate change predicted it. Climate science has evolved to predict all possible outcomes—so no matter what happens, they can claim they predicted it.

In order for the global warming movement to convince policy makers and the public that man-made climate change is a crisis that requires immediate action, a shift in tactics was necessary. And so now every bad weather event is somehow proof of man-made climate change. It's a kind of pseudoscience, ginning up fears of man-made climate change by what noted

Japanese scientist Kanya Kusano of the Japan Agency for Marine-Earth Science and Technology has compared to "ancient astrology."[4]

And Kusano is not the only one to make that comparison. "Whether the ice caps melt, or expand—whatever happens—the anthropogenic global warming theorists claim it confirms their theory. A perfect example of a pseudo-science like astrology," said mathematical physicist Frank J. Tipler of Tulane University in 2009, explaining how the "climate change" tactic works.[5]

In 2007, on a trip to a UN climate conference in Bali, Indonesia, in my job working for the U.S. Senate Environment and Public Works Committee, I had a heated debate with one of Senator John McCain's aides about this switch in tactics. McCain's climate guy was convinced that linking bad weather to climate was the key to influencing the public.

President Obama utilized this tactic often, once telling the members of the audience at an event at a Walmart that they had seen climate change. "Here in California, you've seen these effects firsthand," Obama said. "Climate change is not some far off problem in the future, it's happening now."[6] The president of the United States had officially told Americans that they can see global warming at their local Walmart.

Penn State Professor Michael Mann was even more explicit. "We can see climate change, the impacts of climate change, now, playing out in real time, on our television screens, in the 24-hour news cycle," Mann said at a Democrat platform draft hearing in 2016. "The signal of climate change is no longer subtle, it is obvious."[7]

Back in 2000, when it was still "global warming," David Viner, a senior research scientist at the climatic research unit (CRU) of the University of East Anglia (the institution that would be at the epicenter of Climategate), was featured in a news article in the UK newspaper the *Independent* with the headline, "Snowfalls are now just a thing of the past." Viner predicted that within a few years winter snowfall would become "a very rare and exciting event. Children just aren't going to know what snow is." Another researcher, David Parker, of the UK's Hadley Centre for Climate Prediction and Research, even went as far as to predict that British children would have only "virtual" experience of snow via films and the Internet.[8]

The predictions of less snow by global warming scientists were ubiquitous—and dead wrong. The current decade, from 2010 forward, is now the snowiest decade ever recorded for the U.S. East Coast, according to meteorologist Joe D'Aleo.[9] Talk about an inconvenient truth.

How did the warmist scientists explain record snow after predicting less snow? Easy. More snow is now caused by "climate change."

By 2013, after "global warming" had become "climate change," snow at unusual times was evidence for the supposed man-made crisis. Senator Barbara Boxer, the chair of the Senate Environment and Public Works Committee claimed. "Yeah, it's gonna get hot, but you're also gonna to have snow in the summer in some places."[10] Boxer seems to think any weather event can be made to fit the climate change narrative.

Environmentalist George Monbiot had already tried to explain away the then record cold and snow in a column titled, "That snow outside is what global warming looks like." Monbiot did his best to square the circle: "I can already hear the howls of execration: now you're claiming that this cooling is the result of warming! Well, yes, it could be." Monbiot asked, "So why wasn't this predicted by climate scientists? Actually it was, and we missed it."[11]

We missed it? Predictions of less snow were ubiquitous by global warming scientists. But once that prediction failed to come true, the opposite of what they predicted instead became—what they expected. How did global warming scientists explain record snow after prediction less snow? Easy. More snow is now caused by global warming.

"Snow is consistent with global warming, say scientists" blared a UK *Telegraph* headline in 2009.[12] The *Financial Times* tried to explain "Why global warming means...more snow" in 2012.[13] The December 26, 2010, *New York Times* featured an op-ed with the headline "Bundle Up, It's Global Warming," claiming, "Overall warming of the atmosphere is actually creating cold-weather extremes."[14]

Even former Vice President Al Gore, who had claimed in his Oscar-winning film in 2006 that all the snow on Mount Kilimanjaro would melt "within the decade,"[15] got into the act. Never once in *An Inconvenient Truth* had Gore warned of record cold and increasing snowfalls as a consequence of man-made global warming. As late as 2009, the Environmental News Service was reporting on Gore's hyping the lack of snow as evidence for man-made global warming: "Gore Reports Snow and Ice Across the World Vanishing Quickly."[16] But then, after massive snowstorms hit the United States in 2010, Gore claimed that "increased heavy snowfalls are completely consistent with...man-made global warming."[17]

UN IPCC lead author and Princeton University physicist Michael Oppenheimer had also exploited years of low snowfall totals to drive home the global warming narrative. He was quoted in a 2000 *New York Times* article: "'I bought a sled in '96 for my daughter,' said Michael Oppenheimer, a scientist at the nonprofit Environmental Defense Fund. 'It's been sitting in the stairwell, and hasn't been used. I used to go sledding all the time. It's one of my most vivid and pleasant memories as a kid, hauling the sled out to Cunningham Park in Queens.'... Dr. Oppenheimer, among other ecologists, points to global warming as perhaps the most significant long-term factor" explaining why, in the words of the *New York Times* reporter, "Sledding and snowball fights are as out-of-date as hoop-rolling."[18]

When I confronted Oppenheimer about his sled comment following his appearance at a 2014 Congressional hearing, my interview was cut short. I asked, "In 2000 *New York Times*, you mentioned you bought your daughter a sled, but she hadn't been able to use it..."

★ ★ ★
He Got the Memo

NBC weatherman Al Roker obviously got the "climate change" memo. "This is global warming even though it's freezing?" Larry King asked Roker in 2015.

"Right, well, that's why I don't like the phrase 'Global Warming.' I like 'Climate Change,'" the weatherman explained.

The message went from *global warming causes less snow* to *climate change causes more snow.*

"So Boston at this point, is in number two snowiest winter," Larry King asked just before Boston broke the record for it snowiest winter on record, in 2015. "Is this all part of Climate change?"

Roker did not flinch. "I think it is," he answered.[19]

Oppenheimer's aide intervened to say, "I'm sorry, but Dr. Oppenheimer has to testify."

Perhaps next time we can ask Oppenheimer about his daughter's sled.[20]

MSNBC's Ed Schultz also sees cold and snow as hard evidence of what used to be called global warming. "Every day we are getting new evidence of man-made climate change. Today the Northeast plains and lower Midwest are digging out from another round of snow and freezing rain," Schultz announced to viewers in 2015.[21]

"Bill Nye the Science Guy," for his part has been on television claiming that any weather event "is consistent with what you'd expect" with climate change.[22]

But virtually any kind of weather can be made to fit the climate narrative. A 2007 *Telegraph* article warned that global warming would mean less ice around Antarctica. "Global warming is threatening one of the most endearing symbols of Antarctica—the penguin.... The environmental conservation group WWF is warning that rising temperatures and the resulting loss of sea ice is robbing the emblematic birds of the nesting grounds they need to breed successfully."[23] But in 2008, in order to explain the record growth in sea ice, the media and some scientists started claiming that a colder Antarctic with more sea ice was caused by—global warming. "A cold Antarctica is just what calculations predict," stated a February 12, 2008, post on Real Climate titled "Antarctica is Cold? Yeah, We Knew That."[24]

As the Antarctic sea ice continued to grow, the excuses kept coming. "Global warming has led to more ice in the sea around Antarctica and could help insulate the southern hemisphere from atmospheric warming," claimed a 2013 article.[25]

The Australian *Herald Sun*'s Andrew Bolt ridiculed the heads I win, tails you lose approach to climate science in 2013, noting, "Less ice, more ice, whatever. It's global warming."[26]

Professor Roger Pielke Jr., professor in the environmental studies program at the University of Colorado, mocked the warmists' contradictory claims. "So a warming Antarctica and a cooling Antarctica are both 'consistent with' model projections of global warming. Our foray into the tortured logic of 'consistent with' in climate science raises the perennial question, what observations of the climate system would be inconsistent with the model predictions?"[27]

Climatologist Patrick Michaels also ridiculed government scientists who blamed "global warming" for both low and record high sea ice. "Yes, well of course. You know the problem is that what's happened in climate change is every anomaly can find an expert. An expert to say that this is consistent with climate change," Michaels noted.[28]

Climate change also causes more hurricanes—except when it causes fewer hurricanes. In 2007, it was claimed that hurricanes "have doubled due to global warming."[29] But it turned out that the United States was at the beginning of what would be a twelve-year period—between Hurricane Wilma in 2005 and Hurricanes Irma, Harvey, and Maria in 2017—in which no category 3 or higher hurricane would make landfall.

★ ★ ★

Heads They Win, Tails We Lose

In 2011 the No Tricks Zone website, run by Pierre Gosselin, collated a sampling of climate studies predicting opposite results. The analysis found "more than 30 contradictory pairs of peer-reviewed papers":[30]

Amazon dry season greener[31]	Amazon dry season browner[32]	Indian rice yields to decrease[55]	Indian rice yields to increase[56]
Avalanches may increase[33]	Avalanches may decrease[34]	Latin American forests may decline[57]	Latin American forests have thrived in warmer world with more CO_2[58]
Bird migrations longer[35]	Bird migrations shorter[36]		
Boreal forest fires may increase[37]	Boreal forest fires may continue decreasing[38]	Leaf area index reduced[59]	Leaf area index increased[60]
		Malaria may increase[61]	Malaria may continue decreasing[62]
Chinese locusts swarm when warmer[39]	Chinese locusts swarm when cooler[40]	North Atlantic cod to decline[63]	North Atlantic cod to thrive[64]
Columbian spotted frogs decline[41]	Columbian spotted frogs thrive in warming world[42]	North Atlantic Hurricane frequency to increase[65]	North Atlantic Hurricane frequency to decrease[66]
Coral island atolls to sink[43]	Coral island atolls to rise[44]	North Atlantic Ocean less salty[67]	North Atlantic Ocean more salty[68]
Earth's rotation to slow down (increase length of day)[45]	Earth's rotation to speed up (decrease length of day)[46]	Northern Hemisphere ice sheets to decline[69]	Northern Hemisphere ice sheets to grow[70]
East Africa to get less rain[47]	East Africa to get more rain[48]	Plants move uphill[71]	Plants move downhill[72]
Great Lakes less snow[49]	Great Lakes more snow[50]	Sahel to get less rain[73]	Sahel to get more rain[74]
Gulf stream slows down[51]	Gulf stream speeds up a little[52]	San Francisco less foggy[75]	San Francisco more foggy[76]
Indian monsoons to be drier[53]	Indian monsoons to be wetter[54]	Sea level rise accelerated[77]	Sea level rise decelerated[78]

Soil moisture less[79]	Soil moisture more[80]	UK may get more droughts[85]	UK may get more rain[86]
Squids get smaller[81]	Squids get larger[82]	Wind speed to go up[87]	Wind speed slows down[88]
Swiss mountain debris flow may increase[83]	Swiss mountain debris flow may decrease[84]	Winters may be warmer[89]	Winters may be colder[90]

So no matter what happens, the activists can claim with confidence the event was a predicted consequence of global warming. There is now no way to ever falsify global warming claims.

In 2008, when no major hurricane had hit the United States for over two years, a study from meteorologist Tom Knutson, working for the federal government, found that "warmer temperatures will actually reduce the number of hurricanes in the Atlantic and those making landfall," according to the Associated Press.[91]

In the same year, hurricane expert Kerry Emanuel of Massachusetts Institute of Technology "reconsidered" his views on hurricanes and climate and claimed that "even in a dramatically warming world, hurricane frequency and intensity may not substantially rise during the next two centuries."[92]

In 2015, after the hiatus in which no category 3 or higher hurricanes made landfall in the United States had stretched to ten years, climate researchers were still asserting that climate change would mean fewer hurricanes. As one study found, "Global warming means fewer—but more powerful—hurricanes."[93]

But once Irma and Harvey threatened the United States, reporters began hyping claims that climate change could make hurricanes more frequent. As ABC News in Baltimore reported, "Climate Change Might Make Intense Hurricanes like Harvey More Common."[94]

Emergency! The Government Must Fix the Weather!

UN climate chief Christiana Figueres urged governments around the world to "do something" about extreme weather. "We have had severe climate and weather events all over the world and everyone is beginning to understand that is exactly the future we are going to be looking at if they don't do something about it," Figueres explained at the opening of the annual UN climate summit in 2012.[95]

IPCC AR5 coordinating lead author Professor Oliver C. Ruppel blamed global warming for earthquakes, volcanic eruptions, and tsunamis: "There is wide scientific consensus that the increased number and intensity of climate change induced natural disasters, such as earthquakes, volcano eruptions, tsunamis and hurricanes, is of alarming concern."[96]

Not to be outdone, Bill McGuire, professor emeritus at University College London, released a book titled *Waking the Giant: How a Changing Climate Triggers Earthquakes, Tsunamis, and Volcanoes*.[98] *Newsweek*, which could not resist the temptation to blame earthquakes on climate change, featured McGuire in a 2015 article. "A growing body of scientists" are now concerned "that climate change can affect the underlying structure of the Earth," *Newsweek*'s Alex Renton wrote.[99]

But wait! Not only does climate change allegedly cause earthquakes—earthquakes also cause climate change, according to a 2013 study published in the journal *Nature Geoscience*. The study claimed that earthquakes can contribute to climate change by causing the release of the potent greenhouse gas methane from the ocean floor.[100]

On the morning after the 2013 Moore, Oklahoma, tornado outbreak that killed twenty-four people, Barbara Boxer, at that time the chair of the U.S. Senate Committee on Environment and Public Works, took to the Senate floor to blame the deadly tornadoes on global warming. "This is climate change. We were warned about extreme weather," Boxer said as cleanup crews were only just beginning to deal with the tornado wreckage. "Not just hot weather. But extreme weather. When I had my hearings, when I had the gavel years ago. It's been a while—the scientists all agreed that what we'd start to see was extreme weather.... You're gonna have terrible storms. You're going to have tornados and all the rest. We need to protect our people."

Boxer also took the opportunity to plug her own carbon tax bill, co-sponsored with Senator Bernie Sanders, during the aftermath of the tornado tragedy. "Carbon could cost us the planet," she said. "The least we could do is put a little charge on it so people move to clean energy."

Democratic Senator Sheldon Whitehouse of Rhode Island used the same tornado outbreak as an opportunity to blame Republicans—his exact words were "polluters and deniers"—for helping cause bad weather.[102]

Senator Whitehouse posed a question to himself: "Why do you, Sheldon Whitehouse, Democrat of Rhode Island, care if we Republicans run off the climate cliff like a bunch of proverbial lemmings and disgrace ourselves? I'll tell you why. We're stuck in this together. We are stuck in this together—when cyclones tear up Oklahoma and hurricanes swamp Alabama and

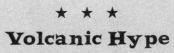

★ ★ ★

Volcanic Hype

The media has also hyped volcanoes and climate change. A 2015 headline at my Climate Depot website read, "Which Is It? Time Mag. Goes Both Ways on Volcanoes! TIME Magazine: 'Climate Change Leads to Volcanoes'—OR Flashback Time Magazine 2014: 'Volcanoes May Be Slowing Down Climate Change.'"[101]

★ ★ ★

Nostradamus Does Climate

Extreme weather expert Roger Pielke Jr. of the University of Colorado wrote of the mystical qualities of the climate change claims in an analysis titled "It Has Been Foretold": "Because various unsupportable and just wrong claims are being advanced by leading scientists and scientific organizations, it would be easy to get the impression that on the issues of extreme events and climate change, IPCC science has a status similar to interpretations of Nostradamus and the Mayan calenders [sic]."[105]

wildfires scorch Texas, you come to us, the rest of the country, for billions of dollars to recover. And the damage that your polluters and deniers are doing doesn't just hit Oklahoma and Alabama and Texas."[103]

Undercutting this blame game was the simple fact that in recent years, as we shall see in the next chapter, tornadoes have been at or near record lows and tornadoes in the region known as "tornado alley" are not proof of man-made climate change. But record low tornado and declining strength does not fit the narrative that "polluters" are making the weather worse. As climatologist Roger Pielke Sr. has explained, "To try to attribute a given weather event, due to added CO_2 or whatever, is impossible...and I think we're misleading the public by telling them that we know why climate is behaving the way it is."[104]

But the lack of evidence doesn't stop politicians from warning that extreme weather will get much worse unless Congress "changes course" to alter storms.

Retired Democrat Representative Henry Waxman of California blamed carbon dioxide for wildfires. "It's time to stop denying science. Extreme events like the wildfires in Colorado and the floods in Florida are going to get worse unless Republican-controlled Congress changes course soon," Waxman explained.[106]

Can the government really fix the weather?

The media and the global warming establishment either pretend to believe or actually do believe that the government can legislate the weather. "The scientists at NASA say we can slow the earth's warming if we cut pollution and have higher carbon emission standards," reported NBC News.[107] "It's our choice how fast the seas rise and that gives us time to prepare and protect our communities in a smart way," claimed Brenda Ekwurzel of the Union of Concerned Scientists.[108]

I have debated Bill Nye, "the Science Guy," several times in prime time on both CNN and Fox News. It's no wonder warmist Randy Olson has said that I "pretty much chewed up Bill Nye the Science Guy on CNN with Piers Morgan a couple years ago," given the statements Nye has made.[111] In a 2014 debate with me on John Stossel's program, Nye made fixing climate change sound as easy as filling potholes. "Is it conceit when politicians claim they're going to fix potholes in the street?" Nye asked. "This is the same thing on a much, much larger scale."[112]

Comments like Nye's have created a backlash among scientists. Atmospheric scientist Dr. Hendrik Tennekes, former director of research at the Netherlands' Royal National Meteorological Institute, rejected such simpleminded claims about how to "fix" or control the climate. "I protest vigorously the idea that the climate reacts like a home heating system to a changed setting of the thermostat: just turn the dial, and the desired temperature will soon be reached," Tennekes wrote.[113]

But scientifically rational comments like these are often ignored. UK warmist professor Mark Maslin is supremely confident in our ability to engineer the global climate. "We are now at the point where we can decide

★ ★ ★
Not Like a Yo-Yo Schoolboy

In 1933, Syrians banned the yo-yo because they thought it caused drought.

A January 23, 1933, headline in the *Barrier Miner* read, "Yo-Yo banned in Syria. Blamed for Drought by Muslims." As the article reported, "The Muslim chiefs at Damascus have attributed the wrath of the heavens to the recent introduction of the yo-yo. . . . The chiefs interviewed the Prime Minister and exposed the evil influence of yo-yos, so they were immediately banned. Today the police paraded the streets and confiscated the yo-yos from everyone they saw playing with them."[109]

Today, global warming activists blame Syrian drought on man-made global warming. In 2013, PBS's Bill Moyers claimed, "Climate change in Syria helped spark the civil war there. Which country is next?"[110]

★ ★ ★

He Called It

During the 2008 presidential election campaign, Barack Obama famously said, "…I am absolutely certain that generations from now, we will be able to look back and tell our children that…this was the moment when the rise of the oceans began to slow and our planet began to heal…"[114]

And lo and behold, in 2011, two years after Obama was sworn in, the European Space Agency's Envisat monitoring of global sea level revealed that a "two year long decline was continuing at a rate of 5mm per year."[115]

NASA also announced in 2011 that global sea level was dropping and was "a quarter of an inch lower than last summer."

Vote
Yourself
Better
Weather!

Obama called it on sea level. Promise kept. He may have literally had powers and abilities far beyond those of mortal men.

Fresh from his command of the seas, President Obama went on to make the case for his re-election by claiming that Americans could "do something" about the weather at the ballot box:

"Climate change is not a hoax. More droughts and floods and wildfires are not a joke. They're a threat to our children's future. And in this election, you can do something about it."

—President Obama in 2012[116]

Really? We can vote ourselves better weather? Who knew?

Ironically, President Obama was unable to control the weather at his acceptance speech for the Democratic nomination; a forecast of possible rain made him move the event indoors. As the *Hill* reported, "Pelosi: "Obama Cannot Control the Weather."[117] Representative Nancy Pelosi, apparently unaware that Obama had already kept his promise to stop the seas from rising, insisted that the weather was under the control of "a higher power": "There are some decisions that are made from a different place and whether it rains or not is not in the president's control."

Tony Heller, who runs the skeptical climate science website Real Science, had some fun with the cancellation: "Obama says he can stop droughts and floods, but he was unable to stop a 30% probability of rain which convinced him to move his speech indoors. Can someone explain how that works?"[118]

how the climate of the future will look," Maslin claimed in 2015. "When we as a collective world community, all nations working together, are able to really prevent global warming, that would be fantastic. That would be the first time that the climate doesn't control us, but rather us controlling it. We could make sure that all future generations will have a stable climate."[119]

Senator Whitehouse, when asked "what impact EPA regulations have on the climate?" responded, "Very, very positive, very positive."[120]

President Obama promoted the scientifically baseless notion that congressional legislation could alter the climate, declaring on June 24, 2009, that the congressional cap-and-trade bill would have huge climate impacts: "A long-term benefit is we're leaving a planet to our children that isn't four or five degrees hotter."[121]

But just a few weeks after President Obama made those remarks, his own EPA admitted the Waxman-Markey cap-and-trade bill would not even detectably reduce atmospheric CO_2 levels—let alone cool the planet. "U.S. action alone will not impact world CO_2 levels," Obama's then–EPA chief Lisa Jackson said before a U.S. Senate committee.[122]

Despite the purely symbolic climate impact, on the day the Trump administration announced it was going to repeal the EPA regulations, filmmaker Michael Moore warned, "Historians in the near future will mark today, March 28, 2017, as the day the extinction of human life on earth began, thanks to Donald Trump."[123]

All the World's Ills

Climate change has been blamed for causing or worsening a range of problems including prostitution, airplane turbulence, murder, rape, car thefts, and barroom brawls.

Media outlets endlessly hype stories on how every aspect of our lives is being made worse by global warming. "Your morning cup of joe could

It's Not Just Coffee Drinkers

"In fact, anyone who eats is under threat from climate change."

—UN IPCC lead author Michael Oppenheimer[125]

become a thing of the past.... a new study is dark with no sugar. It says climate change has the wild Arabica coffee bean plant headed for extinction."[124]

In 2009, climate activists and the UN declared that global warming was causing women in the developing world to become hookers. Yes, global warming causes prostitution.

"Climate change pushes poor women to prostitution," claimed UN official Suneeta Mukherjee, a country representative of the United Nations Food Population Fund in 2009. "Climate change could reduce income from farming and fishing, possibly driving some women into sex work," Mukherjee explained.[126]

A 2013 Democrat Congressional resolution also warned that global warming could send women into prostitution. Several House Democrats called on Congress to recognize that climate change is hurting women more than men, and could even drive poor women to "transactional sex" for survival. The resolution was sponsored by Democrat Representative Barbara Lee of California.[127]

I asked climatologist Judith Curry, the former head of the atmospheric science department at Georgia Tech University, "As a woman, does it worry you that unchecked global warming will lead to prostitution?"

"That's pretty ludicrous," Curry responded.[128]

But what about airplane turbulence?

Michigan senator Debbie Stabenow claimed in 2009, "Global warming creates volatility and I feel it when I'm flying. The storms are more volatile."[129]

NBC News also promoted the scare, in an article whose actual title was "Fasten Seat Belts: Climate Change Could Mean More Airline Turbulence." The 2014 report claimed that "computer models have predicted that climate

★ ★ ★

Climate Change Causing Child Marriages?

Statistician Bjorn Lomborg has dismissed claims that children are in danger of child marriages because of "global warming: "Consider the recent assertion by Unicef's Bangladesh head of mission that climate change leads to an increase in child marriages.... In Bangladesh, nearly 75% of women between the ages of 20 and 49 reported that they were married before they turned 18, giving the country the second-highest rate of child marriage in the world. As the Unicef head tells it, climate change has been a major cause, as warmer weather has worsened the flooding, pushing people to the cities, leading to more child marriages. This entire string of logic is wrong. The frequency of extreme floods in Bangladesh has increased, it's true, but studies show their magnitude and duration have in fact decreased. And Bangladesh is far better at adapting today than it was a generation ago.... Given the weak links between warming, flooding, urbanization and the contrary link between urbanization and child marriage, climate policies would be the least effective in addressing the problem."[130]

change and increased carbon dioxide levels will speed up the jet stream, leading to more serious episodes by 2050."[131]

Climatologist Curry shakes her head in disbelief.

"I don't know who's paying for these studies? Or who, you know—where do these people get their salaries? I mean, this stuff is so ludicrous but there's this whole cottage industry," Curry said.

When I asked, "Will a carbon tax prevent airline turbulence?" Curry said, "No," and laughed.[132]

But supposedly turbulence is only the beginning, when it comes to the effects of climate change on air travel. *Mother Jones* magazine reported in 2014 that "one reason it may be harder to find Flight 370," the missing Malaysia Airlines flight was "climate change." The article explained, "Scientists say man-made climate change has fundamentally altered the currents."[133]

Climate change was even blamed for downing a jet. "Did global warming help bring down Air France flight 447?" asked Russia Today in 2009: "Russian climatologist believes global warming played a significant part."[134]

If you don't buy that, maybe you can believe that global warming will cause...more barroom brawls and rapes?

"Global warming isn't just going to melt the Arctic and flood our cities—it's also going to make Americans more likely to kill each other." *Mother Jones* magazine featured a 2014 study claiming that rape, murder, and car thefts will all increase because of man-made global warming. As the article explains, to prevent crime, we must "rein in the greenhouse gas emissions that are causing global warming in the first place." And just how is climate change going to cause this crime wave? "Police have long operated with the understanding that 'the summer is more dangerous than the winter. To the extent that climate change causes people to be out and interacting more, there will be more crime.'"[135]

In 2013, UN IPCC scientist Solomon Hsiang, a professor at the University of California at Berkeley, did a study claiming, "Global warming sparks fistfights and war." According to Bloomberg News, the study, published in the journal *Science*, found that climate change "'will systematically increase the risk of many types of conflict' ranging from barroom brawls and rape to civil wars and international disputes."[136] A year later the *Los Angeles Times* reported on a similar study published in the *Journal of Environmental Economics and Management* under the headline "Climate Change Brings More Crime." The 2014 study found that "between 2010 and 2099, climate change can be expected to cause an additional 22,000 murders, 180,000 cases of rape, 1.2 million aggravated assaults...a 0.8% increase in cases of vehicle theft."[137]

Salon magazine reported that linking more crime to climate change "could go a long way toward convincing people that climate change can directly affect their lives."[138] Which appears to be the prime motive to do these studies—to lobby for action on "climate change."

★ ★ ★
Godwin's Law

We scoff at medieval witch hunts and Aztec sacrifices, but can we put any crazy theory about the weather past our modern "experts"? In 1941, Clarence A. Mills, a professor of experimental medicine at the University of Cincinnati, claimed that warmer temperatures "may produce a trend toward dictatorial governments" and blamed global warming for—the rise of Hitler. "The rise to power of Adolf Hitler in Germany and Benito Mussolini in Italy may be due in part to the gradually warming temperature of the world," Mills explained. "People are more docile and easily led in warm weather than in cold."[139]

Interestingly, global warming may have *saved* Hitler's life and regime. Very hot temperatures on the day of the nearly successful assassination attempt on the brutal dictator's life may have saved him. Hitler survived the assassination attempt of July 20, 1944, in part because the heat had forced a change of venue for a meeting. "The briefing conference was normally held in a bunker" but on that day "it was held in a flimsy hut owing to the heat," according to Peter Hoffman's *The History of the German Resistance, 1933–1945*. Hitler might not have survived if the briefcase containing the explosives had not been moved and the bomb had been detonated in the original location in the bunker.[140]

So global warming helped bring Hitler to power by making people "more docile and easily led" and then global warming saved Hitler by forcing a change in the critical meeting place.

During the 1940s, some blamed World War II itself for causing extreme weather. An August 19, 1941, article in the *Barrier Miner* stated referred to the "ancient belief that years of war tend to bring abnormal vagaries in the weather" and reported, "There have been remarkable extremes to weather since the war began, and these have continued all through the year."[141]

The reason behind the shift in tactics by the climate science community has become increasingly transparent. Instead of allowing temperature data or other climate metrics to prove or disprove their claims of man-made global warming, the global warming activists have now shifted the goal posts, so that anything from airplane turbulence to rape and murder statistics can be used as some sort of "proof" of man-made global warming.

★ ★ ★

Warmer Is More Stable

A 2011 study from the Center for Strategic and International Studies titled "The Climate Wars Myth" found, "Since the dawn of civilization, warmer eras have meant fewer wars." As author Bruno Tertrais explained, "History shows that 'warm' periods are more peaceful than 'cold' ones. In the modern era, the evolution of the climate is not an essential factor to explain collective violence. Nothing indicates that 'water wars' or floods of 'climate refugees' are on the horizon. And to claim that climate change may have an impact on security is to state the obvious but it does not make it meaningful for defense planning." For this reason: "all things being equal, a colder climate meant reduced crops, more famine and instability. Research by climate historians shows a clear correlation between increased warfare and cold periods."[142]

And of course it's always heads they win, tails we lose. Global warming is causing more crime—and less crime would cause more global warming. The *New York Times* published an article titled, "How Lowering Crime Could Contribute to Global Warming." As the *Times* explained, "Inmates generally consume less than an average citizen in the country, so fewer prisoners might mean higher overall energy consumption. Additionally, the money saved from reducing crime would go into the government's budget and people's pockets. All that money could be spent in other ways— infrastructure, buildings or goods—that may require more energy to produce or operate, possibly adding more greenhouse gases to the atmosphere."[143]

Are you keeping this all straight? Global warming causes more crime. Reducing crime causes more global warming. And if you disagree with any part of this of this crazy scenario…you might just be a "climate denier"— and belong in jail yourself!

Believe it or not, the same dynamic is at work in…moose populations.

"Climate Change Threatens Norway's Moose," blared the headline in the Norwegian newspaper *Aftenposten* on May 15, 2008. "The popular

Norwegian moose now faces another threat: Global warming" because the animals are "[t]hreatened by higher temperatures in spring and early summer that can upset their food supplies." Bernt-Erik Sæther of the Norwegian University explained, "We're not in any doubt. The moose is extremely vulnerable to climate change."[144]

And yet a 2007 *Der Spiegel* article had said Norwegian moose "is harming the climate…through its belching and farting," calling it "more destructive to environment than cars."[145]

The Norwegian moose is threatened by climate change—and also causing climate change! This is settled science. It may be time to list the moose as an endangered species on account of its own "emissions."

CNN reported that climate change was going to cause "changes in disease trajectories, all kinds of implications that we really can't even fully fathom."[146]

Are you so sick of the global warming debate that you look forward to dying to escape it? Sorry, there is no escape. Global warming is even affecting…the dead.

★ ★ ★

Climate Change Cannibals

"Basically none of the crops will grow, most of the people will have died and the rest of us will be cannibals and civilization will have broken down. The few people left will be living in a failed state like Somalia or Sudan."

—**CNN founder Ted Turner explaining the consequences of global warming to Charlie Rose on PBS**[147]

Why do people like Turner make such wacky comments? Professor Roger Pielke Jr. offered up this observation: "I am coming to the conclusion that there is something about the climate issue that makes people—especially but not limited to academics and scientists—completely and utterly lose their senses."[148]

Retired MIT climate scientist Richard Lindzen agreed. "You look at the polls, ordinary people see through this, but educated people are very vulnerable," Lindzen said.[149]

"White Man" Blamed

Aborigines in nineteenth-century Australia blamed the bad climate on the arrival of the white man. A March 11, 1846, article in the *Maitland Mercury* explained that "great changes have taken place in the climate of Australia," citing "heavy rains" and "deluging floods" and noted, "The aborigines say that the climate has undergone this change since the white-man came in country."[152]

In 2013, the white man was once again being blamed for climate change. Climate activist Bill McKibben lamented, "White America has fallen short"—by voting for "climate deniers." In a March 14, 2013, *Los Angeles Times* op-ed, McKibben complained, "Election after election, native-born and long-standing citizens pull the lever for climate deniers."[153]

In a series on "chasing a climate deal in Paris," the *New York Times* reported that a UN climate deal was needed to win the "race to save frozen mummies." Melting mummies, you see, are "what climate change looks like." The *Times* noted, "As global temperatures have warmed, the ice covering the tombs has melted, putting many of the frozen corpses at risk of thawing and rotting."[150]

Not to be outdone, *Science* magazine reported that climate change was causing South American mummies to turn to "black ooze." The "world's oldest mummy" burial sites are now "experiencing higher humidity levels due to climate change" according to Harvard University scientists. "Tests by Harvard's Alice DeAraujo and Ralph Mitchell show that microbes that flourish in an increasingly humid climate are turning the preserved remains of Chinchorro hunter-gatherers into 'black ooze.'"[151]

Is there anything that can't be blamed on global warming? UK scientist John Brignell's list at Number Watch contains hundreds of events that are supposedly the fault of climate change.[154] There's also the impressive "(Not Quite Complete) List of Things Supposedly Caused by Global Warming" at the What Really Happened website.[155]

Who You Gonna Call?

The dire claims of the climate change activists are reminiscent of an iconic scene from the 1984 comedy blockbuster Ghostbusters. But while the film featured "three unemployed parapsychology professors," the climate change industry is about fully employed and very well-funded professors who set up shop as unique sages who can save the planet from the alleged scourge of man-made climate change.

In Ghostbusters, the main characters, played by Dan Aykroyd, Harold Ramis, and Bill Murray, warn the mayor about the alleged supernatural apocalypse about to arrive.

The iconic Ghostbusters scene:

Dr. Peter Venkman: This city is headed for a disaster of biblical proportions.

Mayor: What do you mean, "biblical"?

Dr. Ray Stantz: What he means is Old Testament, Mr. Mayor, real wrath of God type stuff.

Dr. Peter Venkman: Exactly.

Dr Ray Stantz: Fire and brimstone coming down from the skies! Rivers and seas boiling!

Dr. Egon Spengler: Forty years of darkness! Earthquakes, volcanoes...

Winston Zeddemore: The dead rising from the grave!

Dr. Peter Venkman: Human sacrifice, dogs and cats living together...mass hysteria![156]

The similarities are striking.

Ghostbusters	Climate Change Fears
"a disaster of biblical proportions"	Al Gore: "a nature hike through the book of Revelations"[157]
"Old Testament, Mr. Mayor, real wrath of God type stuff."	Huffington Post: "God's wrath caused East Coast earthquake.... God is very disappointed with humanity for leaving the gas on in our home planet. [I]t's dangerous, wasteful, and wrong."[158]
"Fire and brimstone coming down from the skies! Rivers and seas boiling!"	Former NASA lead global warming scientist James Hansen: "The Oceans will begin to boil...."[159]

"Forty years of darkness! Earthquakes, volcanoes..."	UK government advisor Bill McGuire: "Global warming is causing earthquakes and giant landslides, and could bring about an age of 'geological havoc' including 'volcano storms'...a top academic told the Telegraph...."[160]
"The dead rising from the grave!"	*Daily Mail*: "Will Climate Change Bring Back Smallpox? Siberian Corpses Could Ooze Contagious Virus if Graveyards Thaw Out, Claim Scientists"[161]
"Human sacrifice, dogs and cats living together...mass hysteria!"	Princeton physicist Will Happer: "Like the Aztecs, many scientists believe that sacrificial offerings are necessary to stabilize climate."[162]

CHAPTER 12

Not So Extreme

In August and September 2017 three category 4 hurricanes hit the United States—Harvey in Texas, Irma in Florida, and Maria in Puerto Rico. These hurricanes prompted the man-made global warming promotion machine to go into high gear, reminding everyone that they had predicted bad storms and, lo and behold, they came!

The climate change claims heard in the media included, "Harvey's intensity and rainfall potential tied to global warming" and "Because of climate change, hurricanes like Harvey are probably going to become more common." The *Independent* asked, "Will Hurricane Harvey show Trump that climate change exists?" Penn State professor Michael Mann's op-ed was titled: "It's a fact: climate change made Hurricane Harvey more deadly."[1]

Interestingly, 2017 was the first time that a Category 3 or larger storm had made landfall in the United States since 2005. The 2017 hurricanes broke an unprecedented long streak of good-weather fortune in the United States.[2]

Nevertheless, even before Harvey, the climate activists were claiming that "climate change" was causing more extreme weather events. From floods to droughts, from wildfires to hurricanes and tornadoes, the media and climate campaigners try to blame virtually every natural disaster on

Did you know?

★ Landfalls of major hurricanes in the United States have declined over the past 140 years

★ Instances of F3 or larger tornadoes have been in decline since the 1970s

★ Droughts and floods are not at historically high levels

★ ★ ★
Climate Heretics, Repent!

The *Washington Post* featured Susan Brooks Thistlethwaite, a professor of theology and a past president of the Chicago Theological Seminary, literally urging Americans to "repent" for our "sin" of causing 2013's Typhoon Haiyan. Thistlewaite's article was subtitled, "Suffering and the Sin of Climate Change Denial." She warned, "These 'superstorms' aren't an 'act of God,' but an act of willful disregard for God's creation, and the neglect of the human responsibility to care for the planet."

Thistlewaite was just warming up. Her sermon continued, "There is the moral evil of continuing to pump fossil fuels into the atmosphere, producing global warming. Second, however, is the moral evil of climate change denial, that is, those who would continue to deny, in the face of mounting evidence, that violent climate change is upon us and it is accelerating."[5]

man-made global warming. Only the scientific evidence for such claims is severely lacking. The weather is not cooperating with the global warming activists' attempts to persuade us that man-made "global warming" is endangering the planet.

Normal Variation

In 2017, Bloomberg News declared, "It's not your imagination. The weather has been weird. So weird, in fact, it's had an almost biblical feel."[3] Actually, it is your imagination. The weather has in fact been normal.

Extreme weather expert Roger Pielke Jr., a former professor in the Environmental Studies Program at the University of Colorado and author of the 2014 book *The Rightful Place of Science*: *Disasters and Climate Change*, has repeatedly testified to Congress, mostly recently in 2017, that there is "no evidence" that hurricanes, floods, droughts, tornadoes are increasing.[4] "The world is presently in an era of unusually low weather disasters. This holds for the weather phenomena that have historically caused the most damage: tropical cyclones, floods, tornadoes and drought.," Pielke said in July 2017.[6]

The "extreme weather" we're having now is not any worse than the weather we were having during the late twentieth century and earlier, when there was much less CO_2 in the atmosphere. "It's not a theoretical issue what the weather would be like below 350 ppm. Until 1988, the weather was below 350ppm. What we see is all the extreme weather we had back then," as Real Climate Science website founder Tony Heller has pointed out. "If you look through the historical record the weather was just as bad or worse under 350ppm."[7]

Climatologist John Christy testified before Congress in 2012, "There is a lack of evidence to blame humans for an increase in extreme events. One cannot convict CO_2 of causing any of these events, because they've happened in the past before CO_2 levels rose." As Christy pointed out, "There are innumerable types of events that can be defined as extreme events— so for the enterprising individual (unencumbered by the scientific method), weather statistics can supply an unlimited, target-rich environment in which to discover a 'useful' extreme event."[12]

★ ★ ★

It's Always What Climate Change Looks Like

2017: "Harvey Is What Climate Change Looks Like"
    —*Politico*[8]

2015: "Hurricane Matthew Looks a Lot like the Future of Climate Change"
    —**CNN**[9]

2012: "Superstorm Sandy Is 'What Global Warming Looks Like'"
    —**Environment News Service**[10]

2005: "The hurricane that struck Louisiana yesterday was nicknamed Katrina by the National Weather Service. Its real name is global warming."
    —*Boston Globe*[11]

Christy explained why the extreme weather claims are unscientific: "The non-falsifiable hypotheses can be stated this way, 'whatever happens is consistent with my hypothesis.' In other words, there is no event that would 'falsify' the hypothesis. As such, these assertions cannot be considered science or in any way informative since the hypothesis' fundamental prediction is 'anything may happen.' In the example above if winters become milder or they become snowier, the non-falsifiable hypothesis stands. This is not science."

As climatologist Roy Spencer noted in 2016, "Global warming and climate change, even if it is 100% caused by humans, is so slow that it cannot

★ ★ ★

"Weather-Porn"

Climate skeptic Jo Nova called the claims about extreme weather in Al Gore's 2017 *An Inconvenient Sequel: Truth to Power* "primal weather-porn." As Nova wrote, "Cherry-picked extremes. The long tenuous chain of cause and effect was glossed over repeatedly with handwaving. The system was the most complex on earth, yet somehow scientists know what causes what.... This is a never ending game for Gore. Until we get perfect weather on Earth, on all 150 million square kilometers terra firma, he will always be able to say 'boo.'"[13]

be observed by anyone in their lifetime. Hurricanes, tornadoes, floods, droughts and other natural disasters have yet to show any obvious long-term change. This means that in order for politicians to advance policy goals (such as forcing expensive solar energy on the masses or creating a carbon tax), they have to turn normal weather disasters into 'evidence' of climate change."[14]

Meteorologist Tom Wysmuller, formerly with NASA, explained that severe weather has actually decreased: "Precipitation over the last hundred years, precipitation is relatively flat world wide.... If you match CO_2 and tornado counts, they are down. If you plot CO_2 versus hurricane strikes, as you increase CO_2, hurricane strikes go down. Does that mean more CO_2 means less hurricanes? No because correlation is not necessarily causation. But all these claims about more hurricanes, more intense hurricanes, they're false."[15]

Tropical Cyclones and Hurricanes

In 2017, the National Oceanic and Atmospheric Administration stated that it was "premature to conclude that human activities—and particularly greenhouse gas emissions that cause global warming—have already had a detectable impact on Atlantic hurricane or global tropical cyclone activity." The United States went from Hurricane Wilma in 2005 to Hurricane Harvey in 2017 with no Category 3 or larger hurricanes making landfall. That twelve-year period was the longest hiatus since at least 1900 and possibly even back to the U.S. Civil War.[16]

★ ★ ★

New Lyrics to an Old Tune

In the 1970s and 1980s, extreme weather used to be blamed on "global cooling."

According to a 1981 *Chicago Tribune* article, "Climatologists now blame recurring droughts and floods on a global cooling trend that could trigger massive tragedies for mankind."[17]

In 1974, NOAA linked extreme weather events to a cooling Earth: "Many climatologists have associated this drought and other recent weather anomalies with a global cooling trend and changes in atmospheric circulation which, if prolonged, pose serious threats to major food-producing regions of the world."[18]

A 1975 *Newsweek* article titled "The Cooling World" explained how extreme weather was being caused by colder temperatures. The piece claimed that cooling "causes an increase in extremes of local weather such as droughts, floods, extended dry spells, long freezes...."[19]

NOAA's Hurricane Research Division has been tracking hurricanes that have struck the United States since 1851. The drought in major hurricane activity was a very unusual event that "likely" occurs only once every 177 years, according to a study by NASA's Goddard Institute for Space Studies (GISS).[20]

Hurricane Harvey in August 2017, broke the twenty-five-year streak of no Category 4 to 5 hurricanes making landfall in the United States since 1992 (Hurricane Andrew) and the twelve-year record-breaking streak of no Category 3 or larger Hurricanes making landfall since 2005 (Wilma). Both landfalls and hurricane strength were "down by ~20% since 1900," according to Pielke. The worst decade for major (Category 3, 4 and 5) hurricanes was the 1940s, according to National Oceanic and Atmospheric Administration data.[21]

The UN IPCC Fifth Assessment Report in 2013 found, "Current datasets indicate no significant observed trends in global tropical cyclone frequency over the past century.... No robust trends in annual numbers of tropical

★ ★ ★

Cyclone, Typhoon, or Hurricane?

"The only difference between a hurricane, a cyclone, and a typhoon is the location where the storm occurs," NOAA explains. "Hurricanes, cyclones, and typhoons are all the same weather phenomenon; we just use different names for these storms in different places. In the Atlantic and Northeast Pacific, the term 'hurricane' is used. The same type of disturbance in the Northwest Pacific is called a 'typhoon' and 'cyclones' occur in the South Pacific and Indian Ocean."[25]

storms, hurricanes and major hurricanes counts have been identified over the past 100 years in the North Atlantic basin."[22]

In 2016 atmospheric scientist Philip Klotzbach released an analysis showing the "global Accumulated Cyclone Energy" having a "statistically insignificant downward trend since 1985."[23] A 2014 study published in the *Journal of Climate* found that strong hurricanes were more frequent than had previously been thought during the lower-CO_2 period from 1851 to 1898.[24]

In a 2016 analysis, Kenneth Richard found thirty peer-reviewed scientific papers revealing the lack of connection between hurricanes and global warming. "It has been determined from paleoclimate analyses that cooler ocean temperatures are associated with more tropical storms and major hurricanes than warmer ocean temperatures," Richard wrote.[26]

A 2017 paper published in the journal *Natural Hazards*, found "that the annual number of SCS-G TC [South China Sea–generated tropical cyclones] landfalls over China and Vietnam presented a decreasing linear trend in 66 years from 1949 to 2014."[27]

Also in 2017, the Southern hemisphere was having "one of its quietest cyclone seasons on record." Andrew Watkins, the manager of climate prediction services at Australian Bureau of Meteorology, asked, "Where have all the cyclones gone?" It was the second consecutive quiet season in Australia.[28] An October 2017 analysis by Paul Homewood noted, "All of the attention on hurricanes this summer has been in the Atlantic. It may come as a surprise to some then that global cyclone activity has been pretty much normal so far this year.... In fact, the North Atlantic is the only region where ACE (accumulated cyclone activity) is above average, with the Western Pacific and Southern Hemisphere being particularly low."[29]

"Hurricane Sandy" in 2012 had been used as a prime example of man-made global warming. Democrat Barbara Boxer of California, chair of the U.S. Committee on Environment and Public Works, declared, "Hurricane Sandy has shown us all what the scientists sitting right in this room said the day I got the gavel, and they told us exactly what would happen and it's all happening."[30]

In fact, Superstorm Sandy was not even a Category 1 hurricane at the time when it made landfall in Brigantine, New Jersey, in 2012.[31] "Hurricane Sandy was not a hurricane when it came ashore. It was a gigantic atmosphere low pressure system more consistent with a very strong winter type storm," climatologist Patrick Michaels pointed out. "Look, the fact [of] the matter is, when you warm up the planet from greenhouse gases and you decrease

★ ★ ★

Moving the Goalposts

The lack of category 3 or larger "major" hurricanes making landfall from 2005 until 2017, was becoming something of an embarrassment to climate activists—so much so that they advocated to recategorize hurricanes, to revise the statistics to make them look worse.

Andrew Freedman, the warmist science editor at the news outlet Mashable, suggested in 2016 that "it's time to face the fact that the way we measure hurricanes and communicate their likely impacts is seriously flawed." Freedman argued, "We need a new hurricane intensity metric that more accurately reflects a storm's potential to cause death and destruction well inland."[32]

For climate activists, it would be convenient if hurricane measurements could be effectively altered with a new metric redefining storms that did not even meet Category 1 standards with new ranking criteria that inflated storm measurements.

temperature contrast between the polar regions and the tropics—that's what drives the jet stream—you make weather actually lazier. There's less energy to impart to spin tornadoes, you have less cold temperatures."[33]

In 2016, climate activists attempted to blame "climate change" for causing Hurricane Hermine, which hit Florida. Hurricane Hermine was a Category 1 storm that finally broke a record eleven-year span of no land-falling hurricanes at all hitting Florida. As meteorologist Joe Bastardi noted, "portraying Hermine as some kind of climate change demon is either ignorance as to the history of hurricanes or deceit." Climate activists are "warning us about something that occurred much more frequently in the past, yet trying to blame it on an agenda-driven issue."[34]

In 2017, as we have seen, the nearly twelve-year record hurricane streak of no category 3 or larger storms hitting the United States came to a crashing halt with Hurricane Harvey, which made landfall near Rockport, Texas, as a Category 4 hurricane with winds of 130 mph on August 25. Harvey will go down in the record books as the top rainfall-producing hurricane, dropping 60.58 inches in Nederland, Texas, from August 24 to September 1. The previous top rain producer was Tropical Storm Amelia, which produced 48 inches in Medina, Texas, in 1978. [35]

As Hurricane Harvey was battering the coast, the climate activists and the media began the incessant drumbeat of how man-made "global warming" had made Harvey worse.

But Climatologist Judith Curry refuted such claims: "Anyone blaming Harvey on global warming doesn't have a leg to stand on. The huge amounts of rain are associated with Harvey's stalled movement."[36]

Hurricane Irma made its Florida landfall on September 10 as a category 4 hurricane. Hurricane Maria also made landfall in the United States as category 4 hurricane. Neither storm was unprecedented. Colorado State University meteorologist Philip Klotzbach's analysis of Irma found that it made landfall in Florida at 929 millibars (mb) of landfall pressure, tying it for the seventh most powerful storm to hit the mainland since recordkeeping began in the 1850s. Hurricane Harvey, ranked eighteenth at 938 mb, placing it in a three-way tie with an 1898 Georgia hurricane and Hurricane Hazel in 1954.[37] Hurricane Maria had the eighth-lowest landfall pressure (917 mb) on record in the Atlantic Basin.[38]

★ ★ ★

Fake News Makes Landfall

Associated Press climate reporter Seth Borenstein was quick to pounce with his usual promotion of man-made climate fears after Hurricane Harvey made landfall. "With four big hurricanes, a powerful earthquake and wildfires, it seems that nature recently has just gone nuts," Borenstein wrote in a September 7, 2017, AP article. "Some say the recent global increase in powerful hurricanes fits perfectly with global warming."[42]

Meteorologist Anthony Watts noted, "With Irma ranked 7th, and Harvey ranked 18th, it's going to be tough for climate alarmists to try connecting these two storms to being driven by CO_2/global warming. But they'll do it anyway."[39]

And they did. Climate activists immediately claimed man-made climate change was driving these hurricanes, despite the historical evidence. A Bloomberg News article claimed, "Hurricane Irma Made Worse by Climate Change."[40] The UN IPCC's Michael Mann, a Penn State professor embroiled in the Climategate scandal, couldn't restrain himself from engaging in climate porn after Harvey hit Texas. "We're starting to talk about conditions that will literally force us to relocate the major coastal cities of the world, to relocate the better part of the billion people. We're talking about a planet with a larger global population and great competition for diminishing land, water and food. That is a recipe for a catastrophe," Mann said on September 7, 2017.[41]

But the scientific data didn't support those claims. On September 13, 2017, research meteorologist Ryan Maue wrote, "Now almost 50-years of global hurricane data. No trends in frequency in number of named storms or those that reach hurricane-force."[43] He pointed out, "During the past half-century tropical storms and hurricanes have not shown an upward

★ ★ ★

The Horror! The Horror!

After Hurricane Irma hit Florida, the UK newspaper the *Independent* went full warmist, featuring a September 7, 2017 article by Andrew Griffin titled, "Hurricane Irma Likely to be Followed by More Extreme Weather Events So We Should Prepare for Horror of Global Warming Now, Say Experts."[44] I published a point-by-point rebuttal to the article's overwrought claims at my website:

The *Independent* claims:

"Hurricane Irma likely to be followed by more extreme weather events."

Climate Depot responds:

A meaningless statement. Extreme weather events have always happened and will always happen.

The *Independent* claims:

"The world is going to be hit by more horrifying weather events like the hurricanes Irma and Harvey."

Climate Depot responds:

Yes. That is true. The world has always been hit by horrific storms and extreme natural events. Climate activists are basically saying "many bad things will happen because of global warming" and then when a bad thing happens, they tout "we predicted it!" But if you look at the history of major landfalling hurricanes, you can see a declining trend as CO_2 has risen.

The *Independent* claims:

"Global warming is likely to trigger a run of extreme weather events. . . . "

Climate Depot responds:

Concern about "global warming" has coincided with unusually low extreme weather so far, despite the recent hurricanes.

The *Independent* claims:

"The rapid pace of climate change is set by government policies in the U.S. and many other countries."

Climate Depot responds:

Wow. Governments set the "pace of climate change." About as likely as witches controlling the weather.

The *Independent* claims:

"Planet Earth's climate is in upheaval and we know exactly what is causing it: right now."

Climate Depot responds:

Earth's climate is not in any more "upheaval" than past geologic history. It's medieval witchcraft to claim that "we know exactly what is causing" bad weather.[45]

trend in frequency or accumulated energy. Instead they remain naturally variable from year-to-year. The global prevalence of the most intense storms (Category 4 and 5) has not shown a significant upward trend either."[46]

Maue mocked the claims of "extreme weather getting more extreme." That "isn't a scientific statement but anecdotal. In a warmer world, some events may become weaker," Maue wrote. "It's actually been fairly quiet in the Atlantic since the 2004 and 2005 seasons. Curious how two storms changes that so quickly?"[47]

Climatologist Roger Pielke Sr. wrote about the media claims linking Hurricanes Harvey and Irma to about the climate change link, "Amazing to see claims of how hurricane activity will change in future based on climate models, yet same models cannot predict past activity."[48]

Tony Heller of Real Climate Science wrote of the Irma claims, "Florida has had 119 hurricanes since 1850, but the last one was due to climate change."[49]

Statistician Bjorn Lomborg pointed out: "Harvey and Irma are terrible, but.... Major landfalling US hurricanes [are] trending downwards over [the] past 140 years." As he explained, "We need perspective: Despite Harvey and Irma catastrophes, death risk from hurricanes is declining 1900–2016."[51]

The No Tricks Zone website featured a graph showing that "as CO_2 has risen, major landfalling US hurricanes declined over the past 140 years."[52]

★ ★ ★

"Stress, Anxiety, Post-Traumatic Stress Disorder"

A 2014 report by the American Psychological Association (APA) and ecoAmerica, an environmental advocacy group, titled "Beyond Storms and Droughts: The Psychological Impacts of Climate Change," claimed that extreme weather fueled by man-made climate change "will mean more stress, anxiety, PTSD in the Future" as well as growing "substance abuse" and "broad psychological impacts." According to Norman B. Anderson, the APA's CEO, "That means a future with heightened levels of stress, anxiety, post-traumatic stress disorder and depression, as well as a loss of community identity—if nothing is done to stop or slow emissions of industrial-produced greenhouse gases into the atmosphere.... as well as increases in violence and crime rates as a result of higher temperatures or competition for scarce resources."[50]

★ ★ ★

Fossil Fuels Saved Thousands

Fossil fuel advocate Alex Epstein of the Center for Industrial Progress had this take on Hurricane Harvey: "Those in the industry should feel proud of producing the affordable, reliable energy we need to cope with an inherently dangerous climate. Before industrial development, and in the underdeveloped world today, storms like Harvey routinely claimed tens of thousands of lives. That equates to thousands upon thousands of lives saved in Texas thanks to fossil fuels and the development they make possible. The opponents of fossil fuels trying to use this tragedy to promote their anti-energy, anti-development agenda should be exposed as advocating policies that would make storms like Harvey more dangerous and more deadly."[53]

Al Gore declared in his usual hyperbolic style, "We are departing the familiar bounds of history as we have known it since our civilization began." The former vice president asked, "And why? Because today like all days we will put another 110 million tons of man-made heat-trapping pollution into the atmosphere, using the sky as an open sewer."[54]

In September 2017 climate scientist Roy Spencer wrote a fifty-page e-book titled *Inevitable Disaster: Why Hurricanes Can't Be Blamed on Global Warming* just to address these over-the-top claims. Spencer ridiculed the notion that Hurricane Harvey or Irma was "what climate change looks like." As Spencer explained, "This isn't what human-caused climate change looks like. It's what weather looks like."

Spencer dismissed Gore's attempt to explain hurricane science as "some sort of pseudo-meteorological gobbledygook." As he pointed out, "There have been many years with multiple Cat 4 hurricanes in the Atlantic, but there is nothing about global warming theory that says more of those will make landfall. While the official estimate is that this was the first time two Cat 4 storms hit the U.S., since Florida was virtually unpopulated before 1900, we probably don't really know."[55]

Spencer also rejected the claims that warmer sea surface temperatures drive hurricanes. "Major hurricanes in the tropical Atlantic, Caribbean, and Gulf of Mexico are not limited by sea surface temperatures, which are warm enough every hurricane season to support catastrophic hurricanes.... major hurricanes don't really care whether the Gulf is above average or below average in temperature. Why is that? It's because hurricanes require a unique set of circumstances to occur, and sufficiently warm SSTs is only one. The Gulf of Mexico is warm enough every summer to produce a major hurricane." Spencer wrote. He noted, "I did my Ph.D. dissertation on the structure and energetics of incipient tropical cyclones, and have published a method for monitoring their strength from satellites."[56]

Spencer explained why Harvey's rainfall totals were so high. "What made Harvey rain totals exceptional was the system stalled next to the coast, which was due to a very temporary weakening of atmospheric steering currents." As he pointed out, "the 12-year hurricane drought" was much "more unusual." Statistics would predict such a hiatus only "once every 250–300 years, but so far nobody is attributing that fortunate happenstance to climate change."[57]

Professor Roger Pielke Jr. addressed the Hurricane Harvey and Irma hype. "If you predict something bad will occur in 2080–2100 (worse hurricanes!) and you then claim to see it in 2017 (Harvey, Irma! Told you so!), that does not prove you 'right'—it actually says your prediction is wildly off base," Pielke wrote on September 18, 2017. "Neither tropical cyclones globally, Atlantic hurricanes overall, U.S. landfalls nor U.S. normalized damage has gotten worse (that is more frequent or intense) over climate time scales. (Don't take it from me, this is straight out of the IPCC and US government's National Climate Assessment)."[58]

Tornadoes

CBS newsman Dan Rather warned ominously during a 2001 CBS *Evening News* segment that climate change was going to make tornadoes much worse: "The forecast from Hell! Why America may see more killer tornadoes."[59]

★ ★ ★

Misleading Statistics

Meteorologist Michael Mogil is outraged at climate activists' misuse of hurricane statistics. Mogil wrote in September 2017: "We all know that statistics themselves don't lie, but the people who use statistics may intentionally or unintentionally do so.... It would be easy for some people to draw an incorrect conclusion from [climate activist Eric] Holthaus' data—i.e., that intense hurricane activity is escalating. But.... hurricanes are not becoming more intense. Then, one must recognize that there has been a dramatic change in global observing and forecasting systems since the mid 19th century. In fact, it wasn't until the latter part of the 1800's that hurricane warning offices were established and it wasn't until the mid 20th century before the National Hurricane Center was created. Hurricane hunter aircraft were not employed until the 1940's and the first weather satellite didn't arrive on the scene until 1960. Since 1960, satellite observation systems have evolved to be highly powerful, high frequency, and high resolution observing tools. These satellites can now see entire ocean basins; in earlier years, point ship and island reports were all that meteorologists had available. To say that 'There was likely undercounting pre-1960,' would be an understatement."[60]

But tornadoes do not back up climate change claims any more than hurricanes do. There is "no scientific consensus or connection between global warming and tornadic activity," emphasized Greg Carbin, tornado warning coordination meteorologist at NOAA's Storm Prediction Center in Norman, Oklahoma.[61] Climate statistician Caleb Rossiter of American University noted, "What is happening to global temperature and global ice and water levels and all sorts of things, little individual tornadoes, come from long trends, long before there were human beings running around in SUV's."[62]

And in any case big tornadoes have seen a drop in frequency since the 1950s. "There has been a downward trend in strong (F3) to violent (F5) tornadoes in U.S. since 1950s," climatologist Roy Spencer explained. "Obviously, the conclusion should be that warming causes fewer strong tornadoes, not more. (Or, maybe a lack of tornadoes causes global warming!)."[63]

According to Carbin, "NOAA statistics show that the last 60 years have seen a dramatic increase in the reporting of weak tornadoes, but no change in the number of severe to violent ones."[64] And as extreme weather expert Roger Pielke Jr.'s analysis of the data reveals, "Tornadoes have not increased in frequency, intensity or normalized damage since at least 1950, and there is some evidence to suggest that they have actually declined."[65] Pielke found, "Over the past six decades, tornado damage has declined after accounting for development that has put more property into harm's way."[66] The bottom line: "Recent years have seen record low tornadoes."[67]

Tony Heller of the Real Science website has pointed out, "The worst tornado outbreak in recent history occurred on April 3–4, 1974 at the peak of the 1970's ice age scare."[68]

The National Academy of Sciences' review of its draft 2017 "Climate Science Special Report" said that "there is, at best, scant evidence that tornadoes are exhibiting changes linked to climate change."[69]

Meteorologist Bastardi noted the "extreme lack of tornadoes" in 2016,[70] which NOAA tornado data revealed to be one of the quietest years since record keeping started in 1954. Tornadoes were below average for the fifth year in a row, with several recent years at or near record low tornado activity, according to NOAA data.[71]

Still somehow the federal agency managed to hype statistics to show an alleged increasing number of tornadoes. Climate analyst Paul Homewood explained how NOAA tried to spin global warming fears in 2017 by inflating the tornado statistics. "According to NOAA, the number of tornadoes has been steadily growing since the 1950s, despite a drop in numbers in the last five years. But with increased National Doppler radar coverage, increasing population, and greater attention to tornado reporting, there has been an increase in the number of tornado reports over the past several decades. This can create a misleading appearance of an increasing trend in tornado frequency," Homeward pointed out. "The bottom line is that the NOAA headline graph is grossly dishonest," he explained. "NOAA themselves know all of this full well. Which raises the question—why are they perpetuating this fraud?"[72]

Drought

Droughts are also not following the predictions of the activists' warning of man-made climate change.

But readers of the *New York Times* don't get that information. Instead they are told that Americans are to blame for killing babies by causing droughts in places like the island of Madagascar off southern Africa. As Nicholas Kristof wrote in a 2017 *Times* op-ed, of an impoverished African desperately seeking medical aid for her infant son, "We Americans may be inadvertently killing her infant son. Climate change, disproportionately caused by carbon emissions from America, seems to be behind a severe drought that has led crops to wilt across seven countries in southern Africa."[73]

But the scientific reality is quite different.

"Globally, and I quote from a recent paper in *Nature*, there has been little change in drought in the past 60 years.... Drought has, and here I quote the IPCC, 'for the most part, become shorter, less frequent, and cover a smaller portion of the U. S. over the last century,'"[74] noted Roger Pielke Jr. in 2014 Congressional testimony. "U.S. Midwestern drought has decreased in past 50+ years? That is not skepticism; that's according to the UN Intergovernmental Panel on Climate Change."[75]

In 2017 the federal government released yet another key piece of scientific data that counters the man-made global warming narrative. The federal U.S.

Drought Monitor report revealed that droughts in the United States were at record lows. "Drought in the U.S. fell to a record low this week, with just 6.1% of the lower 48 states currently experiencing such dry conditions, federal officials announced Thursday. That's the lowest percentage in the 17-year history of the weekly U.S. Drought Monitor report," *USA Today* reported on April 27.[76]

The National Academy of Sciences' review of its draft 2017 "Climate Science Special Report" revealed that "analysis of global and continental-scale trends indicates that drought severity and other statistics have actually declined."[79]

A 2015 study published in the journal *Science Advances* found that droughts in the past two thousand years were both more severe and longer-lasting than modern-day droughts.[80] "A new atlas shows droughts of the past were worse than those today—and they cannot have been caused by man-made CO_2. Despite the claims of 'unprecedented' droughts, the worst droughts in Europe and the US were a thousand years ago," noted Australian climate analyst Jo Nova of the new investigation.[81]

A 2016 study published in the *International Journal of Climatology* found, "For most of the CONUS [Continental U.S.] drought frequency appears to have decreased during the 1901 through 2014 period."[82]

Much more severe California droughts occurred in times of lower and thus allegedly safe CO_2 levels. According to the data, some droughts lasted for over two centuries.[83]

A Book You're Not Supposed to Read

The Rightful Place of Science: Disasters and Climate Change by Roger Pielke Jr. (Consortium for Science, Policy, & Outcomes, 2014).

★ ★ ★
Thirst-Quenching

Climate activists had declared California to be in a permanent drought, "probably forever."[77] But California's statewide drought literally disappeared in 2017, as the state added more than 350 billion gallons of rain to their reservoirs as the rainy season ranked highest in 122 years of record keeping.[78]

★ ★ ★

New Lyrics to an Old Tune

Before global warming caused drought—global cooling caused drought. In 1974, the *NOAA Magazine* published an article by Patrick Hughes warning that the global cooling trend was going to increase droughts around the world. "Some climatologists think that if the current cooling trend continues, drought will occur more frequently in India—indeed, through much of Asia, the world's hungriest continent," Hughes wrote.[86]

In 1976, the *New York Times* featured climatologists who "believe that the earth's climate has moved into a cooling cycle, which means highly erratic weather for decades to come." According to the *Times*, "The weather seems to have gone berserk lately. The tennis courts at Wimbledon in England have not been as parched since the 1920's. The same is true for croplands in northern France, the Soviet Union, Minnesota and the Dakotas. It's so dry, brush fires have started several weeks early in California, and water is being rationed."[87]

A 2014 analysis of the California drought found, "Researchers have documented multiple droughts in California that lasted 10 or 20 years in a row during the past 1,000 years—compared to the mere 3-year duration of the current dry spell. The two most severe mega-droughts make the Dust Bowl of the 1930s look tame: a 240-year-long drought that started in 850 and, 50 years after the conclusion of that one, another that stretched at least 180 years."[84]

California's recent record drought also fails the man-made global warming attribution test.

U.S. government scientists have admitted that recent droughts hyped by the media and climate campaigners are not due to climate change. "This is not a climate change drought," said Robert Hoerling, a NOAA research meteorologist, who served as the lead author of the U.S. Climate Change Science Plan Synthesis and Assessment Report. Hoerling was referring to a 2011 Texas drought. "The good news is that this isn't global warming. This is not the new normal in terms of drought."[85]

A 2016 study published in *Journal of Climate* found that "climate change" had made California drought "less likely." According to the abstract of the article, "The results thus indicate the net effect of climate change has made agricultural drought less likely, and that the current severe impacts of drought on California's agriculture has not been substantially caused by long-term climate changes."[88]

Climatologist Dr. David Legates of the University of Delaware testified to the U.S. Senate in 2014, "My overall conclusion is that droughts in the United States are more frequent and more intense during colder periods. Thus, the historical record does not warrant a claim that global warming is likely to negatively impact agricultural activities."[89]

Floods and Heavy Rains

Peer-reviewed studies and current data refute the claims that global warming is making floods worse.

Extreme weather expert Roger Pielke Jr. asked, "Are US floods increasing? The answer is still 'No,' A new paper...in the *Hydrological Sciences Journal* shows that flooding has not increased in U.S. over records of 85 to 127 years. This adds to a pile of research that shows similar results around the world," Pielke wrote.[90] Pielke explained that the UN IPCC had found, "[n]o gauge-based evidence...for a climate-driven, globally widespread change in the magnitude and frequency of floods." He asked, "How about IPCC SREX authors on floods?...'A direct statistical link between anthropogenic climate change and trends in the magnitude/frequency of floods has not been established.'"[91] Pielke pointed out, "Flood disasters are sharply down. U.S. floods not increasing either." Admittedly, "Floods suck when they occur. The good news is U.S. flood damage is sharply down over 70 years."[92]

The National Academy of Sciences' review of its draft 2017 "Climate Science Special Report" said that "within the existing literature, few locations show statistically significant changes in flooding nor have they [changes in flooding] been clearly linked to precipitation or temperature." The report added that flooding was exhibiting "no clear national trend."[93]

Tony Heller of Real Science explained, "The world's ten deadliest floods all occurred before 1976." In other words, "all of the world's deadliest floods occurred with CO_2 well below 350 PPM."[94]

A 2011 U.S. government study titled "Has the Magnitude of Floods across the USA Changed with Global CO_2 Levels?" found no evidence that man-made climate change had caused more severe flooding during the past hundred years in the United States. Currently, "we do not see a clear pattern that enables us to understand how climate change will alter flood conditions in the future," U.S. Geological Survey (USGS) scientist Robert Hirsch explained. The report, published in the *Hydrological Sciences Journal*, found "In none of the four regions [of the U.S.] defined in this study is there strong statistical evidence for flood magnitudes increasing with increasing global mean carbon dioxide concentration. One region, the southwest, showed a statistically significant negative relationship between [rising CO_2] and flood magnitudes."[95] Nevertheless, the USGS pleaded for more study (and money!). The *Hill* newspaper reported that, according to the authors of the report, "More research is necessary to better understand the relationship between climate change and flooding."[96]

In 2015, Robert Holmes, USGS's National Flood Hazard Coordinator, reported, "USGS research has shown no linkage between flooding (either increases or decreases) and the increase in greenhouse gases. Essentially, from USGS long-term streamgage data for sites across the country with no regulation or other changes to the watershed that could influence the streamflow, the data shows no systematic increases in flooding through time."[97]

In September 2017, after Hurricane Harvey's massive flooding in Houston, yet another peer-reviewed study revealed that flooding was not on the rise. The findings, published in the *Journal of Hydrology*, "provide a firmer foundation and support the conclusion of the IPCC (Hartmann et al., 2013) that compelling evidence for increased flooding at a global scale is lacking." According to the study, "the number of significant trends was about the number expected due to chance alone" and the "changes in the frequency of major floods are dominated by multidecadal variability."[98]

The media and climate activists like to hype individual rare storm events, but these claims don't hold up to scrutiny, either. South Carolina's "thousand-year flood" in 2015 turned out to be wildly overhyped as "the majority of USGS streamgages had flood peaks that were less than 10-year floods," according to the USGS analysis, which also found "no linkage between flooding and increase in greenhouse gasses" in the United States. In fact, a USGS report found, "The majority of USGS streamgages had flood peaks that were less than 10-year floods" and the "analysis show[ed] NO indication that a 1000-year flood discharge occurred at any USGS streamgages." [99]

In 2017 Al Gore overcranked the hype when he claimed that Hurricane Harvey's rainfall totals constituted a "once-in-25,000-year event" and in some parts of Texas, a "once-in-500,000-year event."[100]

Climatologist Roy Spencer pointed out that those claims were unsubstantiated. "Unfortunately, there seems to be a trend toward classifying events as '1 in 1,000 years,' when there is no way of knowing such things. This is especially true for floods, where paving of urban and suburban areas causes increasing runoff, making river flooding worse for the same amount of rainfall. This is a big reason why flood events have gotten worse in the last 100

★ ★ ★

New Lyrics to an Old Tune

Ironically, heavy rains were said to be caused by "global cooling" in the 1970s. *Time* magazine in an article titled "Another Ice Age?" warned that a cooling climate was responsible for such bad weather events in 1974: "During 1972 record rains in parts of the U.S., Pakistan and Japan caused some of the worst flooding in centuries."[101]

> ★ ★ ★
> # The Weather Lottery
>
> Your chance of the winning the lottery is very low, but the chance of someone, somewhere winning the lottery are very high. The climate campaigners and the media essentially hype the "winners" of the extreme weather lottery, wherever they are, and attempt to imply these events are happening everywhere. Extreme weather always strikes somewhere at some time, and it always will, so there is no shortage of examples of "record" storms. Lotteries and casinos do the same thing in their ads—showing the winners, and implying that you are just one ticket or spin away from joining them.

years.... it has nothing to do with 'climate change.'" Spencer also explained, "Remember, it is perfectly normal to have a 1 in 100 year event every year…as long as they occur in different locations. That's how weather records work."[102]

Meteorologist Topper Shutt explained the misuse of the term 100-year flood after Hurricane Harvey hit Houston in 2017. "A 500 year flood does not mean that an area will see a flood of that magnitude once in 500 years. It means that in any given year there is a .2% chance of a 500 year flood and likewise a 1% chance every year for a 100 year flood," Shutt wrote. "Remember, we are talking about billions of years of climate and usually just a hundred years of actual, observational data. Secondly, urban development reduces the surface of the ground that allows the rain to permeate into the ground. Adding parking lots, more roads and driveways create more runoff. Thirdly, at least in the case of Houston 1000s of homes have been built close to streams, creeks and bayous that should have never been built in the first place."[103]

Wildfires

The mainstream media seems to be very sure that wildfires are getting worse because of man-made global warming. *ABC World News Tonight* warned in 2014 that "here in America, more wildfires, intense burns" have arrived courtesy of climate change. *CBS This Morning* featured climate fear promoter Michio Kaku, predicting "hundred-year droughts, hundred-year forest fires" and claiming that "something is very dangerously happening with the weather."[104]

Al Gore also thinks he knows all about wildfires. "All over the West we're seeing these fires get much, much worse," Gore said in 2017, adding, "the underlying cause is the heat."[105]

But the science tells a very different story. A 2016 study published by the Royal Society reported, "There is less fire in the global landscape today than centuries ago" and the "global area burned" has seen a "slight decline over past decades." The study, by Stefan Doerr and Cristina Santín of Swansea University in Wales, noted that "many consider wildfire as an accelerating problem, with widely held perceptions both in the media and scientific papers of increasing fire occurrence, severity and resulting losses. However, important exceptions aside, the quantitative evidence available does not support these perceived overall trends." The study also found that the data for the western U.S. indicates "little change overall, and also that area burned at high severity has overall declined compared to pre-European settlement. Direct fatalities from fire and economic losses also show no clear trends over the past three decades." The researchers concluded, "The data available to date do not support a general increase in area burned or in fire severity for many regions of the world. Indeed there is increasing evidence that there is overall less fire in the landscape today than there has been centuries ago, although the magnitude of this reduction still needs to be examined in more detail."

According to the study, "fire is a fundamental natural ecological agent in many of our ecosystems and only a 'problem' where we choose to inhabit these fire-prone regions or we humans introduce it to non-fire-adapted ecosystems. The 'wildfire problem' is essentially more a social than a natural one."[106]

Researchers from the Universidad Carlos III de Madrid found that "climate change" is not to blame for increased forest fires in Mediterranean basin. "The change in the occurrence of fires that are recorded in the historical research cannot be explained by the gradual change in climate." The fires instead "correspond to changes in the availability of fuel, the use of sources of energy, and the continuity of the landscape."[107]

In the United States, wildfires are also due in part to a failure to thin forests or remove dead and diseased trees. In 2014, forestry professor David B. South of Auburn University testified to the U.S. Senate Environment and Public Works Committee that "data suggest that extremely large megafires were four-times more common before 1940," adding that "we cannot reasonably say that anthropogenic global warming causes extremely large wildfires." As he explained, "To attribute this human-caused increase in fire risk to carbon dioxide emissions is simply unscientific."[108]

The evidence is so strong that even the *Los Angeles Times* featured an article rebuking Governor Jerry Brown for his claims that California's 2015 wildfires were "a real wake-up call" to reduce carbon dioxide emissions, which he claimed were "in many respects driving all of this." The *Times* article noted, "But scientists who study climate change and fire behavior say their work does not show a link between this year's wildfires and global warming, or support Brown's assertion that fires are now unpredictable and unprecedented. There is not enough evidence, they say."[109]

Dominick DellaSala, chief scientist at the Geos Institute in Ashland, Oregon, has conducted research on fires in the western United States and found them declining. "If we use the historical baseline as a point in time for comparison, then we have not seen a measurable increase in the size or the severity of fires," DellaSala said. "In fact, what we have seen is actually a deficit in forest fires compared to what early settlers were dealing with when they came through this area."[110]

A 2014 study found that Colorado wildfires have not become more severe since the 1900s. "The severity of recent fires is not unprecedented when we look at fire records going back before the 1900s," said research scientist Tania

Schoennagel. The study, "one of the largest of its kind ever undertaken in the western United States," was published in the journal *PLOS ONE* and funded by the National Science Foundation.[111]

Ocean Acidification

One of the mainstays of climate fear is that rising carbon dioxide concentrations will cause the oceans to acidify. But the data refutes these claims.

A Climate Science Report You're Not Supposed to Read

"Why Scientists Disagree About Global Warming" by the NonGovernmental International Panel on Climate Change (NIPCC), published by the Heartland Institute as an alternative to the UN IPCC reports.[112]

As Christopher Monckton, a former advisor to British prime minister Margaret Thatcher, explained, "They have lost that argument so now they're saying but no never mind but even if global warming doesn't happen we still mustn't put CO_2 into the atmosphere because it will melt the oceans by making them acid. And there is no basis for this in science at all."[113]

Geologist Robert Giegengack also rejected ocean acidification claims. "We have had periods in earth's history when the CO_2 concentration in the atmosphere was 15 times higher than it is now and the organisms survived. They are still here," Giegengack said. "The level of acidity that people are talking about prevailing in the surface ocean as a consequence of 400 ppm of CO_2 in the atmosphere is trivial compared to what it has been in the past when that concentration was five or six thousand ppm."[114]

Australia's Great Barrier Reef—supposedly threatened by ocean acidification—has been a kind of poster child for the global warming cause. But a 2011 study found that the reef's overall health was not impacted by climate change. The researchers from the Australian Institute of Marine Science found that the "overall regional coral cover was stable (averaging 29% and ranging from 23% to 33% across years) with no net decline between 1995 and 2009." They found "no evidence of consistent, system-wide decline in coral cover since 1995."[115] In 2016, the chairman of the Great Barrier Reef Marine Park Authority in Australia publicly called out climate activists for exaggerating surveys, maps, and data to hype coral-bleaching on the reef.[116]

In 2015 Greenpeace co-founder Patrick Moore did an analysis on ocean acidification and concluded, "There is no solid evidence that ocean acidification is the dire threat to marine species that many researchers have claimed. The entire premise is based upon an assumption of what the average pH of the oceans was 265 years ago when it was not possible to measure pH at all, never mind over all the world's oceans. Laboratory experiments in which pH was kept within a range that may feasibly occur during this century show a slight positive effect on five critical factors: calcification, metabolism, growth, fertility and survival. Of most importance is the fact that those raising the alarm about ocean acidification do not take into account the ability of living species to adapt to a range of environmental conditions. This is one of the fundamental characters of life itself."[117]

CHAPTER 13

The Ever-Receding Tipping Point

There's not much scientific evidence for catastrophic man-caused global warming. The link between climate and CO_2 is tenuous, the earth has been warming since the end of the last ice age (and entered an indefinite pause nearly two decades ago), and the catastrophic polar melting, sea-level rise, and increased extreme weather aren't actually happening. So what's a global warming fear promoter to do? Kick the can down the road—over and over again.

The climate change scare campaign has always relied on arbitrary deadlines, dates by which we must act before it's too late. Global warming advocates have drawn many lines in the sand, claiming that we must act to solve global warming—or else.

"We are running out of time. We have to get an ambitious global agreement," warned then–UN climate chief Christiana Figueres at the 2014 People's Climate March. "This is a huge crisis."[1]

At the UN climate summit in Copenhagen in 2009, Al Gore sought UN climate agreement—immediately. "We have to do it this year. Not next year, this year," he demanded. "And of course the clock is ticking because Mother Nature does not do bailouts."[2]

Did you know?

★ Prince Charles claimed that we had only eight years to stop global warming—in 2009

★ In the '80s, the UN warned that the world had to act by the year 2000 or "entire nations could be wiped off the face of the Earth" by rising seas

★ Since 2001, at least seven different climate conferences have been touted as the "last chance" to stop global warming

★ ★ ★

Differing Deadlines

It's difficult to keep up with whether we have years, months, days, or merely hours to avert catastrophe. Here's a small collection of contrasting "tipping points."[8]

Years	January 2009	NASA's James Hansen said Obama had "four years to save Earth."[9]
Months	July 2009	Prince Charles claimed in July 2009 that we had just ninety-six months to stave off "irretrievable climate and ecosystem collapse, and all that goes with it."[10]
Days	October 2009	UK prime minister Gordon Brown warned we had only fifty days to prevent climate "catastrophe."[11]
Hours	March 2009	"We have hours" to prevent climate disaster, declared Elizabeth May of the Canadian Green Party.[12]

Gore has warned repeatedly of the coming tipping point. Climate change "can cross a tipping point and suddenly shift into high gear," the former vice president claimed in 2006.[3] Laurie David, the producer of Gore's film *An Inconvenient Truth*, said in 2007 that "we have to have action we have to do something right now to stop global warming."[4]

Prince Charles has also warned that time is running out. "We should compare the planet under threat of climate change to a sick patient," urged the heir to the British throne.[5] "I fear there is not a moment to lose."[6]

"The clock is ticking.... Scientists believe that we have ten years to bring emissions under control to prevent a catastrophe," reported ABC News.[7]

But these "tipping points" and "last chance" claims now have a long history. The United Nations alone has spent more than a quarter of a century announcing a series of ever-shifting deadlines by which the world must act or face disaster from anthropogenic climate change.

Deadlines Come and Go

Recently, in 2014, the United Nations declared a climate "tipping point" by which the world must act to avoid dangerous global warming. "The world now has a rough deadline for action on climate change. Nations need to take aggressive action in the next 15 years to cut carbon emissions, in order to forestall the worst effects of global warming, says the Intergovernmental Panel on Climate Change," reported the *Boston Globe*.[13]

But way back in 1982, the UN had announced a two-decade tipping point for action on environmental issues. Mostafa Tolba, executive director of the UN Environment Program (UNEP), warned on May 11, 1982, that the "world faces an ecological disaster as final as nuclear war within a couple of decades unless governments act now." According to Tolba, lack of action would bring "by the turn of the century, an environmental catastrophe which will witness devastation as complete, as irreversible as any nuclear holocaust."[14]

In 1989, the UN was still trying to sell that "tipping point" to the public. According to a July 5, 1989, article in the *San Jose Mercury News*, Noel Brown, the then-director of the New York office of UNEP was warning of a "10-year window of opportunity to solve" global warming. According to the Herald, "A senior U.N. environmental official says entire nations could

Eighteen-Sixty-Four Tipping Point Warns of "Climatic Excess"

"As early as 1864 George Perkins Marsh, sometimes said to be the father of American ecology, warned that the earth was 'fast becoming an unfit home for its "noblest inhabitant,"' and that unless men changed their ways it would be reduced 'to such a condition of impoverished productiveness, of shattered surface, of climatic excess, as to threaten the depravation, barbarism, and perhaps even extinction of the species.'"

—**MIT professor Leo Marx**[15]

be wiped off the face of the Earth by rising sea levels if the global warming trend is not reversed by the year 2000. Coastal flooding and crop failures would create an exodus of 'eco-refugees,' threatening political chaos."[16]

But in 2007, seven years after that supposed tipping point had come and gone, Rajendra Pachauri, then the chief of the UN IPPC, declared 2012 the climate deadline by which it was imperative to act: "If there's no action before 2012, that's too late. What we do in the next two to three years will determine our future. This is the defining moment."[17]

UN Secretary-General Ban Ki-moon announced his own deadline in August 2009, when he warned of "incalculable" suffering without a UN climate deal in December 2009.[18]

And in 2012, the UN gave Planet Earth another four-year reprieve. UN Foundation president and former U.S. Senator Tim Wirth called Obama's re-election the "last window of opportunity" to get it right on climate change.[19]

Heir to the British throne Prince Charles originally announced in March 2009 that we had "less than 100 months to alter our behavior before we risk catastrophic climate change." As he said during a speech in Brazil, "We may yet be able to prevail and thereby to avoid bequeathing a poisoned chalice to our children and grandchildren. But we only have 100 months to act."[20]

To his credit, Charles stuck to this rigid timetable—at least initially. Four months later, in July 2009, he declared a ninety-six-month tipping point.[21] At that time the media dutifully reported that "the heir to the throne told an audience of industrialists and environmentalists at St James's Palace last night that he had calculated that we have just 96 months left to save the world. And in a searing indictment on capitalist society, Charles said we can no longer afford consumerism and that the 'age of convenience' was over."[22]

At the UN climate summit in Copenhagen in 2009, Charles was still keeping at it: "The grim reality is that our planet has reached a point of crisis and we have only seven years before we lose the levers of control."[23]

As the time expired, the Prince of Wales said in 2010, "Ladies and gentlemen we only—we now have only 86 months left before we reach the tipping point."[24]

By 2014, a clearly exhausted Prince Charles seemed to abandon the countdown, announcing, "We are running out of time. How many times have I found myself saying this over recent years?"[25]

In the summer of 2017, Prince Charles's one-hundred-month tipping point finally expired.[26] What did Charles have to say? Was he giving up? Did he proclaim the end times for the planet? Far from it. Two years earlier, in 2015, Prince Charles abandoned his hundred-month countdown and gave the world a reprieve by extending his climate tipping point another thirty-five years, to the year 2050!

A July 2015 interview in the *Western Morning News* revealed that "His Royal Highness warns that we have just 35 years to save the planet from catastrophic climate change."[27] So instead of facing the expiration of his tipping point head on, the sixty-nine-year-old Charles kicked the climate doomsday deadline down the road until 2050 when he would be turning the ripe age of 102. (Given the Royal Family's longevity, it is possible he may still be alive for his new extended deadline.)

Former Irish President Mary Robinson issued a twenty-year tipping point in 2015, claiming that global leaders have "at most two decades to save the world."[28]

Al Gore announced his own ten-year climate tipping point in 2006 and again in 2008, warning that "the leading experts predict that we have less than 10 years to make dramatic changes in our global warming pollution lest we lose our ability to ever recover from this environmental crisis."[29] In 2014, with "only two years left" before Gore's original deadline, climatologist Roy Spencer mocked the former vice president, saying "in the grand tradition of prophets of doom, Gore's prognostication is not shaping up too well."[30]

★ ★ ★

New Lyrics to an Old Tune

Newsweek magazine weighed in with its own tipping point: "The longer the planners delay, the more difficult will they find it to cope with climatic change once the results become grim reality."[36] That warning appeared in an April 28, 1975, article about global cooling! Same rhetoric, different eco-scare.

Penn State Professor Michael Mann weighed in with a 2036 deadline. "There is an urgency to acting unlike anything we've seen before," Mann explained.[31] Media outlets reported Mann's made a huge media splash with his prediction, noting "Global Warming Will Cross a Dangerous Threshold in 2036."[32]

Other global warming activists chose 2047 as their deadline,[33] while twenty governments from around the globe chose 2030 as theirs, with Reuters reporting that millions would die by 2030 if the world failed to act on climate: "More than 100 million people will die and global economic growth will be cut by 3.2% of GDP by 2030 if the world fails to tackle climate change, a report commissioned by 20 governments said on Wednesday. As global avg. temps rise due to ghg emissions, the effects on planet, such as melting ice caps, extreme weather, drought and rising sea levels, will threaten populations and livelihoods, said the report conducted by humanitarian organisation DARA."[34]

As we saw in chapter five, top UK scientist Sir David King warned in 2004 that that by 2100 Antarctica could be the only habitable continent.[35]

Tipping point rhetoric seems to have exploded beginning in 2002. An analysis by *Reason* magazine's Ron Bailey found that tipping points in environmental rhetoric increased dramatically in that year.[37]

The Last Chance

Michael Mann warned that the 2015 UN Paris summit "is probably the last chance" to address climate change.[38] But the reality is that every UN climate summit is hailed as the last opportunity to stop global warming.

Here, courtesy of the great research published at Climate Change Predictions, is a sampling of previous "last chance" deadlines that turned out to be—well—not the last chance after all.[39]

Bonn, 2001: "A Global Warming Treaty's Last Chance" —*Time* magazine, July 16, 2001

Montreal, 2005: "Climate campaigner Mark Lynas warned 'with time running out for the global climate, your meeting in Montreal represents a last chance for action.'" —*Independent*, November 28, 2005

Bali, 2007: "World leaders will converge on Bali today for the start of negotiations which experts say could be the last chance to save the Earth from catastrophic climate change." —*New Zealand Herald*, December 3, 2007.

Poznan, Poland, 2008: "Australian environmental scientist Tim Flannery warned, 'This round of negotiations is likely to be our last chance as a species to deal with the problem.'" —*Age*, December 9, 2008

Copenhagen, 2009: "European Union Environment Commissioner Stavros Dimas told a climate conference that it was 'the world's last chance to stop climate change before it passes the point of no return.'" —Reuters, February 27, 2009

Cancun, 2010: "Jairem Ramesh, the Indian environment minister, sees it as the 'last chance' for climate change talks to succeed." —*Telegraph*, November 29, 2010

Durban, 2011: "Durban climate change meeting is "the last chance." Attended by over 200 countries, this week's major UN conference has been described by many experts as humanity's last chance to avert the disastrous effects of climate change." —UCA News, November 28, 2011[40]

"Serially Doomed"

Perhaps the best summary of the tipping-point phenomenon comes from UK scientist Philip Stott. "In essence, the Earth has been given a 10-year survival warning regularly for the last fifty or so years. We have been serially doomed," Stott explained. "Our post-modern period of climate change angst can probably be traced back to the late-1960s, if not earlier. By 1973, and the 'global cooling' scare, it was in full swing, with predictions of the imminent collapse of the world within ten to twenty years, exacerbated by the impacts of a nuclear winter. Environmentalists were warning that, by the year 2000, the population of the US would have fallen to only 22 million. In 1987, the scare abruptly changed to 'global warming', and the IPCC (the Intergovernmental Panel on Climate Change) was established (1988), issuing its first assessment report in 1990, which served as the basis of the United Nations Framework Convention on Climate Change (UNFCC)."[41]

Why do the warmists continue to hype ever-changing tipping points? Author Donna Laframboise has explained, "The experts often don't know any more than you and I about what's going to happen in the future. So the idea that climate scientists have this crystal ball and they know what is coming, I find that very hard to believe.... And we should also understand that psychologically there's obviously something in us as human beings where we are perhaps, we've been primed to always be worried about our survival and our existence. So we are very predisposed to a narrative that says, you know, 'we are all going to die, we are all going to die.'"[42]

As Laframboise noted, "There was global cooling. Prior to that there was the population bomb, ya know. Millions of people were going to starve to death because we could not possibly feed this many humans. There's always something."

Greenpeace co-founder Patrick Moore thinks it is all about fear: "I think people are just playing on people's fear that there will be a catastrophe or calamity that you cannot recover from."[43]

Geologist Don Easterbrook agrees: "It instills fear. If you think that your house is going to be under water in the next 10 years you are going to be frightened and you're going to be willing to accept things that are being proposed by other people."[44]

Comedians Penn & Teller explain how the tipping point works. Their segment on their TV show *Bullshit*! featuring global warming activist Ross Gelbspan is priceless: "Ross Gelbspan has asserted that 'Unchecked, global warming will bankrupt the global economy by 2065.'" But Penn Jillette was having none of it: "Where did Gelbspan get that data, how did he choose that date of 2065? I will tell you how, this asshole figures he will be dead by then and not have to own up."[45]

In 2016, just after President Donald Trump's election, *New York Times* columnist Eduardo Porter wrote an article featuring the headline "Earth Isn't Doomed Yet. The Climate Could Survive Trump."[46]

It may be that the only authentic climate "tipping point" we can rely is this one, issued in 2007: New Zealand atmospheric scientist Augie Auer, former University of Wyoming professor of atmospheric science, said "We're all going to survive this. It's all going to be a joke in five years."[47]

CHAPTER 14

Controlling Climate...or You?

Global warming is not about the science. There's very little scientific evidence to back up the threats of catastrophic and man-made climate change, and copious data refutes the fears. So what is it really about? Is there a larger agenda at play? (Hint: Of course there is.)

MIT climate scientist Richard Lindzen has laid out the real agenda behind the global warming scare. "Controlling carbon is a bureaucrat's dream. If you control carbon, you control life," Lindzen said in 2007.[1]

The climate change panic is all about central planning, global governance, planned recessions, and redistributing wealth. It's just the most recent in a long chain of eco-scares—overpopulation, deforestation, the ozone hole, resource scarcity, and so forth—for which the solution is always the same: global regulation by central planners. At the moment, the UN IPCC is the leader of this agenda. The UN "experts" openly say they want to redistribute wealth by climate policy. They are using the science as tool in a partisan political campaign effort for centralized government planning through the United Nations and the European Union.

Former EU Climate Commissioner Connie Hedegaard gave the game away when she said that global warming policy is right even if science is wrong. In

Did you know?

★ Numerous politicians have admitted that they would be pushing the very same policies in the absence of climate change

★ A top UN IPCC official said climate policy is to "redistribute the world's wealth"

★ The EPA has said that its regulations will have no effect on global CO_2 levels—let alone the global temperature

★ ★ ★
Different Environmental Scare, Same Solution

In 1974, future Obama science czar John Holdren proposed "redistribution of wealth" to battle environmental degradation. Holdren testified to the U.S. Senate Commerce Committee, "The neo-Malthusian view proposes conscious accommodation to the perceived limits to growth via population limitation and redistribution of wealth in order to prevent the 'overshoot' phenomenon. My own sympathies are no doubt rather clear by this point. I find myself firmly in the neo-Malthusian camp."[6]

2013 the *Telegraph* reported her as saying that "regardless of whether or not scientists are wrong on global warming, the European Union is pursuing the correct energy policies even if they lead to higher prices." Hedegaard asked, "Let's say that science, some decades from now, said 'we were wrong, it was not about climate,' would it not in any case have been good to do many of things you have to do in order to combat climate change?…I think we have to realize that in the world of the 21st century for us to have the cheapest possible energy is not the answer."[2]

Hedegaard was echoing former Obama White House science czar John Holdren, who had claimed in an essay in the *Windsor Star*, "The U.S. is threatened far more by the hazards of too much energy, too soon, than by the hazards of too little energy, too late."[3] Holdren had also warned that the United States already had too much energy—in 1975. "Mounting evidence suggests that the United States is approaching (if not beyond) the level where further energy growth costs more than it is worth."

Former Democratic U.S. Senator Timothy Wirth said something similar—but even more revealing—in the 1990s: "We've got to ride the global warming issue. Even if the theory of global warming is wrong, we will be doing the right thing, in terms of economic policy and environmental policy."[4]

So, according to the admissions of its own leadership, the point of the campaign against global warming is not really the supposed climate science. No, the real agenda is the "solutions" to the alleged crisis. Those "solutions," which the members of the global governance lobby want in any case for other reasons, are what is driving the movement.

Spreading the Wealth, Governing the Globe

In November 2010 a top UN IPCC official revealed the real agenda. "One must say clearly that we redistribute de facto the world's wealth by climate policy. Obviously, the owners of coal and oil will not be enthusiastic about this," said Ottmar Edenhofer, who was co-chair of the IPCC's Working Group III and a lead author of the IPCC's Fourth Assessment Report, released in 2007. Edenhofer's candid revelations about the motivations for climate change policy were eye-opening. "One has to free oneself from the illusion that international climate policy is environmental policy. This has almost nothing to do with environmental policy anymore, with problems such as deforestation or the ozone hole."[5]

At the Bali UN climate summit in 2007, climate activists advocated the transfer of money from rich to poor nations, purportedly in order to fight global warming. "A climate change response must have at its heart a redistribution of wealth and resources," said Emma Brindal, a climate justice campaigner coordinator for Friends of the Earth.[7]

If, as Lindzen says, controlling carbon dioxide is a bureaucrat's dream, then the global warming advocates are in a field of dreams as they envision centrally planned energy economics and regulations over ever-increasing aspects of our lives.

Former Vice President Al Gore even declared in July 2009 that the congressional climate bill that passed the House would help bring about

"Eyes Bulge" at Climate Taxation

"Global warming, climate change, all these things are just a dream come true for politicians. The opportunities for taxation, for policies, for control, for crony capitalism are just immense, you can see their eyes bulge."

—MIT professor emeritus Richard Lindzen[8]

★ ★ ★

Different Environmental Scare, Same Solution

Amherst College professor Leo Marx warned in 1970 about the "global rate of human population growth. All of this is only to say that, on ecological grounds, the case for world government is beyond argument."[13]

"global governance." Gore explained, "one of the ways it will drive the change is through global governance and global agreements."[9] UN Secretary General Ban Ki-moon trumpeted that same concept in a 2009, *New York Times* op-ed. "A [climate] deal must include an equitable global governance structure," he wrote.[10] Also in 2009, it was announced that the Obama State Department wanted to form a global "Ecological Board of Directors."[11] In 2000, then French President Jacques Chirac had said the UN's Kyoto Protocol on global warming represented "the first component of an authentic global governance."[12]

The much-touted "carbon tax" is another step toward both wealth distribution and a central global governance structure dictating every aspect of our lives from our light bulbs to transportation to home energy to our diets. In December 2007, the UN climate conference in Bali urged the adoption of a global tax on CO_2 emissions that would represent "a global burden-sharing system, fair, with solidarity and legally binding to all nations."[14]

Two years later, a top German climate advisor proposed the creation of a CO_2 budget for every person on planet." Hans Joachim Schellnhuber told *Der Spiegel* that this internationally monitored "budget" would apply to "every person on the planet, regardless whether they live in Berlin or Beijing." According to Schellnhuber, the "industrialized nations have already exceeded their quotas if you take into account past emissions." Describing a form of climate reparations, he explained, "The West would give back part of the wealth it has taken from the South in the past centuries and be indebted to countries that are now amongst the poorest in the world."[15]

A 2008 *Daily Mail* article reported on a proposal to create a "personal carbon-trading scheme" in which "every adult in U.K. should be forced to use 'carbon ration cards.'" According to the report, "Everyone would be given an annual carbon allowance to use when buying oil, gas, electricity and flights—anyone who exceeds their entitlement would have to buy top-up credits from individuals who haven't used up their allowance."[16] Under this energy rationing scheme, the UK government would have the authority to impose fines, "monitor employees emissions, home energy bills, petrol purchases and holiday flights."[17]

The *London Times* accurately called the scheme "Rationing being reintroduced via workplace after an absence of half a century" and pointed out, "Employees would be required to submit quarterly reports detailing their consumption."[18]

Also in 2008, the California state government stunned the nation when it sought to control home thermostats remotely. Even the *New York Times* appeared to be shaken by this proposal, comparing it to the 1960s science-fiction show *The Outer Limits*. "California state regulators are likely to have the emergency power to control individual thermostats, sending temperatures up or down through a radio-controlled device," the *Times* reported.[19]

★ ★ ★
Unholy Alliance

In 2015, Pope Francis released his climate change encyclical, *Laudato Si': On Care for Our Common Home*. At Climate Depot, I did a special report titled "'Unholy Alliance'—Exposing the Radicals Advising Pope Francis on Climate," in which I pointed out, "The Vatican relied on advisors who can only be described as the most extreme elements in the global warming debate…many of the Vatican's key climate advisors have promoted policies directly at odds with Catholic doctrine and beliefs. The proceedings of the Vatican climate workshop included activists like Naomi Oreskes, Peter Wadhams, Hans Joachim Schellnhuber, and UN advisor Jeffrey Sachs…. The Vatican advisors can only be described as a brew of anti-capitalist, pro–population control advocates."[20]

Dictator Envy

Global warming activists appear to be enchanted with China and its one-party rule. "One party can just impose politically difficult but critically important policies needed to move a society forward," gushed *New York Times* columnist Tom Friedman in 2009.[21]

The China fixation seems widespread.

In 2014 then UN climate chief Christiana Figueres lamented U.S. democracy as "very detrimental" in the global warming battle and lauded China for "doing it right" on climate change. Bloomberg News reported, "China is also able to implement policies because its political system avoids some of the legislative hurdles seen in countries including the U.S., Figueres said."[22]

In fact, to truly get an understanding of the global warming agenda, listen to what Figueres revealed to the *Guardian* in 2012 about the UN's plan. "What we are doing here is we are inspiring government, private sector, and civil society to [make] the biggest transformation that they have ever undertaken. The Industrial Revolution was also a transformation, but it wasn't a guided transformation from a centralized policy perspective. This is a centralized transformation that is taking place because governments have decided that they need to listen to science. So it's a very, very different transformation and one that is going to make the life of everyone on the planet very different."

It bears emphasizing: The UN climate chief has publicly stated that the goal of the UN climate policies and pacts is to seek "a centralized transformation" that is "going to make the life of everyone on the planet very different."[23]

Five years earlier, Democratic House Speaker Nancy Pelosi told audiences in China that "every aspect of our lives must be subjected to an inventory" in order to combat global warming.[24] In the United States "every aspect" of our lives already includes regulations on our dishwashers, dryers, SUVs, home thermostats, energy bills, and even our light bulbs.

There have even been calls to drug us—and to use bioengineering to shrink humans as a species. Professor Matthew Liao of the Center for Bioethics at New York University has promoted a "solution of human engineering" for climate change. It involves the "biomedical modification of humans." Liao proposes "genetically engineered humans" with "pharmacological enhancements" to combat global warming. "I call this human engineering. It involves biomedical modification of humans to better deal with" climate.[25]

With human engineering "we can make humans smaller," he noted. "By reducing the height of average man in U.S. by just 15 cm, it would mean a 23% reduction in metabolic reduction," he calculated. "Another possible and more dangerous possibility is to consider hormone treatments to close growth hormone earlier than normal," he explained. "How can height reduction be achieved? One possibility is to use pre-implantation genetics to select shorter children," Liao noted.[26]

★ ★ ★
I'll Have the Vegetable Plate

Do you like eating meat but worry about its climate impact? Not anymore, with Liao's solutions. He has argued that we should "make ourselves allergic to those proteins" in meat so that we suffer an "unpleasant reaction" when we attempt to eat it. "The way we can do that is to create some sort of meat patch," he explained, "Kind of like a nicotine patch where you put it on before you go to dinner go out to restaurant and this will curb your enthusiasm for eating meat."[27]

Liao's "human engineering" would reduce the literal footprint of the human race—and thus our carbon footprint as well. But it would also have other benefits. As Liao explained in 2012, his genetic modifications "could also make behavior solutions more likely to succeed. For example pharmacologically induced altruism and empathy could increase the likelihood we can adopt necessary market solutions to combat climate change."

You don't care about global warming? You don't support a UN climate treaty or a carbon tax? Here is a pill that will make you care!

Yes, Professor Liao is suggesting that medications would alter people who don't care enough about global warming, so that they would fall in line. "Having the option to use pharmacological means to increase altruism and empathy to allow this person volunteer to overcome his weakness of will and do the right thing," Liao said. Mind-altering drugs will make you want to act on climate change. "Interventions affecting the interventions in neurosystems could increase the willingness to cooperate with social goals."

Liao claimed that he would not force his solutions on the public. This is "meant to be a voluntary activity, possibly supported with taxes, nobody is being coerced," he insisted. But even fellow climate alarmists find Liao's ideas alarming. Climate activist Bill McKibben, has rejected Liao's proposals as the "worst climate change solutions of all time."[28]

Voluntary Poverty

Human engineering. Carbon taxes, "budgets," and rationing. Climate-friendly bedtimes. Inventories of every aspect of our lives. The cure sounds worse than the disease. And here's yet another proposed solution that has been floated for global warming: "de-growth" of the economy and "planned recessions."

In 2013 Kevin Anderson, the deputy director of the Tyndall Centre for Climate Change Research in the UK, called for "a planned economic recession" to reduce emissions and battle climate change. Anderson asserted that economic "de-growth" was needed to fight climate change. "Continuing with economic growth over the coming two decades is incompatible with meeting our international obligations on climate change."[29]

Many climate activists find the idea of shrinking the world economy to lower emissions appealing. The Sierra Club has also touted economic "de-growth." A 2014 report from the

★ ★ ★

What's That I Smell?

Professor Anderson is also bathing less to fight climate change. "I've cut back on washing and showering—but only to levels that were the norm just a few years back," Anderson told me during an impromptu hallway debate at the UN climate summit in Poland in 2013. "That is why I smell, yes," Anderson joked to me. But his reasons are dead serious. "I think it's extremely unlikely that we wouldn't have mass death at 4 degrees" global temperature rise.[30]

environmentalist organization urged, "We have to de-grow our economy" to "temper climate disruption, and foster a stable, equitable world economy."[31]

University of Manchester climate researcher Alice Bows-Larkin has also called for "planned recessions" to fight global warming. "Economic growth needs to be exchanged" for "planned austerity," Bows-Larkin wrote. She advocates "whole system change."[32]

In 1974, Stanford University Professor Paul Ehrlich, author of *The Population Bomb*, predicted the United States would "move to a no-growth" economy. "I think that what is not realized, and it's going to be one of the hardest things to be accepted by the Americans in general, is that the onset of the age of scarcity essentially demolishes current models of economists. We are going to move to a no-growth" economy, Ehrlich testified to the U.S. Senate Committee on Commerce and Committee on Government Operations. He recommended that we move to economic austerity "intelligently through the government by planning as rapidly as possible."[33]

The climate agenda becomes quite clear. When you try to restrict energy access and manage and plan every aspect of the economy in the name of global warming—you suddenly have the power to create "planned recessions."

Prince Charles has urged a "fundamental transformation" of capitalism to tackle climate change.[34]

Naomi Klein, the author of *This Changes Everything: Capitalism vs. The Climate* is clearly thinking along the same lines. She claims that capitalism "is irreconcilable with a livable climate." According to Klein, "Facing climate change head-on means changing capitalism." Her book lays out in detail how solving climate change is incompatible with our current economic system. "Core inequalities need to be tackled through redistribution of wealth and technology," she explains.[35]

"If climate can be our lens to catalyze this economic transformation that so many people need for other even more pressing reasons, then maybe that's a winning combination," Klein, who also serves as an advisor to Pope

Francis on his climate campaign, explained.[36] She is calling for a whole new system. "I think we might be able to come with a system that is ecologically rooted that is better than anything that we have tried before."

During the panel discussion at an event at the People's Climate March in New York City in 2014, I asked Klein, "Even if the climate change issue did not exist, you would be calling for the same structural changes?" Klein responded, "Yeah."[37] Once again we see how the "science" that we are all supposed to accept without question is actually irrelevant in the minds of the leaders of the climate movement.

Others are even more out there than Klein. "The only way to stop runaway climate change is to terminate industrial civilization," according to Professor Emeritus Guy McPherson of the University of Arizona. McPherson was a natural resources, ecology, and evolutionary biology instructor, but he has since turned into a sort of grief counselor. "I think we are walking around to save our own funeral expenses at this point," he explained.[38]

And McPherson is not alone. In 2009 NASA then–lead global warming scientist James Hansen endorsed a book called *Time's Up! An Uncivilized Solution to a Global Crisis*, which ponders "razing cities to the ground, blowing up dams and switching off the greenhouse gas emissions machine" as possible solutions to global warming. "The only way to prevent global ecological collapse and thus ensure the survival of humanity is to rid the world of Industrial Civilization," the book explains. "A future outside civilization is a better life; one in which we can actually decide for ourselves how we are going to live." Hansen declared on the book's Amazon page that author Keith Farnish, "has it right: time has practically run out, and the system is the problem. Governments are under the thumb of fossil fuel special interests—they will not look after our and the planet's well-being until we force them to do so, and that is going to require enormous effort."[39] Remember, Hansen was NASA's lead global warming scientist, in charge of the temperature dataset, endorsing a book suggesting the solution of calling for ridding the world of industrial civilization. Hansen is a hardcore activist who has been arrested multiple times protesting climate change. [40] He has accused climate skeptics of "crimes against humanity and nature."[41]

The climate activists are waging war on modern energy, and it seems at times they are winning. As the *Daily Telegraph* reported in 2011, the "Era of Constant Electricity at Home is Ending." Because of the increasing use of wind power instead of gas, "the days of permanently available electricity may be coming to an end.… Families would have to get used to only using power when it was available, rather than constantly, said Steve Holliday, chief executive of National Grid." Holliday explained that consumers of electricity would "have to change their behavior."[42]

The Carbon Tax

A carbon tax remains one of the most persistent "solutions" for the dangers climate change supposedly threatens us with. But proponents seem to be ascribing almost magical powers to the tax. In 2013 the *Los Angeles Daily News* featured the headline, "Can a Tax Stop Global Warming? The Citizens Climate Lobby thinks so."[43]

In an August 25, 2012, *New York Times* op-ed, Cornell University economics professor Robert Frank urged a "steep" carbon tax—one that "might raise gasoline prices by 70 cents a gallon"— for the sake of "climate stability" and to prevent an "end [to] life as we know it." Frank called this a "relatively modest cost" to "insulate ourselves from catastrophic risk."[44]

Former President George W. Bush's treasury secretary Hank Paulson believes a carbon tax can prevent "natural disasters." According to Paulson, "What I've said about a carbon tax is some people that oppose it are opposing it because they don't like the government playing a big role. And, you know, the perverse aspect of that is, frankly, those that are resisting taking action now are guaranteeing that the government will be playing a bigger

★ ★ ★

Increasing Our Freedom, or Their Revenue?

Yoram Bauman, a professor at Florida State University, has claimed that a carbon tax—which he dubbed "the most sensible tax of all"—would lower taxes and "increase personal freedom."[46]

But perhaps the real reason a carbon tax has been proposed has nothing to do with "personal freedom," or even with stopping global warming.

As the *Washington Post* noted in 2012 when it endorsed a carbon tax, it "could bring in serious money"—raising "$125 billion a year." The *Post* editorial argued, "A carbon tax would make sense regardless of revenue it would raise. But the policy could bring in serious money."[47]

role, because we're seeing now and we're going to see an increasing number of natural disasters, Mother Nature acts."[45]

But as we saw in chapter twelve, the scientific data show there has been no increase in extreme weather. And of course the Earth's climate will not notice a carbon tax one way or another.

Former Vice President Al Gore has promoted a carbon tax as well. In 2017 Gore's newly formed Energy Transitions Commission issued a document calling for carbon tax. "Gore would start the tax at $50 per ton, which would increase to $100 per ton over time, essentially destroying the market for continued robust development of the world's fossil-fuel base," according to Heartland Institute energy analyst Fred Palmer. "The all-in estimated cost to re-engineer humanity is only a mere $15 trillion—enough money to give every man, woman, and child in the United States more than $46,000," Palmer wrote.[48]

Gore's 2017 carbon tax plan represented a complete reversal from his 1992 testimony to the Senate Foreign Relations Committee, when he declared that "no government mandated requirements would be necessary of any kind" to control carbon dioxide emissions, instead touting "purely voluntary measures."

Back then, Gore had testified, "You said for example that you were convinced that the only way to stabilize CO_2 emissions at 1990 levels is with a large tax. Quite to the contrary, the Bush Administration itself produced a rather comprehensive study of the options available to us to stabilize CO_2 emissions and found that not only would a tax be unnecessary but no government mandated requirements would be necessary of any kind, that we could all but meet the goal with purely voluntary measures if we had leadership from the executive branch with a series of new initiatives that are outlined in that study."[49]

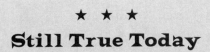

★ ★ ★

Still True Today

The late ABC News anchor Peter Jennings reported in 1998: "Al Gore genuinely believes that if he does not prevail, the apocalypse is coming. His opponents think he is the danger."[50]

Just Turn Down the Heat

The warmists seem to believe that international agreements can act as thermostatic controls for the earth's climate. A G8 Summit in 2009 announced a "historic breakthrough" when Western leaders "agreed to the goal of keeping the world's temperatures from rising more than 2C."[51]

Philip Stott, professor emeritus of Biogeography at the University of London ridiculed the idea that politicians can control the Earth's climate with regulations. "Angela Merkel, the German chancellor, and my own good Prime Minister [Tony Blair], for whom I voted—let me emphasize —arguing in public two weeks ago as to who in 'Annie get the gun style' could produce the best temperature. 'I could do two degrees C,' said Angela. 'No, I could only do three,' said Tony. Stand back a minute. Those are politicians telling you that they can control climate to a degree Celsius," Stott said in disbelief.[52]

★ ★ ★

There's a New Sheriff in Town

Given the absurd faith in the power of climate regulations, President Trump's reversal of President Obama's regulations has caused lots of hand-wringing. Even before the elections, climate activist and Penn State professor Michael Mann warned that Trump was a "threat to the planet," whose future "could quite literally lie in the balance."[53]

The pervasive belief that politicians can control the weather by agreements to curtail our lifestyle has not gone away.

But the Climategate emails leak revealed the elusive 2-degree C regulatory cap of the Earth's temperatures has no basis in science—according to a former top UN scientist. Former UN lead author Phil Jones admitted that the 2-degree C limit was "plucked out of thin air." In 2007, Jones emailed, "The 2 deg C limit is talked about by a lot within Europe. It is never defined though what it means. Is it 2 deg C for globe or for Europe? Also when is/was the base against which 2 deg C is calculated from? I know you don't know the answer, but I don't either! I think it is plucked out of thin air."[54]

"Two degrees is not a magical limit—it's clearly a political goal," says Hans Joachim Schellnhuber, director of the Potsdam Institute for Climate Impact Research (PIK).

Professor Roger Pielke Jr. explained in 2017 that the 2-degree goal "is an arbitrary round number that was politically convenient. So it became a sort of scientific truth. However, it has little scientific basis but is a hard political reality." He noted that the goal "originated in a local, long-term record of temperature variation in England, which was adapted by an economist in a 'what if?' exercise."[55]

Ironically, the EPA had admitted that its climate regulations would not only have no measurable effect on global temperatures, they would not even impact global CO_2 levels. Even if the UN Paris pact were fully enacted, it would have no measurable impact on global temperatures. And that's on the assumption that the UN is correct on the science.[56] The entire campaign is pure symbolism on climate—but the more centrally planned economy would be a reality.

When the UN Paris climate pact was agreed to by President Obama in December 2015, to exuberant celebration, I penned this response: "Now that the United Nations has officially 'solved' man-made global warming, does this mean we never have to hear about 'global warming' fears again!? Does this mean we can halt the endless supply of federal tax dollars funding 'climate change' studies? Does this mean we can stop worrying about 'global warming's' ability to end civilization and cause wars, and increase prostitution, bar room brawls, rape, airline turbulence, etc.? Can we finally move on to other issues?" I added, "If climate activists celebrating truly believe the UN has solved 'climate change,' even skeptics should rejoice! Now that the UN treaty has 'solved' global warming, can we all just move on to something else?"[57]

Of course, that was before President Trump announced his intention to withdraw from the UN Paris pact in June of 2017. But it is obvious that the activists have no intention of ever admitting that climate change has been solved because if they did they would jeopardize all the other agendas wrapped up in the climate cause.

Former Czech president Václav Klaus, who lived under a totalitarian regime, perhaps summed it up best. "As someone who lived under communism for most of his life, I feel obliged to say that I see the biggest threat to freedom, democracy, the market economy and prosperity now in ambitious environmentalism," Klaus warned. "This ideology wants to replace the free and spontaneous evolution of mankind by a sort of central (now global) planning."[58]

I travelled to Prague in 2016 to interview Klaus. He explained to me the forces that were behind the climate change movement. "Today, the green

A Book You're Not Supposed to Read

Scare Pollution: Why and How to Fix the EPA by Steven J. Milloy (Bench Press, 2016).

★ ★ ★

Pope or UN Climate Lobbyist?

Pope Francis urged prayers for the passage of a UN climate treaty, specifically exhorting Catholics "to ask God for a positive outcome" for the Paris UN agreement. "We believers cannot fail to ask God for a positive outcome to the present discussions, so that future generations will not have to suffer the effects of our ill-advised delays," Francis wrote in his encyclical, *Laudato Si'* in 2015.[59]

Climate statistician William Briggs was not impressed by the Pope's climate claims. "Most of the scientific claims cited in Pope's encyclical are not true," Briggs wrote.[60]

agenda and the environmentalist agenda and the fighting climate agenda are trying to basically influence the economic system. It's trying to basically stop the existence of the free market system which we were fighting for and were dreaming about in the communist era for decades."[61]

British professor Philip Stott echoes that sentiment, warning, "Global warming 'has become the grand political narrative of the age, replacing Marxism as a dominant force for controlling liberty and human choices.'"[62]

CHAPTER 15

Climbing onto the Climate Change Gravy Train

The popular perception, encouraged by the media, is that the battle over climate change features David against Goliath. Selfless defenders of the environment are pitted against greedy multi-billion-dollar corporations and their lavishly funded stooges who only care about profits and don't give a hoot about the Earth or future generations.

Thus in 2007 NBC Nightly News breathlessly reported a claim by the environmental pressure group Union of Concerned Scientists that Exxon-Mobil had "given almost $16 million" over seven years to "denier groups."[1]

To the mainstream media, this $15,837,873 was the smoking gun—proof that climate change skeptics were in the pay of the world's largest oil company. Case closed.

Or was it?

Left unreported by NBC News was that just a few years earlier Exxon-Mobil had given a $100 million grant to Stanford University to combat global warming. The oil company pledged the $100 million in 2002 to "research on ways to meet growing energy needs without worsening global warming," according to the *New York Times*.[2]

Did you know?

★ Global warming promoters receive "3,500 times as much money as anything offered to skeptics"

★ The world is spending nearly $1 billion *a day* to prevent climate change

★ Gore took millions of dollars in Qatari oil money for his Current TV network

★ ★ ★

The "Exxon Knew" Myth

Climate campaigners have claimed that "Exxon knew" about "climate change" risks in the 1970s, but suppressed the information and instead promoted skeptical "misinformation" instead.[3] But Breakthrough Institute president Mike Shellenberger, a man *Time* magazine has called a "hero of the environment," declared these claims "false." Exxon in "many cases advocated for climate policy!" Shellenberger noted. "The picture painted of Exxon seeking out and funding 'climate change deniers' to mislead public and prevent climate policy is false." He pointed out that the charges that "Exxon was paying people to lie about climate while acknowledging it privately" were not accurate. "In reality, Exxon funded conservative think tanks that were mostly *not* 'climate deniers' and in many cases advocated climate policy!" He added that even if all the money Exxon "spent was on 'deniers'—and it mostly wasn't—[it was a] drop in the bucket compared to green money."[4]

The media finds outrage that $15,837,873 was allegedly given to skeptical organizations over seven years, but is silent on the over $100 million grant to Stanford to fight global warming.

Deep Pockets

Let's put this $16 million of alleged funds to skeptics in some perspective. A single 2007 U.S. Department of Agriculture (USDA) grant of $20 million to study how "farm odors" contribute to global warming exceeded all of the money that skeptical groups reportedly received from oil giant Exxon-Mobil.[5] The USDA awarded "nearly $20 million in Conservation Innovation Grants to fund 51 research projects across the country designed to refine new technologies helping dairy and other agricultural producers cut back on their greenhouse emissions and cash in on governmental incentives for the research." USDA reports have stated "that when you smell cow manure, you're also smelling greenhouse gas emissions."

In comparison to the *hundreds of billions* of dollars in U.S. government funding of activist climate change research, the few million dollars Exxon allegedly gave the skeptics is a drop in the bucket.

A 2017 report from the Capital Research Center revealed how climate activists have repeatedly tried to inflate the financial clout of groups affiliated with skepticism of climate change. "In a widely cited 2014 study, the sociologist Robert Brulle purportedly exposed a network of nonprofit groups executing 'a deliberate and organized effort to misdirect the public discussion and distort the public's understanding of climate change,'" reported Steven Allen, the vice president of the Capital Research Center.[6] The media ran with the story, portraying skeptical groups as lavishly funded, with endless resources at their disposal.

But the analysis by the Capital Research Center found "that no more than 6%" of the spending by the ninety-one conservative groups "engaged the public on climate science."

Biologist turned filmmaker Randy Olson excoriated his fellow climate activists in 2017 for what he called their inaccurate claims about skeptics' funding. "There is so much money in this climate community, in the foundations, the gargantuan sums of money over a billion dollars as of 2011 poured into this issue of climate and energy," Olson said. "Matt Nisbet documented that in his Climate Shift report." Olson, who made the 2008 climate film *Sizzle* and wrote the 2009 book *Don't Be Such a Scientist: Talking Substance in an Age of Style*, pointed out, "It isn't right for them to be crying poor or somehow Exxon outspends them. That's a myth."[7]

And, according to Allen at Capital Research Center, government spending on climate issues is massive. "The best estimate, based on Office of Management and Budget data, is that from 1993 to 2014, federal expenditures exceeded $166 billion in 2012 dollars," Allen wrote. "Who really has the power?" he asked.

Another analysis, by the Science and Environmental Policy Project (SEPP), a group of scientists skeptical of man-made climate change fears, came to similar conclusions, which were published in a 2014 report: "Based on US government reports, SEPP calculated that from Fiscal Year (FY) 1993 to FY 2013 total US expenditures on climate change amount to more than $165 Billion. More than $35 Billion is identified as climate science. The White House reported that in FY 2013 the US spent $22.5 Billion on climate change. About $2 Billion went to US Global Change Research Program (USGCRP). The principal function of the USGCRP is to provide to Congress a National Climate Assessment (NCA).... Much of the remaining 89% of funding goes to government agencies and industries claiming they are preventing global warming/climate change, even though they do not understand the natural causes of climate change and, likely, far overestimate the influence of CO_2. These entities have a vested interest in promoting the fear of global warming/climate change."[8]

But even more revealing is the work of Australian researcher Joanne Nova, who examined funding of just U.S. climate skeptics versus the proponents of man-made global warming. What Nova's research revealed is that global warming promoters in the United States receive "3,500 times as much money as anything offered to skeptics." As Nova discovered, "Skeptics are fighting a billion dollar industry aligned with a trillion dollar trading scheme."[9] Her 2009 financial analysis uncovered how the "global warming science machine" had been financed—up to that time—to the tune of "$79 billion and counting." The report revealed a "well-funded, highly organized climate monopoly."[10]

And what does all this money buy? According to Nova, it "buys a bandwagon of support, a repetitive rain of press releases, and includes PR departments of institutions like NOAA, NASA, the Climate Change Science Program and the Climate Change Technology Program. The $79 billion figure does not include money from other western governments, private industry, and is not adjusted for inflation. In other words, it could be…a lot bigger." The report's bottom line: "Big Oil's supposed evil influence has been vastly outdone by Big Government."

In the fall of 2013, the White House reported to Congress that there were a total of eighteen federal agencies involved in the global warming debate, spending an estimated $22.2 billion in 2013, and $21.4 billion in 2014.[11] And in 2017 Bloomberg News reported that the Obama administration had "stashed" $77 billion in "climate money" across various agencies to elude budget cuts. "Obama aides spread money across the government to elude cuts," the news agency reported. The goal was to make "programs hard to disentangle" by "integrating climate programs into everything the federal government did."[12]

At the National Science Foundation, the geosciences program almost doubled to $1.3 billion. The budget for NASA's Earth Science program increased 50 percent, to $1.8 billion.

Feds awarded $1 billion through its Community Development Block Grant program to projects protecting against climate change-related natural disasters.

In 2012, the Federal Highway Administration made climate-adaptation projects eligible for federal aid.

The Bureau of Indian Affairs created the Tribal Climate Resilience Program.

The range of climate programs is vast, stretching across the entire government.

The Congressional Research Service estimated total federal spending on climate was in 2013. It concluded 18 agencies have climate-related activities, and calculated $77 billion in spending from fiscal 2008 through 2013 alone. But that figure could well be too low.[13]

★ ★ ★

Hot Rodding

The hidden climate funding has spread to almost every aspect of the federal government, with sometimes wacky results. The Department of Transportation, for example, has studied the alleged link between climate change and fatal car crashes. The agency asked in a study: "How might climate change increase the risk of fatal crashes in a community?"[14]

The Trump administration's Office of Management and Budget director Mick Mulvaney declared that there would be no more of the "crazy stuff the previous administration did." Mulvaney said in 2017, "What I think you saw happen during the previous administration is the pendulum went too far to one side, where we were spending too much of your money on climate change and not very efficiently."[15]

Aside from government funding of the climate change issue, environmental groups have massive private financial funding to promote climate fears. A 2014 report from the Center for the Defense of Free Enterprise (CDFE) revealed just how deep Big Green's pockets actually are. "U.S. environmental activist groups are a $13-billion-a-year industry—and they're all about PR and mobilizing the troops. Their climate change campaign alone has well over a billion dollars annually, and high-profile battles against drilling, fracking, oil sands and Keystone get a big chunk of that," explained CDFE executive vice president Ron Arnold. "The liberal foundations that give targeted grants to Big Green operations have well over $100 billion at their disposal."[16]

Making Out like a Bandit

And it's not just universities, professors, and green organizations that have reaped financial benefits from the climate panic. Former vice president Al Gore has done quite well for himself, too.

As Bloomberg News reported, "In the last personal finance report he filed as vice president, Gore disclosed on May 22, 2000, that the value of his assets totaled between $780,000 and $1.9 million."[17] But by 2007 Gore's wealth had skyrocketed. By that point he had a net worth "well in excess" of $100 million, including pre–public offering Google stock options, according to an article at *Fast Company*.[18] MIT scientist Richard Lindzen declared that Gore wanted to become the world's first "carbon billionaire."[19] After the Obama administration bloated climate and energy stimulus packages, Gore was on the path to that achievement. By 2008, Gore was so flush that he announced a $300 million campaign to promote climate fears and so-called solutions.[20]

And he just kept raking it in. According to a 2012 *Washington Post* report, "14 green-tech firms in which Gore invested received or directly benefited from more than $2.5 billion in loans, grants and tax breaks, part of Obama's historic push to seed a U.S. renewable-energy industry with public money." The *Post* explained that Gore "benefited from a powerful resume and a constellation of friends in the investment world and in Washington. And four years ago, his portfolio aligned smoothly with the agenda of an incoming administration and its plan to spend billions in stimulus funds on alternative energy. The recovering politician was pushing the right cause at the perfect time.... Gore's orbit extended deeply into the administration, with several former aides winning senior clean-energy posts."[21]

Republican Congressman Fred Upton of Michigan, the chair of the Energy and Commerce Committee, has been a critic of Gore's profiting off the taxpayer funds using his government connections. Gore's portfolio "is reflective of a disturbing pattern that those closest to the president [Obama] have been rewarded with billions of taxpayer dollars...and benefited from the administration's green bonanza in the rush to spend stimulus cash."[22]

Gore was essentially either a founder, a member, or a partner in a whole wide range of groups that were profiting or poised to profit from a green energy stimulus and federally mandated carbon trading schemes if they became law. Gore would have personally benefited if the carbon cap-and-trade bill he supported had become law. The media never

★ ★ ★

Not So Smart

Warren Buffett's vice chairman Charlie Munger told a small meeting of investors in 2017 that Gore is "not very smart" and "an idiot" but he was still able to amass a personal fortune in the investment world. "Al Gore has hundreds of millions [of] dollars in your profession. And he's an idiot. It's an interesting story." Munger added, "he's not very smart. He smoked a lot of pot as he [coasted] through Harvard with a gentleman's C."[24]

treated his Congressional testimony in support of the climate bills for what it actually was—a former vice president supporting legislation that would make him richer. These reports prompted one sarcastic skeptic to suggest, "Maybe Al Gore Should Be the Subject of a RICO Investigation."[23]

The power of carbon trading schemes to enrich politicians and corrupt politics is one reason that environmental guru James Lovelock has slammed carbon trading, declaring, "Most of the 'green' stuff is verging on a gigantic scam. Carbon trading, with its huge government subsidies, is just what finance and industry wanted. It's not going to do a damn thing about climate change, but it'll make a lot of money for a lot of people."[25]

In 2013, Gore sold his Current TV network to the Qatar-funded Al Jazeera for a reported $100 million. The sale inspired this headline at my Climate Depot website: "AlGorjeera—It's Official: Al Gore Is by Far the Most Lavishly Funded Fossil Fuel Player in the Global Warming Debate Today."[26] I asked if the media would now accurately label Gore an industry-funded activist every time they reported on him. Gore had literally sold out to big oil and gas: Al-Jazeera "received its initial funding through a decree from Emir of Qatar, and Qatar gets its wealth from its vast oil and natural gas reserves."[27]

The freshly laid off staffers from Current TV did not hesitate to lash out at Gore. "Gore's supposed to be the face of clean energy and just sold [the channel] to very big oil, the emir of Qatar! Current never even took big oil advertising—and Al Gore, that bulls***ter sells to the emir?" declared one former staffer, according to the New York Post.[28] Another staffer commented, "He [Gore] has no credibility."

They Get More Money—Even from Gas Producers

Strangely, the fossil fuel industry itself appears to be giving generously to the green movement these days.

In 2012, the *New York Times* reported that the Sierra Club was guilty of the "secret acceptance of $26 million in donations" from the natural gas industry, raising what the paper termed an "uncomfortable debate among environmental groups about corporate donations and transparency" and prompting accusations from fellow greens of "sleeping with the enemy." The money was used by the Sierra Club as part of a "campaign to block new coal-fired power plants and shutter old ones."[29] The natural gas industry paid the Sierra Club to help crush natural gases' competitor: coal. Thus, when, on CNN, Sierra Club executive director Michael Brune accused climate skeptics of taking funding from the fossil fuel industry—"The only folks who are arguing this are the occasional climate skeptic or the people who are paid for by the fossil fuel industry"—I retorted —"The Sierra Club took $26 million from natural gas and Michael has the audacity to try to imply that skeptics are fossil-fuel funded."[30]

It's important to put this $26 million figure in perspective. Just how big a funding advantage do global warming advocates have over skeptics?

The $26 million funding from the natural gas industry to the Sierra Club exceeded the entire combined "total revenue" of three of the most prominent U.S. climate skeptic groups in 2015, the most recent year for which records are available. The Competitive Enterprise

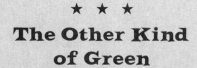

★ ★ ★
The Other Kind of Green

"Al was always lecturing us about green. He kept his word about green all right—as in cold, hard cash!" said another staffer.

★ ★ ★
Busting a Myth

Greenpeace co-founder Patrick Moore: "Well it's a common myth that the skeptic community is receiving all this money from the fossil fuel sector from coal and oil, etc. but in fact hundreds times more money is going from the fossil fuel industry into the green movement."[31]

Institute ($7.4 million), the Heartland Institute ($4.4 million), and the Committee for a Constructive Tomorrow (CFACT, which is the parent organization of my Climate Depot website, $2.2 million).[32] That adds up to total combined budgets of $14 million.

Fourteen million dollars was the *total* budgets of three combined skeptics groups in 2015, versus the $26 million the Sierra Club received just from natural gas. Sierra Club's 2015 revenue was over $109 million.[33]

A 2014 analysis by climate skeptic Steven Hayward noted that the skeptical organizations and bloggers that the media often accused of being "well-funded" actually have very small financial footprints. "These are boutique operations next to the environmental establishment: The total budgets for all of these efforts would probably not add up to a month's spending by just the Sierra Club. And yet we are to believe that this comparatively small effort has kept the climate change agenda at bay," Hayward wrote. "Rather than reflect, they deflect, blaming the Koch brothers, the fossil fuel industry, and Republican 'climate deniers' for their lack of political progress. Yet organized opposition to climate change fanaticism is tiny compared with the swollen staffs and huge marketing budgets of the major environmental organizations, not to mention the government agencies around the world that have thrown in with them on the issue."[34]

Nonetheless, this is one of the favorite talking points of global warming proponents. The Koch Brothers are supposed to be the most powerful financing force on the planet, funding the "deniers" and personally causing global warming. Democrat Senate majority leader Harry Reid asserted in 2014 of the Koch Brothers, "They are one of the main causes of this. Not a cause, the main cause."[35] Even the *Washington Post* had to give Reid "three Pinocchios" on that one, noting that using emissions as the gauge, the Koch Industries ranks twenty-seventh in the study Reid attempted to cite—putting out about six one-thousandths of 1 percent of global carbon dioxide emissions.[36]

A 2016 election-cycle analysis of the biggest donors in American politics compiled by OpenSecrets reveals the truth. Koch Industries came in at thirty-eighth, with only $11.2 million in total contributions in the 2016 cycle. San Francisco billionaire and climate activist Tom Steyer's NextGen Climate Action ranked ninth on the list, with $34.6 million in total contributions, and George Soros's group ranked fourteenth, with $26 million in contributions. The lobbying group Environment America also ranked ahead of Koch Industries, with $11.7 million in contributions.[37]

The media like to portray skeptical scientists as bought and paid for by the fossil fuel industry. But they don't dig into the lavish funding of some of the highest-profile climate-panic-promoting scientists—and the possible conflicts of interest involved.

Take lead global warming scientist James Hansen, now retired from NASA. In 2001, Hansen, who at that time was the director of NASA's Goddard Institute for Space Studies, was awarded $250,000 by the Heinz Foundation, run by Teresa Heinz Kerry. Hansen won the Heinz Award in the Environment for "his exemplary leadership" in climate change.[38] And then, lo and behold, in 2004 Hansen publicly endorsed Teresa Heinz Kerry's husband John Kerry for president over President George W. Bush. A presidential political endorsement is considered to be highly unusual for a NASA scientist. In short, Hansen got $250,000 in 2001 from Teresa Heinz Kerry and three years later endorsed her husband for president.[39]

Senator James Inhofe was critical of Hansen's conduct.[40] "The foundation's money originated from the Heinz family ketchup fortune," Inhofe said, noting that the media accuses skeptics of being funded by big oil. "So it appears that the media makes a distinction between oil money and ketchup money." The *Congressional Quarterly* story was headlined, "Inhofe Complains the Media Failed to Report Climate Change 'Ketchup Money' Grant."[41]

Hansen's colleagues at NASA were surprised at the apparent conflict of interest. Former NASA scientist Roy Spencer noted that "Hansen even campaigned for John Kerry, and received a $250,000 award from Theresa Heinz-Kerry's charitable foundation—two events he maintains are unrelated. If I had done anything like this when I worked at NASA, I would have been crucified under the Hatch Act. Does anyone besides me see a double standard here?"[42]

UN IPCC lead author and Princeton University professor Michael Oppenheimer is one of the media's most cited proponents of man-made climate change. He is the source of such memorable quotations as the prediction that, once global warming really kicks in, "It will be functionally impossible to be outside, including for things like construction work and farming, as well as recreation."[43]

In a 2006 climate documentary, former NBC newsman Tom Brokaw asked Oppenheimer, "Why do you think there is such a persistent minority of people who just refuse to believe that global warming is a fact?"

"There are some people who have a financial interest in not believing," Oppenheimer responded.[44]

A financial interest? But—fair is fair—what about *Professor Oppenheimer*'s financial interests? Michael Oppenheimer is put forth as some sort of objective scientist to present the claims of the global warming establishment. But what is almost never noted is that he was a paid partisan of the Environmental Defense Fund (EDF) and still serves as a scientific advisor for the green group.[45] Oppenheimer "was the holder of the 'Barbra Streisand Chair of Environmental Studies'" at EDF.[46]

Streisand, who once declared a "global warming emergency," explained her support of Oppenheimer this way. "My Foundation started supporting climate change work in 1989, when I donated a quarter of a million dollars to support the work of environmental scientist Dr. Michael Oppenheimer at EDF. Since then, I, and others have spent countless millions on this issue."[47] Oppenheimer was on the payroll of "big Hollywood."[48] And in 2010, Oppenheimer was also awarded a cash prize of $100,000 from the Heinz foundation, run by John Kerry's wife—six years after Oppenheimer belonged to a group which campaigned to help get Kerry elected president in 2004.[49]

Just before Professor Oppenheimer testified on Capitol Hill in 2014, I asked him about his own apparent conflict of interest:

> Morano: "In a 2006 interview with Tom Brokaw you'd said that global warming skeptics had a financial interest in promoting it. Yet you yourself were the Barbra Streisand chair at Environmental Defense Fund.... is it credible to say you also had a financial interest given that a quarter million dollars came from Hollywood, Barbra Streisand to fund you?"
>
> Oppenheimer: "So let me put that in perspective. What I think I was asked was what are the reasons beyond the science itself why people might be skeptical about climate change. Then I gave a variety of reasons including, for instance, it might be in some people's or some organizations' financial interests—like coal companies—to oppose measures to deal with climate change.... I think everyone's not only financial, political, and other interests and biases ought to be examined closely, and people are to be transparent about them—as I have always been."

Will the media take Professor Oppenheimer's advice and examine the financial ties of the global warming promoters?

Unfortunately, they continue to focus on the supposedly deep pockets funding the "deniers." Occasionally a reporter will go too far even for her colleagues. Journalist Sharon Begley, who has written for *Newsweek* and the *Boston Globe*, was rebuked by her *Newsweek* colleagues in 2007 for her "highly contrived" and "discredited" reporting claiming skeptics were better funded than warmists.[50]

In fact the biggest climate funding of all comes from governments around the world, which have spawned an entire industry cashing in on climate fears. The flood of money going to promote the belief in man-made climate change and the policies to address it continues unabated, with the European Union spending $7 trillion on climate policies—which, as we shall see in more detail in chapter eighteen, have no impact on the climate but are building a huge constituency of researchers, scientists, bureaucrats, rent seekers, and other government employees.[51] The more the green plans fail, the more the green planners plan.

The World Economic Forum seeks to spend $14 trillion, or $700 billion a year until 2030, to "green" the global economy in order to "keep a lid on global warming," the *Independent* reported. The article noted that according to a report from the organization, "Only a sustained and dramatic shift to infrastructure and industrial practices using low-carbon technology can save the world and its economy from devastating global warming."[52]

Global spending on global warming is skyrocketing. The world is spending $1 billion per day in an attempt to tackle global warming, according to a 2013 report by the Climate Policy Initiative.[53]

But this almost $1 billion a day is still just a small fraction of the $5 trillion that the International Energy Agency wants to spend for alternative energy projects—to limit global temperatures from rising more than 2 degrees Celsius. And if you eat candy bars, you are helping spend another $1 billion in efforts to fight climate change, as the Mars candy company announced their climate initiative in 2017 after signing a petition urging President Trump not to withdraw from the UN Paris climate pact.[54]

A 2011 UN report admitted—years before the Paris climate pact—that going green will cost $76 trillion over forty years. Dan Gainor of the Media Research Institute has explained, "So let's do the math: That works out to a grand total of $76 trillion, over 40 years—or more than five times the entire Gross Domestic Product of the United States ($14.66 trillion a year). It's all part of a 'technological overhaul' 'on the scale of the first industrial revolution' called for in the annual report. Except that the U.N. will apparently control this next industrial revolution."[55]

Cut Them Off

In 2017, with President Trump seeking to rein in massive spending on climate change, climatologist Roy Spencer urged the new administration to end the "incestuous relationship" between government and science research. Spencer ridiculed the notion that government-funded science is objective. "Since politicians are ultimately in charge of deciding how much money agencies receive to dole out to the research community, it is inevitable that politics and desired outcomes influence the science the public pays for," he wrote.[57]

Spencer pointed out, "All of the scientific institutions are going to jump on the bandwagon, with politically savvy committees agreeing with each other; they are in effect being paid by the government to agree with the consensus through billions of dollars in grants and contracts. If there is no global warming crisis, there would be little congressional funding to study it, and thousands of climate-dependent careers (including mine) simply wouldn't exist."[58]

Glaciologist Dr. Terry Hughes, a pioneer in his field, ended his silence in 2015, declared his dissent, and explained how he had witnessed government funding corrupting climate science. "Too many (the majority) of climate research scientists are quite willing to prostitute their science by giving these politicians what they want," Hughes, a professor emeritus at the University of Maine's Climate Change Institute and School of Earth and Climate Sciences, wrote. "I'm now retired, so I have no scientific career to protect by

★ ★ ★

It's Never Enough

Christiana Figueres, the UN climate chief who described her job as "sacred," has complained that far too little is being spent in the attempt to alter the climate of the Earth. "We need to be investing a trillion every year," Figueres said.[56]

spreading lies." Hughes has trekked to the Arctic ten times and Antarctica thirteen times since 1968 as the principal investigator of National Science Foundation (NSF)–funded glaciological research. Climate fears, he said, help politicians "get electoral visibility by pounding the panic drums."[59]

A key example of how funding is distorting climate science is the way researchers strive for media attention in an effort to motivate climate "action." Professor Camilo Mora, an ecologist at the University of Hawaii, has admitted that the press release for his 2014 study warning of climate doom by 2047 (unless the world takes action) took as long as the actual scientific study. As Mora explained, "The actual writing of the paper took us about two months. It's 40 pages altogether. But it's amazing that the press release on this paper took us two months to prepare. It was a massive investment of time for just two pages of paper. So another limitation for us as scientists is that it's very hard for us to commit that kind of time to prepare for that press release."[60]

Why spend two months on a press release? "My motivation and everybody's motivation whenever we produce these papers is trying to increase the level of awareness of people and politicians to take action on these things.... So as scientists, we are struggling to figure out how we can increase public awareness on this issue," Mora said. In other words, if a climate doomsday prediction study is produced at a university and the media fails to report on it, the funding may dry up.

A former NASA GISS atmospheric scientist broke his silence in 2017 and spoke of money "wasted" in climate research. "The wasted and misspent money at NASA GISS and all climate research institutions is staggering," said Climate Scientist Dr. Duane Thresher, who had worked under James Hansen and current lead NASA climate scientist Gavin Schmidt. "Follow the money. I am going to concentrate on NASA GISS, where I was for 7 years, but it applies to all climate research institutions, of which I have been at several and am familiar with several more." Thresher's solution is simple:

"Start with defunding NASA GISS where this whole global warming nonsense started. It was started by James Hansen, formerly head of NASA GISS and considered the father of global warming. It was continued by Gavin Schmidt, current head of NASA GISS, anointed by Hansen, and leading climate change warrior scientist/spokesperson." "NASA GISS is a monument to bad science that truly should be torn down."[61]

MIT atmospheric scientist Richard Lindzen agreed. "By now, most of the people working in this area have entered in response to this funding. Note that governments have essentially a monopoly over the funding in this area. I would expect that the recipients of this funding would feel obligated to support the seriousness of the problem. Certainly, opposition to this would be a suicidal career move for a young academic," Lindzen explained.[62]

Lindzen urged Trump to seize the opportunity and "cut the funding of climate science by 80% to 90% until the field cleans up." As the now retired MIT scientist pointed out, "Climate science has been set back two generations, and they have destroyed its intellectual foundations." Money has had a corrupting influence. "Groupthink has so corrupted the field that funding should be sharply curtailed rather than redirected," Lindzen explained.[63]

Climate statistics professor Caleb Rossiter of American University also noted how money is driving research. "It is absolutely true that the money available for global warming statements and research is driving academia right now and people line up to get it," Rossiter explained.[65]

In the early days of the Trump administration, OMB Director Mick Mulvaney announced to stunned officials in Washington that global warming funding is "a waste of your money." Mulvaney said at a White

Pigs at the Trough

"Even in 1990 no one at MIT called themselves a 'climate scientist,' and then all of a sudden everyone was. They only entered it because of the bucks; they realized it was a gravy train. You have to get it back to the people who only care about the science."

—MIT atmospheric scientist Richard Lindzen[64]

House briefing, "I think the president was fairly straightforward on that: We're not spending money on that anymore."[66] We shall see if that pledge is kept.

CHAPTER 16

Hypocrisy on Parade

Why don't the climate activists live like they believe in catastrophic climate change?

Hollywood celebrities are some of the most prominent pushers of the global warming panic, ceaselessly urging action to battle climate change. The stars regularly harangue Americans to sacrifice in order to save the planet. They call for more austerity for the rest of us, but are they willing to walk the walk in their own lives? I think you already know the answer.

In 2007, former Vice President Al Gore appeared at the Seventy-Ninth Academy Awards ceremony in Hollywood with fellow climate activist Leonardo DiCaprio.[1] Just before they walked out onstage, the Oscars flashed helpful tips on what people can do to lower their carbon footprints. Columnist Charles Krauthammer wrote in *Time* magazine that his personal favorite tip was, "Ride mass transit."[2] As Krauthammer pointed out, "This to a conclave of Hollywood plutocrats who have not seen the inside of a subway since the moon landing and for whom mass transit means a stretch limo seating no fewer than 10." When was the last time Gore or DiCaprio rode a city bus?

Did you know?

★ Al Gore's house uses twenty times as much energy as the average American home

★ A Greenpeace executive was caught commuting 250 miles to work—by plane

★ Producer James Cameron flew to South America to protest a dam that will bring electricity to poor Brazilians

Gore's 2006 Oscar-winning film *An Inconvenient Truth* asks viewers at the end, "Are you ready to change the way you live?" Gore was confronted with his own question when it was reported that his home energy usage at his residence in Tennessee was twenty times as high as that of the average American household at that time. Asked about his outsized carbon footprint during a March 21, 2007, Senate hearing, Gore refused to take a "Personal Energy Ethics Pledge" to promise that the Gore household would henceforth consume no more energy than the average American household. The pledge was presented to Gore by my former boss, Republican senator James Inhofe of Oklahoma, then the ranking member of the Environment and Public Works Committee.[3]

Personal Energy Ethics Pledge

As a believer:

that human-caused global warming is a moral, ethical, and spiritual issue affecting our survival;

that home energy use is a key component of overall energy use;

that reducing my fossil fuel-based home energy usage will lead to lower greenhouse gas emissions; and

that leaders on moral issues should lead by example;

I pledge to consume no more energy for use in my residence than the average American household by March 21, 2008.

The pledge that Inhofe asked Gore to take. (Gore refused.)

Fast forward ten years and not much has changed. In June 2017 Gore reassured the world that he had mended his lavish lifestyle. "I don't have a private jet. And what carbon emissions come from my trips on Southwest Airlines are offset. I live a carbon-free lifestyle, to the maximum extent possible," he told CNN.[4]

★ ★ ★

Gore-Morano Encounter at Thirty-Seven Thousand Feet over the South China Sea

In 2007 I encountered Al Gore on the flight home from the Bali UN climate conference. This is the account I wrote at the time:

December 14, 2007—37,000 feet over South China Sea—"Former Vice President Al Gore rebuffed Senator James Inhofe's spokesman during the return flight from the Bali UN climate conference. Gore, in front of excited passengers posing with photos, revealed he was not happy with the criticism that he has endured 'You all attack me all the time,' an agitated Gore said. Morano responded, 'Yes. We do.' After a long stare, Gore refused to have [a] photo taken and walked off. Inhofe has been one of Gore's harshest critiques. Inhofe's website mocks Gore for hypocrisy and maintains a running counter of how many days since Gore has refused to pledge to use no more than the average electricity use."

Two years later, in a *New York Times* profile of me, reporter Leslie Kaufman wrote, "Gore had no memory of the encounter. Mr. Morano does not care. He tells the story anyway." But according to Joe Romm of Climate Progress, Gore remembered the incident after all. Romm reported in April 10, 2009, "I happened to be speaking to Gore today and he remarked on this Morano fable and said

he just doesn't remember it happening the way Morano describes." Hmm. Gore told the *New York Times* that he has "no memory of the encounter," yet he told his fellow climate activist Romm that he "doesn't remember it happening the way Morano describes."[5]

In July of 2017, Gore and I had another encounter, this one on captured on video. I met up with Gore in Melbourne, Australia, after his talk to the EcoCity World Summit and offered him a DVD of my film *Climate Hustle*. During the inconvenient encounter, Gore refused to accept the DVD of the film and walked on by to his waiting Lexus RX 450h SUV "hybrid."[6]

Note the horrified face of the lady on the right as she catches sight of the DVD of the skeptical film *Climate Hustle*.

★ ★ ★

The Crying Climate Activist

Laurie David, the producer of Al Gore's *An Inconvenient Truth*, has told the story of her eye-opening conversion to climate activist.

"My concerns about global warming began soon after we had our first child. I was a new mom, feeling very overwhelmed with the realization that I was now irreversibly responsible for this tiny creature. There was no turning back. I remember crying every day at five in the afternoon, the witching hour, my stress level at a breaking point.... I spent a lot of time walking around the neighborhood, pushing a stroller. I started noticing an enormous amount of SUVs on the street. Everyone was driving them.... So, every time you drove somewhere, to the store, the school, the freeway, you were now all of a sudden doubling your personal CO_2 pollution. I panicked, because everyone I knew was driving them. I had had other lightbulb moments in my life—like the first time I tasted good wine and then couldn't drink the cheap stuff any more...."[7]

David came out of climate retirement in 2017, prompted by Donald Trump's presidential victory. "After the election, it took me two weeks to just stop crying. I was just anticipating what was to come and what has come was worse than what I was crying about," David told the *Hollywood Reporter*.[8] She is becoming known as the crying activist. David says she also wept for three days and three nights when John Kerry lost the presidency in 2004. According to the *Guardian*, "she cried because, in her mind, the end of the planet as we know it had just got that much closer.[9]

But in August 2017, when Gore's *An Inconvenient Sequel: Truth to Power* was released, The Daily Caller reported on a comprehensive analysis revealing that Gore's Nashville home (one of three he owns) was now using twenty-one times as much energy as the average American home per year—and in the month of September, thirty-four times as much. The report noted that "the green extremist shells out about $22,000 a year to pay his electric bills" but that Gore can afford it because "he has manipulated environmental concerns into a big business" and grown his net worth from under $2 million to approximately $30 million since 2001. The coup de grâce? "Over the last 12 months, Gore used more electricity just heating his outdoor swimming pool than six typical homes use in a year."[10]

An editorial in the *Richmond Times-Dispatch* in 2017 summed up the problem with Gore as a spokesman for global warming. "So Gore, despite his sermons about how others should live, is quite the energy hog," the paper noted, adding "Maybe climate change isn't so dangerous after all—at least not enough to make any personal sacrifices." The editorial concluded with a warning to "be skeptical of really rich guys (especially if they're former politicians) preaching apocalypse and salvation."[11]

High Fliers

Gore has also been caught flying private jets instead of commercially—Sean Hannity of Fox News revealed footage of Gore flying a B2 aircraft, "one of most fuel inefficient private jets."[14]

Gore's ally Leonardo DiCaprio also suffers from this same one-rule-for-me-and-another-for-thee attitude. "I will fly around the world doing good for the environment," DiCaprio, who currently serves as the United Nations Messenger of Peace, boldly declared in 2013.[15] He seems to be a little unclear on the concept of sacrificing for the planet. A 2017 *Daily Mail* investigation into celebrities' "carbon footprints" found DiCaprio "flew around 87,609 miles on various business trips and jaunts around the world which burned up 14.8 tonnes of carbon dioxide."[16]

★ ★ ★
"Stupendous Wealth"

In the of summer 2017, former Vice President Al Gore released *An Inconvenient Sequel: Truth to Power*. The sequel tanked at the box office.[12] Climate activist Stephen Lacey, editor-in-chief of Greentech Media, blamed Gore's hypocritical lifestyle for the public not warming to his film, noting that his "stupendous wealth complicates his climate message." As Lacey pointed out, "We're in an era of backlash against elites, so Gore, a guy who bought a 6,500-square-foot seafront home in California for $8.8 million, and who hangs around with other celebrities who talk big on climate but who live lavish lifestyles, is the perfect target at this point in time."[13]

But despite his high-consumption lifestyle, DiCaprio told the UN climate summit in Paris in 2015 that everyone needed to make changes. "The solutions we seek require all of us to make real changes in the way we live our lives, operate our businesses, and govern our communities. Our future will hold greater prosperity and justice when we are free from the grip of fossil fuels," DiCaprio said.[17]

In 2016, DiCaprio, who is frequently seen on yachts and owns multiple homes, generated more controversy when he took a private jet an extra eight thousand miles to collect an environmental award in New York City. One analysis found that the amount of fuel he used flying to get the award could power ten thousand cars for a day.[18] Environmental analyst Robert Rapier, who owns a renewable-energy company, said DiCaprio's high-flying "diminishes his moral authority to lecture others on reducing their own carbon emissions." DiCaprio "demonstrates exactly why our consumption of fossil fuels continues to grow. It's because everyone loves the combination of cost and convenience they offer. Alternatives usually require sacrifice of one form or another." Rapier said.[19]

A humbler actor who understands the concept of walking the walk a little better is John Travolta. He has urged his fans to "do their bit" to fight global warming—while owning five planes and his own private runway. But at least Travolta has realized and publicly acknowledged his own hypocrisy. Travolta admitted, "I'm probably not the best candidate to ask about global warming because I fly jets."[20]

Actor Chevy Chase has touted socialism and cited Cuba as a model country. I interviewed the *Vacation* and *Fletch* actor on the Mall in Washington, D.C., on Earth Day 2000 for my Amazon Rainforest documentary *Clear-Cutting the Myths*. A highly agitated Chase took offense at my question about Hollywood hypocrisy in preaching to Americans conservation to ordinary Americans.

I asked, "Skeptics would say that Hollywood has all this wealth and money—"

But Chase interrupted, "I am not from Hollywood! I'm from Upstate New York first of all and second of all, I don't know who you are!"[21]

Later in our testy exchange, which included Chase grabbing my microphone, Chase said, "I would also like to address your skepticism about Hollywood people. A lot of them are airheads. There are also well educated, college-educated people with degrees, not only BA's, but MA's. Intellectuals who read and care about these things. You're talking to one and I don't particularly like being, you know, set upon by a skeptic who thinks he's talking to some guy from Hollywood who just plays tennis all day and spills water all over the place. That isn't the way we live and that isn't the way people who care about the environment live. So take it easy on them!"

So there.

Actress Donna Mills of *Knots Landing* fame has opined that Americans need to cut down on their high living. Mills said at the Earth Day 2000 rally in D.C., "We do have to cut down on consumerism. We're the worst offenders in the world, this country is. We have to realize it's not things that make us happy, it's people and our environment."[22] Mills owns a "nearly 5,000-square-foot home with her longtime boyfriend, Larry."[23]

Another major Hollywood celebrity who has taken up the cause of saving the Earth is actor Harrison Ford. But Ford, who was featured in the 2014 climate series *Years of Living Dangerously* appears to fall short when it comes to his personal habits. He has admitted that he is so passionate about flying that he uses his plane for takeout. "I often fly up the coast for a cheeseburger," Ford said in a 2010 interview in the UK *Daily Mail*. Ford's personal flying habits fly in the face of his climate warnings. In 2015, Ford predicted that "if we don't work together" to stop global warming, "The planet will be ok, there just won't be any damn people on it."[24]

Former GOP California governor and global warming activist Arnold Schwarzenegger came under fire for his near-daily private jet commute from his Brentwood home to the governor's mansion in Sacramento. "The governor's Gulfstream jet does nearly as much damage to the environment in one hour as a small car does in a year," the *Los Angeles Times* reported. But not to worry, because "Schwarzenegger is well aware of this and makes amends by purchasing pollution credits for the carbon dioxide his jet releases."[25] Still, Schwarzenegger's fellow environmentalists were distressed by his hypocrisy. Environmental activist Denis Hayes told the paper, "If you are going to be talking about an issue, you should be living the reality you are trying to embrace."

> ★ ★ ★
> ## It's Not Easy Being Green
>
> Greenpeace decided to take a page from the flying antics of celebrities like Travolta, Ford, and Schwarzenegger. In 2014 the *Telegraph* noted that despite Greenpeace campaigning for limits on "the growth in aviation," which, it says, "is ruining our chances of stopping dangerous climate change," the green group was paying one of its senior executives to commute 250 miles to work by airplane. Pascal Husting, Greenpeace International's international program director, was commuting twice a month between Luxembourg and Amsterdam.[27]

An undaunted Schwarzenegger did not let this stop him from touting "air-drying your clothes" to fight global warming. "These are the kind of things that anyone can do. Did you know that air-drying your clothes for six months saves 700 pounds of carbon dioxide?" he asked in his Governor's Earth Day message in 2009.[26] There were no reports of Schwarzenegger's clothing hanging in the wind at the governor's mansion in Sacramento.

The Hummer-driving Schwarzenegger had all kinds of suggestions for other people.

"Keep the right tire pressure in your car and you will reduce gasoline costs by 4 percent. Now maybe that doesn't seem like much. But listen to this: If everyone in the country took this simple step, America's oil consumption would drop by 800,000 barrels a day. You get it? That's the power the individual has," he said.

In 2017, celebrities appeared at a telethon to raise money for hurricane recovery. They could not miss the opportunity to toss in their opinions on climate change.

"Anyone who believes that there's no such thing as global warming must be blind or unintelligent," said Stevie Wonder during the telethon. Beyoncé added, "The effects of climate change are playing out around the world everyday." She also seemed to link an earthquake in Mexico to climate change as well.[28] The "Hand in Hand" telethon to raise money for victims of Hurricanes Harvey and Irma also featured Justin Bieber, George Clooney, Cher, and Leonardo DiCaprio. A *Daily Mail* investigation of the telethon revealed the telethon "started with a lecture about global warming—then celebrities with multiple homes, cars and private jets starting soliciting much needed money."[29]

Noble Savages

Actresses Cameron Diaz and Drew Barrymore starred in a 2005 MTV environmental series called *Trippin'* that "lauded traditional tribal lifestyles, which lack running water, electricity." The show featured Hollywood celebrities praising the developing world's primitive lifestyles as Earth-friendly—despite those poor nations' high infant mortality rates and short life expectancies. Hollywood celebrity Barrymore was so enthralled by the lack of a modern sanitary facilities that she gleefully bragged in one episode about, well—defecating in the forest. "I took a poo in the woods hunched over like an animal. It was awesome!" Barrymore bragged.[30]

Diaz responded, "I am so jealous right now, I am going—I am going to the woods tomorrow."

Barrymore laughed, repeating, "It was awesome."

Barrymore, who was reportedly earning $15 million a film, found the episode that she spent in a primitive, electricity-free Chilean village inspiring.

"I aspire to be like them more," Barrymore said. Diaz, who was making a reported $20 million a movie, criticized the lifestyles of many Americans after visiting the Chilean village. "It's kinda gotten out of hand how much convenience we think we need," she said. She boasted that the cow-dung slathered walls of a Nepalese village hut were "beautiful" and "inspiring" and called the primitive practice of "pounding mud" with sticks to construct a building foundation "the coolest thing."

Despite the celebrities' praise for the primitive life, the *Trippin'* MTV show featured them flying on multiple airplanes and chartering at least two helicopters and one boat to reach remote locations over the course of the first four episodes. The series also showed the celebrities being chauffeured to the airport in a full-size Chevy SUV—despite several on-screen anti-SUV factoids about how environmentally unfriendly SUVs are.[31]

Actor Alec Baldwin is another celebrity climate activist. Baldwin, who acted in the kids' global warming film *Arctic Justice*, declared in 2016, "I believe climate change denial is a form of mental illness."[32]

But when the actor attended the 2015 UN Paris climate summit, he admitted his hypocrisy to CNN. "Here I am, staying in a fancy hotel, going out to dinner with my friends, hanging out, and we're doing our best to disseminate some information, but I'm on one end of the spectrum, and this is all new to me," Baldwin explained, adding, "I've done a lot of environmental work back home related to climate change."[33]

Super-wealthy Hollywood director Chuck Lorre, who is responsible for such shows as *Two and a Half Men*, waxed nostalgic about the world before the advent of modernity in one of his trademark "Vanity Panels" that air following every one of his programs, noting that the old days were much better. "It was relatively stable. Peace on Earth, if you will. Then, with the advent of large-scale agriculture and the need for ever larger swathes of land to accommodate it, the tribal system collapsed and people began to live huddled together in towns, villages and cities. In short order, the priceless wisdom that taught us who we were and how we could live a happy life was forever lost to mankind," Lorre wrote.[34]

More Hollywood Hypocrites

In 2011, actor Robert Redford declared that if the EPA failed to regulate CO_2 as a pollutant, "I'm going to be devastated.... I think the country will be devastated too."[35] Redford, who was featured on *Time* magazine's list of environmental heroes, is a long-time environmentalist with the Natural Resources Defense Council who has championed global warming as his cause celebre and promoted alternative energy sources like wind and solar.[36] Redford has compared the climate change movement to the 1960s Civil Rights Movement and called global warming the defining issue of our time.[37]

Redford has said, "Fossil fuels are literally cooking our planet."[39] He chastised President Obama for not taking enough action on climate change, and he opposed the Keystone Pipeline.[40] But he appears to have fallen short in his personal life.

Filmmakers Ann McElhinney and Phelim McAleer exposed Redford's shortcomings in their film *Robert Redford—Hypocrite*. The actor, they discovered, "opposes eco-village near his property while quietly selling $2 million lots in the Sundance Preserve."[41] Redford also did promotional voiceovers for a United Airlines commercial

★ ★ ★
He Can Afford It

Redford has gotten some pushback on his civil rights comparison. African American ministers protested the actor in 2009, "linking his environmentalism to racism" and calling him an "enemy of the poor." The Congress on Racial Equality's Niger Innis said Redford's advocacy of limiting cheap affordable carbon-based energy such as oil, gas, and coal, "hurts a lot of low-income families" by keeping electricity rates higher. "Robert Redford can afford to heat his 13,000-square-foot mansion in Utah no matter how high home heating prices get," said Harry R. Jackson Jr., chairman of the High Impact Leadership Coalition. "But grandmothers on a fixed income and single mothers dependent upon public assistance count on energy production in states like Utah to continue so that their home-heating costs stay as low as possible," Jackson explained. "The high energy prices we're going to see this winter are essentially discriminatory."[38]

proclaiming, "It's time to fly," even though he has frequently called on the world to reduce its carbon footprint.[42]

Hollywood producer James Cameron, responsible for such mega hits as *Titanic* and *Avatar*, has also been a huge climate activist. Cameron once challenged skeptics to a public debate using the rhetoric of an Old West gunslinger: "I want to call those deniers out into the street at high noon and shoot it out with those boneheads."[43]

In 2010, Cameron's representatives reached out to me to assemble a skeptical debate team to face off against the producer at a public event. We agreed to the terms; Ann McElhinney and the late Andrew Breitbart were going to be joining me on the skeptical side of the debate.

I was flying to Aspen, Colorado, for the great global warming Wild West showdown when Cameron got cold feet and canceled the debate. At the very last moment, Cameron pulled the plug on a debate he himself had initiated and organized. When my connecting flight landed in Denver, I was informed that the debate was off.[44] The official reason given by Cameron's spokesman was that "Morano is not at Cameron's level to debate, and that's why it didn't happen. Cameron should be debating someone who is similar to his stature in our society." But the real reason had nothing to with "stature in society" and more to do with fear of losing a climate debate. Cameron backed out of the debate at the last minute after environmentalists "came out of the woodwork" to warn him not to engage in a debate with skeptics because it was not in his best interest. I responded to Cameron's last-minute debate ducking with this statement: "Cameron let his friends in the environmental community spook him out of this debate. When he was warned that he was probably going to lose and lose badly, he ran like a scared mouse."[45] Cameron had gone from Wild West gunslinger to chicken of the sea. But Cameron's real failing is not his debate cowardice; it is his indifference to the needs of the developing world.

In 2010, Cameron and actress Sigourney Weaver flew to Brazil to protest a dam that would be one of the world's largest hydroelectric projects. Even

★ ★ ★

Let Them Eat Cake

A UN climate and Earth summit is quite the extravaganza. I attended my first UN environmental summit in 2002 in Johannesburg, South Africa. The *New York Post* detailed how the politicians and the UN behaved while speaking of poverty:

> While starving children line up for handouts in the shantytowns of Johannesburg, South Africa, delegates to the world conference just blocks away have been pigging out on Kilimanjaros of lobster, oysters and filet mignon. Desmond Morgan, the head chef at Johannesburg's five-star Michelangelo Hotel, where world leaders and other VIP delegates are staying during the "save the planet" conference, says he's cooking round the clock. As famine looms across southern Africa, Morgan has stocked a thousand pounds of shellfish, a trove of caviar and pate, and thousands of lobsters. Summit participants from countries including the United States can order up a smorgasbord at a moment's notice. "Whether they want Beluga caviar, foie gras or bacon sandwiches—we have it all. In my experience, heads of state don't decide what they want to eat or drink until the last minute. So I have to make sure I have everything they can possibly want," Morgan told the London Sun.[46]

then Brazilian President Luiz Inácio Lula da Silva, of the leftist Workers' party, objected to Cameron's attempts to keep energy out of the developing world and "argued that the dam will provide clean energy and is needed to meet current and future energy needs."[47] Cameron opposed a dam—now under construction—that will bring vital electricity to Brazilians. Cameron flew to the developing world to campaign against improving the lifestyles of its poor citizens.

But Cameron seems to be guided by his own form of utopian philosophy. "We are going to have to live with less," the fabulously wealthy producer told the *Los Angeles Times* in 2010.[48] Cameron, whose net worth was estimated at around $900 million in 2014,[49] warned that we face "a dying world if we don't make some fundamental changes about how we view ourselves and how we view wealth." He warned against the "consumer society where you buy something and then throw it away when you get the next new thing, filling up huge landfills with plastic and electronics."

Cameron also wants Americans to change their ways. "Honestly, the truth is, we have to revisit almost every part of our lives and our existence over the next few years. Energy consumption, I think, being the biggest one. Energy and global warming are interlinked issues obviously," the producer explained.[50]

But revisiting "almost every part of our lives" did not seem to impact Cameron's personal life. He owned not one but two adjacent eight-thousand-square-foot mansions in Malibu—and a submarine.[51]

CHAPTER 17

Child Propaganda

If you can't convince adults, who is a more willing and pliable audience than children?

John Kerry signed the Paris climate pact at the UN with his two-year-old granddaughter seated on his lap for full effect. Kids from kindergarten through college are prime targets for the climate change fear promoters' propaganda. Politicians, academia, Hollywood, and global warming activists have focused on kids, feeding them a steady diet of fear and doom, using vulnerable children to promote climate fears.

The media amplifies these tactics. ABC News gushed that Kerry's ploy of bringing his toddler granddaughter to the UN signing was "a symbolic photo opportunity designed to remind the public of its duty to protect the health of future generations."[1]

Leonardo DiCaprio, now an official UN climate scare spokesperson, once declared, "We need to get kids young."[2] Get them and depress them. "I want the public to be very scared by what they see. I want them to see a very bleak future," DiCaprio said while promoting his global warming film *The 11th Hour*.[3]

Did you know?

★ A judge in the UK ruled that schools must warn students about scientific inaccuracies in Al Gore's *An Inconvenient Truth*

★ Gore, who has four children, says Africans should have fewer children on account of climate change

★ One climate activist says we should "protect our kids by not having them"

★ ★ ★
DiCaprio "Brainwashed" into Activism by His Parents?

In 2016, a clue to Leonardo DiCaprio's motivation on climate change was revealed by *New York Post* entertainment columnist Kyle Smith in an article titled, "Leo DiCaprio's Parents Brainwashed Him into Becoming an Environmentalist Freak":

If you've ever considered what underlies the DiCaprio Paradox—the way Leo's persona seems divided between partying with half a hundred bikini nymphs and preaching imminent global apocalypse to fellow parishioners of the Church of Doom—his new movie offers a possible key. It turns out DiCaprio's hippie-artist dad, who knew Andy Warhol, Lou Reed and LSD proselytizer Timothy Leary, abused the poor child by putting a picture of one of the most terrifying paintings of all time over young Leo's crib so the kid would nightly fall asleep to a scene of ravagement, despair and damnation. "My first visual memories are of this framed poster above my crib," DiCaprio explains in his new climate-change documentary, *Before the Flood*, out Friday. "I would stare at it every night before I went to bed…I was brought up on all kinds of wacky visuals when I was a kid. So I would stare at this painting over and over—*The Garden of Earthly Delights* by Hieronymus Bosch…. The deadly sins start to infuse their way into the painting," DiCaprio says. "There's overpopulation, there's debauchery and excess."

Smith summed it up: "Nice job warping your kid's still-forming brain cells, George DiCaprio! Most of us put our kids to bed with fluffy bunnies, not the End of Days."[4]

Hollywood activist Laurie David, Gore's co-producer on *An Inconvenient Truth*, co-authored a kids' book with Cambria Gordon for Scholastic Books titled *The Down-to-Earth Guide to Global Warming*. David, in an open letter to her children, stated, "We want you to grow up to be activists." A Canadian high school student named McKenzie was shown Gore's film in four different classes. "I really don't understand why they keep showing it," McKenzie said on May 19, 2007.[5]

Frightening the Children—and Using Them

No climate change campaign is complete without messages to the kids urging them to pester their parents to change their ways. The Hollywood global warming documentary *Arctic Tale* ends with a child actor telling kids, "If your mom and dad buy a hybrid car, you'll make it easier for polar bears to get around."[6]

In November 2016 the Weather Channel released *When Kids are Talking Climate—Maybe It's Time to Listen!*, a climate change video featuring young children attempting to convince their parents of the seriousness of the issue. "Dear Mom and Dad, Science says the impact of climate change could be very catastrophic during my lifetime," the kids warned their parents. "Climate change is real, it's bad, and it's caused by humans," the video announced.[7]

Weather Channel founder and meteorologist John Coleman ripped his old network for what he termed an "immoral" kids' video. "Right or wrong, using children to promote a point of view borders on immoral. Even knowing that climate change is not happening, it is far beneath my values to use children to promote this truth," Coleman said. "I know without a doubt that there is no significant threat to the future climate of Earth from the industrialized civilization we have created and the drastic climate changes

★ ★ ★

Children's Crusade

When Kids are Talking Climate—Maybe It's Time to Listen! was unabashed climate fear.

"Dear Mom and Dad, The science is clear," the kids said.

The children also made these claims:

- Climate change is "a major threat to national security."
- It's the "hottest year on record."
- "This is about our families' health."
- "97% of scientists agree that global warming started decades ago."
- "Rising sea levels would displace millions."

And

- "It rains harder now."

★ ★ ★

Struck Speechless?

Bill Nye claimed that that he had rendered me speechless during our Central Park interview in 2016: "I confronted Marc Morano, another climate denier, and I said, 'What about your kids?' And he was at a loss for words—it's on camera—he was at a loss for words. Because kids are the reason you live, as a parent, to pass your genes on, and if you pass your genes on to an environment that you ruined, you're just not doing a very good job as a parent. So we'll see what happens as the kids and grandkids of deniers come of age."[11]

But as I told Nye during the interview, I was not even sure what his point was in mentioning my kids. My kids frequently travel with me to TV appearances, speeches, trips to climate meetings, and I have spoken to their classrooms to a very positive reception. One inspiration for my work is the hope of leaving my kids a legacy where they are free to question the "consensuses," promote open scientific debate, and be free of politicized science with hidden agendas. During the same interview, Nye suggested jailing climate skeptics for their dissent. If my legacy to my kids is helping to stop Nye's anti-scientific ideas from becoming reality, then I will be content. Now, can we leave my kids out of this?

predicted by the Al Gore clan and the UN's IPCC are not occurring and are based on an invalid theory. But I will not stoop to the use of children to promote my position."[8]

The UN also likes to use kids to promote man-made climate change fears. The UN awarded children "astronauts" in 2016 for a video in which they sang a song claiming, "We don't need no CO_2," and "don't need no bath." The kids sang, "We are astronauts of Mother Earth—we don't need no cars."[9]

Singer Katy Perry produced a UN video claiming that "man-made climate change is hurting children around the world." The celebrity declared, "It's always children who are first to suffer from its impact."[10]

Children were also used to promote Al Gore's 2017 sequel *An Inconvenient Sequel: Truth to Power*, with a promotional video featuring an eleven-year old warning, "People are releasing toxic gases that are ruining the world." In the video, elementary school kids ages seven through thirteen watch rapper and climate activist Prince Ea, also known as Richard Williams, explain the dangers of global warming. The kids are exposed to scary climate claims and then "inspired" to fix the problem. Prince Ea tells the kids, "Storms are stronger than ever before... more drought, wildfires, hurricanes than ever before." One kid reacts, "This is like, making me feel sad." Another boy explains, "I'm like shaking now." The Huffington Post explained. "Participant [Media] acquired the video production firm Soul-Pancake last year, and has used the company to produce YouTube videos to promote 'An Inconvenient Sequel.'"[12]

Schools have been urging and organizing elementary school–age kids to get involved in climate change issues. One fifth grader launched a "youth ambassador program" to teach kids about global warming.[13]

And the relentless campaign of fear and propaganda seems to be working. The *Washington Post* quoted nine-year-old Alyssa Luz-Ricca saying, "I worry about [global warming] because I don't want to die."[14] The same article explained, "Psychologists say they're seeing an increasing number of young patients preoccupied by a climactic Armageddon." According to a June 14, 2007, article in the *Portland Press Herald*, fourth graders from the East End Community School in Portland, Maine, released a frightening report: "Global Warming Is a Huge Pending Global Disaster."[15]

Greenpeace co-founder Patrick Moore fiercely opposes kids' being injected with climate change fears. "It's basically child propaganda, it's not right. We should be teaching children critical thinking. We should be teaching them about carbon cycle, not about carbon pollution. We should be teaching them about how carbon dioxide is the most important food for all life on earth," Moore explained.[16]

Climatologist Judith Curry is also opposed to young kids being pumped full of global warming fears. "I don't see any reason to teach kids about this unless you're trying to brainwash them in some way. Get 'em while they're young kind of thing," Curry said. "I'm afraid that seems to be the motive in some of these educational efforts," she explained. "I would teach it in a political science class more than I would an actual science course."[17]

A judge in the UK finally said enough, ruling that schools must issue a warning before they show Gore's film to children because of its scientific inaccuracies. Gore's film "was described in the High Court as irredeemable, containing serious scientific inaccuracies and 'sentimental mush.'"[18]

And there has been some headway against the propagandizing of children in the United States, as well. In 2015 I presented the following testimony on "Common Core" educational standards to the West Virginia school board: "Even if you are not a global warming skeptic, these changes are basically fostering an open debate and they are against indoctrination. We must not tell kids there is no debate and no dissent is allowed.... There is nothing controversial here except the idea that we should allow open debate."[19]

Several scientists also submitted testimony on the same side of the issue. Physicist Thomas P. Sheahen pointed out, "The science is NOT settled at all, and it would be a great disservice to children to indoctrinate them with one currently-fashionable theory."

Princeton University physicist Will Happer urged, "You should not let yourself be deceived by those who claim that you are opposing 'science.' Science has been badly hurt by the global warming cult, and it will be hard to repair its reputation."[20]

It turned out to be an educational victory for climate skeptics. The West Virginia school board voted "to allow classroom debate on climate change."[21]

In 2017, the *Washington Examiner* reported on the continued efforts to shut down any climate debate in schools. "Three prominent House liberals have called for what amounts to a mass burning of books and DVDs that question global warming and sent to 200,000 K-12 teachers," the paper reported. The "cleansing effort" of skeptical climate science book titled *Why*

Scientists Disagree About Global Warming from the Heartland Institute, was being signed by Congressional Democrats Rep. Raúl Grijalva of Arizona and Reps. Bobby Scott of Virginia and Eddie Bernice Johnson of Texas.[22]

Banning books that question the alleged climate "consensus" seems to be spreading. The Portland, Oregon, public school board unanimously passed a resolution that attempted to ban "climate change-denying materials" in 2016. "It is unacceptable that we have textbooks in our schools that spread doubt about the human causes and urgency of the crisis," declared high school student Gaby Lemieux in testimony to the school board.[23]

Portland's school board decision was controversial, and I engaged in a heated debate on Fox News with a supporter of the ban.

Climate activist Taryn Rosenkranz supported the school board's position. "When it comes down to it, much like smoking causes cancer, we know carbon emissions causes climate change. We need to make sure that our students know the facts about what they can do to help," Rosenkranz said in the May 30, 2016, live debate with me on Neil Cavuto's program. I retorted, "Carbon dioxide is a trace gas that is essential for life on Earth and you are demonizing it as tobacco, as some kind of cancer causing gas—it's absurd.... This comes down to conformity must be enforced in the name of diversity. That is what they are doing here, they cannot allow dissent and it's very sad for kids."[24]

Children versus the Climate

On the one hand, climate change threatens the children. On the other, having children is terrible for the climate. So U.S. environmentalists are taking a page from China's mandatory one-child policy, even as China abandons the policy. It's not enough for the warmists to regulate our light bulbs, coal plants, and SUVs. Now they want to regulate the size of our families.[25]

★ ★ ★

Not Happy to Go Extinct

Warmist meteorologist Eric Holthaus is a passionate and emotional climate activist. After reading the UN IPCC report, Holthaus had a good cry and decided to take personal responsibility for his carbon footprint. "I just broke down in tears in boarding area at SFO while on phone with my wife. I've never cried because of a science report before," Holthaus tweeted in 2013. "I'm thinking of vasectomy," he wrote, adding "no children, happy to go extinct." He also vowed never to fly again, explaining "I realised just now: This has to be the last flight I ever take. I'm committing right now to stop flying. It's not worth the climate.... We all have to do everything we can, every day to reverse CO_2 emission.... All of our energy, each one of us, should be devoted to this issue."[26]

Alas, Holthaus didn't stay devoted. Two years later, he announced to the world, "My wife and I just had a baby, and it's quickly becoming the best decision we ever made.... Our baby has brought us back from the brink. It's impossible to be hopeless with a newborn. Climate change has changed me. And I don't think I'm the only one."[27]

In 2017, "science guy" Bill Nye featured a professor on his Netflix program, *Bill Nye Saves the World*, warning that having too many kids is bad for the planet. Nye asked, "So, should we have policies that penalize people for having extra kids in the developed world?"

Professor Travis Rieder responded, "I do think that we should at least consider it."

And Nye answered, "Well, 'at least consider it' is like 'Do it.'"[28]

An NPR segment titled "Should We Be Having Kids in the Age of Climate Change?" featured philosophy professor Travis Rieder saying, "We should protect our kids by not having them." Rieder proposed "actually penalizing new parents" with a tax that "should be progressive, based on income, and could increase with each additional child. Think of it like a carbon tax, on kids."[29]

Scientist William Briggs mocked Rieder's claims. "Protect our kids by not having them? That's like saying the way to protect your house from fire is by not building it, or that the way to protect against crop failure is to cease farming," Briggs wrote.[30]

In 2014, Al Gore advocated "fertility management" to reduce the number of Africans and help mitigate climate change. Gore made his remarks alongside of Bill Gates at a World Economic Forum. The former vice president explained that "making fertility management ubiquitously available" is "crucial" to reduce resource use. "Africa is projected to have more people than China and India by mid-century—more than China and India combined by end of the century. and this is one of the causal factors that must be addressed," Gore urged, claiming that "contraception is a key in controlling the proliferation of unusual weather they say is endangering the world."[32]

Physicist Lubos Motl, formerly at Harvard, ripped into Gore's comments. "It is immoral for Al Gore to organize 'fertility management' for other nations," Motl wrote. "It is impossible not to think that there's some racism and stunning hypocrisy if a jerk who has produced four children is 'working' on the reduction of the number of newborn babies in a completely different nation."[33]

Want to Have a Child? License Required

"Childbearing [should be] a punishable crime against society, unless the parents hold a government license.... All potential parents [should be] required to use contraceptive chemicals, the government issuing antidotes to citizens chosen for childbearing."
—David Brower, the first executive director of the Sierra Club[31]

★ ★ ★
Al Gore's "Fertility Management" Pitch, in a Nutshell

A wealthy white Western politician is telling the world that there are going to be far too many black people in Africa.

But if these wacky climate activists believed their own literature, they would realize that "global warming" should by itself lead to fewer kids. A 2015 study claimed that "climate change kills the mood" as "economists warn of less sex on a warmer planet."[34]

CHAPTER 18

Bypassing Democracy to Impose Green Energy Mandates

Polling has consistently shown Americans and much of the world are either skeptical or unconcerned about the man-made climate change narrative, ranking "global warming" last among environmental issues and frequently dead last among all issues. So the Obama Administration had to advance climate change policies by moving the climate debate away from public eye and the people's representatives. He bypassed democracy and implemented his agenda through executive order.

The strategy of we-don't-need-no-stinkin'-Congress worked, as Obama's EPA regulations imposed domestic climate regulations on America without a single vote of Congress. (Until President Donald Trump was elected, that is. But more on that in chapter twenty.)

Even Obama's professor Laurence Tribe, a liberal constitutional scholar at Harvard University, declared that Obama's executive orders "raised serious questions under the separation of powers" because "the EPA is attempting to exercise lawmaking power that belongs to Congress and judicial power that belongs to the federal courts." Tribe quipped, "Burning the Constitution cannot be part of our national energy policy."[1]

Did you know?

★ Climate skeptics defeated legislation and international treaties—only to see Obama impose them without the benefit of Congress

★ The Paris climate accord would theoretically postpone global warming by four years—and cost $100 trillion

★ Global warming is not a top concern for 90% of Americans

★ Forty-two percent of U.S. adults are not willing to pay even $12 a year to stop climate change

★ ★ ★

Environmentalism Hijacked

The Obama administration track record was a perfect example of something that Canadian physicist Denis Rancourt has been warning about for years—that concerns about man-made climate change have "hijacked" the environmental movement and let authentic ecological problems fall to the wayside.[5] In 2017, about 40 percent of the country is still in nonattainment for EPA air quality standards. And EPA toxic waste cleanup is also lagging, as over thirteen hundred Superfund sites still have not been cleaned up.[6] The Obama administration was smitten with the nearly sole focus of making carbon dioxide the modern-day boogeyman.

President Obama had laid out his agenda during the 2008 presidential campaign: "So if somebody wants to build a coal-powered plant, they can; it's just that it will bankrupt them because they're going to be charged a huge sum for all that greenhouse gas that's being emitted."[2]

Obama bypassed democracy to impose huge regulatory costs on Americans through his EPA's "Clean Power Plan," and also by signing the UN Paris pact—and not submitting it for Senate approval.[3]

Chinafication

Legislating via executive orders and making an international agreement without Senate ratification were two giant steps toward the Chinafication of America. Many global warming activists have praised China's ability to impose climate and energy regulations without the messiness of democracy.

We have already seen *New York Times* columnist Tom Friedman and former UN climate chief Christiana Figueres, gripped with dictator envy, lauding the eco-policies that China can "just impose," lamenting that U.S. democracy is "very detrimental" in war on global warming, and lauding one-party-ruled China for "doing it right" on climate change.[4]

Under the EPA rules and the Paris climate agreement—now no longer officially in force, thanks to the Trump administration—the American people would have faced an increasingly centrally planned domestic energy economy. And it looked like a case of the skeptics' losing the war—after winning every battle.

Climate change skeptics had a stellar record at holding back regulations and legislation. With more and more prominent scientists—many of them politically left-wing—joining their ranks after reexamining the evidence, skeptics had

- helped ensure the UN's Kyoto protocol was never ratified
- helped beat back multiple attempts to enact cap-and-trade into law
- helped ensure that a carbon tax was never implemented
- won the battle of public opinion, according to multiple methods of measurement

Cap-and-trade legislation, for example, went down to defeat in Congress in 2003, 2005, 2008, and 2010. It passed the House in 2009, but the members of Congress who had voted for it got such an earful from their constituents when they went back to their home districts that Senate Majority Leader Harry Reid never brought the bill up for a vote in the Senate, and it died a quiet death.

But the Obama administration was poised to enforce a climate-regulation scheme that would have huge costs to America's economy, liberty, and sovereignty—and that was scientifically meaningless. In fact, it was based not on science but on a superstition: that government regulations and UN treaties could regulate the climate and storminess of the Earth.

During a live Fox News climate debate in 2017, I pointed out, "The EPA climate plan is the signature Obama executive order—he couldn't get it through Congress so he bypassed democracy. But this plan wouldn't even impact global CO_2 levels, let alone global temperatures or storms. Yet Obama administration officials like John Podesta actually sold the regulations as a way to prevent storms and we need this because the storms are getting worse."

Podesta claimed in 2014 that EPA CO_2 regulations—which, as we have already seen, in actuality would not even impact global CO_2 levels, much less global temperatures, much less the frequency or intensity of storms—were needed to combat extreme weather. "The risk on the downside you're seeing every day in the weather," was, Podesta warned, why we needed to impose the EPA's onerous Clean Power Plan regulations on American industry.[7]

Professor Roger Pielke Jr. of the University of Colorado, who supported the EPA regulations, nonetheless rejected the notion that they would have any impact on climate or weather. "The so-called climate benefits of the regulations are thus essentially nil, though I suppose one could gin some up via creative but implausible cost-benefit analyses," Pielke wrote in 2014. "The US carbon regulations won't influence future extreme weather or its impacts in any detectable way. Hard to believe I felt compelled to write that."[8]

As Pielke pointed out, "These regulations mainly switch electricity from coal to gas and thus do very little to increase the US proportion of carbon-free electricity generation."

Obama EPA chief Gina McCarthy refused to quantify how much the carbon dioxide regulations would impact the climate. During a July 9, 2015, testimony to the U.S. House Science Committee McCarthy was forced to admit that the signature policy of the Obama Administration would have a purely symbolic impact on the climate—and that was even assuming the Al Gore–UN view of the science. "I'm not disagreeing that this action in and of itself will not make all the difference we need to address climate action, but what I'm saying is that if we don't take action domestically we will never get started," McCarthy testified. She dodged questions about the impact of the regulations on the climate.

Here is the key portion of the exchange between House Science Committee Chairman Representative Lamar Smith and McCarthy:

> SMITH: Do you consider one one-hundredth of a degree to be enormously beneficial?
>
> MCCARTHY: The value of this rule is not measured in that way. It is measured in showing strong domestic action which can actually trigger global action to address what's a necessary action to protect....
>
> SMITH: Do you disagree with my one one-hundredth of a degree figure? Do you disagree with the one one-hundredth of a degree?
>
> MCCARTHY: I'm not disagreeing that this action in and of itself will not make all the difference we need to address climate action, but what I'm saying is that if we don't take action domestically we will never get started and we'll never....
>
> SMITH: But if you are looking at the results, the results can't justify the cost and the burden that you're imposing on the American people in my judgement.[9]

Obama EPA Administrator Lisa Jackson made a similar admission in 2009, conceding under questioning, "I believe the central parts of the [EPA] chart are that U.S. action alone will not impact world CO_2 levels."[10] And former Obama Department of Energy Assistant Secretary Charles McConnell was brutally frank about the administration's policies. "The Clean Power Plan has been falsely sold as impactful environmental regulation when it is really an attempt by our primary federal environmental regulator to take over state and federal regulation of energy," McConnell wrote in 2016. "What is also clear, scientifically and technically, is that EPA's plan will not significantly impact global emissions," McConnell explained. "All of the U.S. annual emissions in 2025 will be offset by three weeks of Chinese emissions. Three weeks."[11]

In layman's terms: All of the so-called "solutions" to global warming are purely symbolic. So if we actually faced a climate catastrophe and we had to rely on a UN climate agreement, we would all be doomed! Climate campaigners who tout UN agreements and EPA regulations as a way to control Earth's temperature and storminess are guilty of belief in superstition.

And these toothless but costly regulations are entirely unnecessary. As author Paul Driessen has explained, the use of fossil fuels is compatible with a clean environment.

"Emissions of key air pollutants declined nearly 90% from 1970 to 2010—even as coal-based electricity generation increased 180%, miles traveled rose 170%, and the U.S. population grew by 110 million, according to EPA and other government data," Driessen pointed out. "A big part of the reason is that U.S. coal-fired generators invested over $100 billion in technologies to reduce power plant emissions. Today's air quality is safe, and pollution continues to decline under pre-Obama regulations."[12]

The regulations were wholly unnecessary—and destructive to the U.S. economy. Obama's climate change–driven energy policies took a toll on coal-fired power plants. As Driessen wrote, the EPA rules were "shutting down facilities, preventing new ones from being built, pummeling coal mining communities, reducing the reliability of our power grid, sending electricity prices higher, and threatening millions of manufacturing and other jobs."

Analyses of these regulations from such groups as the Institute for Energy Research and the U.S. Chamber of Commerce calculated that the EPA's CO_2 regulations were on pace to increase energy costs by up to $50 billion per year—with the biggest financial hit falling disproportionately on lower income families and seniors on fixed incomes. Many households would have suffered "a staggering $1,200 reduction in effective annual incomes and spending, with few or no health or environmental benefits in return," Driessen noted. Even with the Trump-era repeal of these regulations, "replacing that lost capacity will take many years and many billions of dollars."[13]

> ### A Book You're Not Supposed to Read
>
> *Cracking Big Green: To Save the World from the Save-the-Earth Money Machine* by Ron Arnold and Paul Driessen (CFACT, 2014).

No Impact

Trying to centrally plan energy economies many decades into the future while factoring in economic growth, population size, technology, and the needs of society in the year 2050 and beyond—is simply not realistic. International efforts like the UN Paris climate accord will also have no detectable impact on the climate—even if you accept UN science claims and models.

UN climate agreements are totally meaningless. University of Pennsylvania geologist Robert Giegengack has noted, "None of the strategies that have been offered by the U.S. government or by the EPA or by anybody else has the remotest chance of altering climate if in fact climate is controlled by carbon dioxide."[14]

Danish statistician Bjorn Lomborg conducted a cost-benefit analysis of the Paris climate accord and found that the $100 trillion cost of the pact buys no significant impact on global temperatures. Lomborg is the President of the Copenhagen Consensus Center. "You won't be able to measure it in 100 years," Lomborg said in 2017, noting that by the year 2100 the pact would postpone warming by less than four years. "The Paris Treaty will be the most expensive global agreement in world history. It is foolhardy and

Two Books You're Not Supposed to Read

The Skeptical Environmentalist: Measuring the Real State of the World by Bjorn Lomborg (Cambridge University Press, 2001).

Cool It: The Skeptical Environmentalist's Guide to Global Warming by Bjorn Lomborg (Alfred A. Knopf, Inc., 2007).

foolish for world leaders to stay fixated on Paris—not only will it likely falter, but it will be hugely costly and do almost nothing to fix climate change," Lomborg explained.[15]

"After hundreds of billions of dollars in annual subsidies, we only get, according to the International Energy Agency, 0.5 per cent of the world's energy needs from wind, and 0.1 per cent from solar PV," he added, noting that President Trump was right to reject the Paris pact.[16]

"Trump's climate plan might not be so bad after all," wrote Lomborg, adding that Trump withdrawing from the UN treaty "will stop the pursuit of an expensive dead end" because even if you accept the climate claims of the UN, the agreement "will matter very little to temperature rise."[17] As the statistician explained, the debate about the UN Paris agreement is "about identity politics. It's about feeling good…but the climate doesn't care about how you feel." The bottom line? "If the U.S. delivers for the whole century on the President Obama's very ambitious rhetoric, it would postpone global warming by about eight months at the end of the century."[18]

"But here is the biggest problem: These minuscule benefits do not come free—quite the contrary. The cost of the UN Paris climate pact is likely to run 1 to 2 trillion dollars every year," Lomborg explained. "This is likely to be among most expensive treaties in the history of the world." The UN Paris pact is estimated to have a $100 trillion price tag.

"We will spend at least one hundred trillion dollars in order to reduce the temperature by the end of the century by a grand total of three tenths of one degree—the equivalent of postponing warming by less than four years. Again, that is using the UN's own climate prediction model," Lomborg wrote.[19]

★ ★ ★

Not a Lot of Bang for the Buck

"Germany spends $110 billion to delay global warming by 37 hours," Bjorn Lomborg. discovered. "The Germans are spending about $110 billion on subsidies for these solar panels," said the statistician. "The net effect of all those investments will be to postpone global warming by 37 hours by the end of the century."[20] Lomborg also reported,

"For every dollar spent, the EU stands to avoid about 10 cents of damage.... Over the course of this century, the ideal EU policy would cost more than $7 trillion, yet it would reduce the temperature rise by just 0.05 degrees Celsius and lower sea levels by a trivial 9 millimeters."[21]

Even former NASA lead global warming scientist James Hansen, who has spent his career warning of a climate crisis, is not a big fan of the UN Paris accord. He has called it "a fraud really, a fake." As Hansen wrote in 2015, "It's just bullshit for them to say: 'We'll have a 2C warming target and then try to do a little better every five years.' It's just worthless words. There is no action, just promises. As long as fossil fuels appear to be the cheapest fuels out there, they will be continued to be burned."[22]

Other climate activists share Hansen's view. Anthony Rogers-Wright, a campaigner at climate website The Leap, expressed his severe angst about Al Gore's 2017 film *An Inconvenient Sequel.*

"I'd probably walk out if the movie celebrates the Paris climate agreement," Rogers-Wright told the *New Republic.*[23]

U.S. greenhouse gas emissions have actually been declining in recent years. But that decline is not due to the heavy hand of regulation.[24] A 2016 U.S. Energy Information Administration (EIA) report attributed the "increased use of natural gas for electricity generation" or fracking, as the reason for declining emissions.[25]

"Global warming crusader Al Gore won a Nobel Prize merely for his profit-making activities as a green activist. Here's an idea: If the Nobel committee geniuses really want to reward those who've done the most to reduce greenhouse gas emissions, they should give Gore's Nobel to the U.S. fracking industry," noted a 2017 editorial in *Investor's Business Daily*. "Ironically, while the U.S. was pilloried for not ratifying the Kyoto Accord (though then–Vice President Al Gore ostentatiously signed it, despite knowing that the Senate wouldn't ratify it) to reduce global greenhouse gas emissions, it is the only major industrial nation actually slashing its output. Since the Kyoto Accord was struck in 1997 (which U.S. did not ratify) Energy Department data show, U.S. output of greenhouse gases plunged 7.3%, even though real U.S. GDP over that time has grown a whopping 52%."[26]

And the good energy news does not stop there. "Fracking can give us centuries of new oil and gas supplies. Natural gas is needed to back up wind turbines and provide petrochemical feed stocks," Driessen wrote.[27]

Not Winning over the Public

The reason President Obama had to bypass Congress was simple. Public support for "climate action" has never been sufficient to support passing big legislation. A July 2017 Bloomberg poll found that for 90 percent of Americans, global warming is not a top concern. Only one in ten Americans say climate change is the most important issue facing the United States. More than three times as many Americans say health care reform is the most important issue facing the United States. Other top issues included terrorism, jobs, and taxes.[28]

According to Gallup, Americans' concern about the climate has not changed much since the 1980s,[29] despite Al Gore's drumming it into our ears that it is "the most serious challenge we face."[30] A 2016 Gallup poll found that environmental issues were the least important to Americans, with only 3 percent citing the environment as the most important issue.[31]

★ ★ ★

Losing and Ducking Debates

In 2007, a high-profile climate debate between prominent scientists ended with global warming skeptics being voted the clear winner by a tough New York City audience. The debate was sponsored by the Oxford-style debating group Intelligence Squared and featured a three-on-three debating format. Before the start of the nearly two-hour debate, the audience polled 57.3% to 29.9% in favor of the proposition that global warming was a "crisis." But following the debate, the numbers had completely flipped to 46.2% to 42.2% in favor of the skeptical point of view, argued by MIT scientist Richard Lindzen, University of London professor emeritus Philip Stott, and the physician-turned-novelist-and-filmmaker Michael Crichton. After the stunning victory, NASA's Gavin Schmidt, one of the scientists on the losing side promoting belief in a climate "crisis," excused the defeat by noting that his debate team was "pretty dull" and at "a sharp disadvantage" against the skeptical scientists. *Scientific American* agreed, saying the warmists "seemed underarmed for the debate and, not surprising, it swung against them." NASA's Schmidt appeared so demoralized that he realized that debating skeptical scientists was not something he would ever want to do again. "So are such debates worthwhile? On balance, I'd probably answer no (regardless of the outcome)," Schmidt wrote.[34] In 2013, Schmidt was true to his word, refusing to even appear alongside skeptical climatologist Roy Spencer on John Stossel's Fox TV program. Schmidt literally walked off the set when Spencer came on to talk.[35]

A 2016 Pew Research survey found that Americans reject the claims of a 97 percent consensus of scientists agreeing on global warming. Only 27 percent of Americans say that "almost all" scientists are in agreement about global warming.[32] Even Al Gore has lamented in 2016 that there is "formidable denial" about climate change by "some smart people."[33]

A 2016 UN poll revealed that climate change ranked the lowest of all global issues, coming in sixteenth out of sixteen.[36] And a survey by the Energy Policy Institute and the Associated Press–NORC Center for Public Affairs Research found "42 percent of respondents are unwilling to pay even $1" per month "to confront the climate challenge." The survey noted

★ ★ ★

To Persuade Texans, What You Need Is "Less Science"

Chief NASA climate scientist Gavin Schmidt thinks he may have the answer on why poll numbers for global warming are not higher. "Now, you know there's some communities I can't talk to because, you know, I'm a liberal, Jewish atheist from New York City, right? So if I go to Texas and try and tell people about climate change, I'm totally the wrong messenger, right? Because we don't have any shared values quite frankly." Schmidt added, "A lot of times we think, 'oh, more science, more science', and really we want to allow less science and more cultural understanding, and that might take us a lot further."[39]

that "Party affiliation is the main determinant of how much people are willing to pay, not education, income, or geographic location. Democrats are consistently willing to pay more than Republicans."[37]

The lack of concern about climate has caused "science guy" Bill Nye to fantasize about the deaths of climate deniers. "Climate change deniers, by way of example, are older. It's generational," Nye explained in an interview with the *Los Angeles Times*. "We're just going to have to wait for those people to 'age out,' as they say. 'Age out' is a euphemism for 'die.'

"But it'll happen, I guarantee you—that'll happen."[38]

Nye's comments reflect the fact that young people tend to believe in the man-made climate "crisis" much more than older people do. Nye himself has had a big influence with young minds with his popular children's show; he is confident he can convert the young people to advocate for climate action. I saw Nye's appeal to the younger generation firsthand during my 2016 outdoor interview of him in Central Park in New York. Within minutes of the start of our interview, dozens of young people swarmed Nye, seeking an autograph from the iconic science educator.

CHAPTER 19

Green Colonialism

As we have seen, attempts to control weather and climate will have no measurable impact on climate, but a huge impact on energy prices and the economy. That's true for the United States and the rest of the developed world. But it matters a lot more to poor nations. The so-called solutions to the supposed threat of man-made climate change would limit economic development and ban many forms of life-saving carbon-based energy. These restrictions and other policies inspired by the global warming panic function as a form of modern day colonialism.

Secretary of State John Kerry asked this question in 2014: "If we make the necessary efforts to address this challenge—and supposing I'm wrong or scientists are wrong, 97 percent of them all wrong—supposing they are, what's the worst that can happen?"[1]

A fair question. Sadly, the answer is poverty, disease, and death for poor people across the developing world. And all for no impact on the climate.

The climate regulatory scheme that the warmists are doing their best to impose on the world—including drastic limits on carbon emissions, severe curtailment of fossil fuels, "planned recessions," and the promotion of primitivism—are frankly immoral. The developed world is denying a

Did you know?

★ Electricity helped change Chinese life expectancy from fifty-nine to seventy-five years

★ One in three Africans still don't have electricity

★ Climate policies are denying life-saving technology to the world's poor billion people of color in the developing world—Asia, Africa, South America—the coal, natural gas, and other carbon-based energy that they need to pull themselves up out of dire poverty, as the wealthy Western world has already done. There is an unmistakable racial component to this: black and brown people are essentially being told they can never be allowed to have the same standard of living as the predominantly white Western world has enjoyed since the industrial revolution. American and European climate activists are telling Africans that we will manage their economies for them so they don't make our "mistakes."

We've Got Ours

At the UN Earth Summit in Johannesburg, South Africa, in 2002 I asked Democratic California governor Jerry Brown whether he didn't think the residents of the poorest nations of the world wanted to develop economically as the United States has done. His answer: "Many do, but it's not viable.... the developed model cannot work without another five planets."[2]

British scientist Philip Stott compared Brown's views to Marie Antoinette's infamous "Let them eat cake." Stott said, "I am deeply worried when I hear a white, Western male start to lecture the developing world on what they should or should not want."

In my Earth Day 2000 interview of Chevy Chase for my *Clear-Cutting the Myths* documentary, the star of *Vacation* and *Fletch* made the case against economic development for the poor:

> Morano: "Is capitalism and development a good way to help the developing world's poor?"
>
> Chase: "No, not necessarily. No, not necessarily."
>
> Morano: "Why not?"
>
> Chase: "Because sometimes socialism works."

Morano: "Socialism works to help people out of poverty?"

Chase: "Yeah."

Chase added, "I think it's conclusive that there have been areas where socialism has helped to keep people at least stabilized at a certain level." He did acknowledge that there is less freedom under socialism but argued that "when you just say capitalism versus socialism, it's too simple."[3]

Chase then offered an example of a country he thought was doing things right: "I think free markets are important, but you know you can do both and I think Cuba might prove that."

Later in the day, an incensed Chase confronted me and claimed that "intentions count as much as anything else. Some people may not know every increment of every decision and every controversy, but the intention to make the earth better is something that simple. It's good."

But we all know the road to hell is paved with good intentions. And hell is exactly what the results of these supposed good intentions feel like to the people trapped in poverty in the developing world. In recent years, the World Bank has estimated that 1.1 billion poor people still don't have electricity. "Around one in seven people across the globe still live without electricity, despite some progress in expanding access, and nearly three billion cook using polluting fuels," the World Bank reported. "The global electrification rate rose to 85 percent in 2012 from 83 percent in 2010, pushing the number of people without access to electric power down to 1.1 billion from 1.2 billion."[4]

As Ugandan activist Fiona Kobusingye has pointed out, "The average African life span is lower than it was in U.S. and Europe 100 years ago. But Africans are being told we shouldn't

Keeping Us Rich and Them Poor

"Global warming, as a political vehicle, keeps Europeans in the driver's seat and developing nations walking barefoot."

—Takeda Kunihiko, vice-chancellor of the Institute of Science and Technology Research at Chubu University in Japan[5]

A Book You're Not Supposed to Read

Eco-Imperialism: Green Power, Black Death by Paul Driessen (Merril Press, 2010).

develop, or have electricity or cars because, now that those countries are rich beyond anything Africans can imagine, they're worried about global warming."

Kobusingye has been waging war against the climate change campaign on moral grounds.

"Al Gore and UN climate boss Yvo de Boer tell us the world needs to go on an energy diet. Well, I have news for them. Africans are already on an energy diet. We're starving!" she declared. "Al Gore uses more electricity in a week than 28 million Ugandans together use in a year. And those anti-electricity policies are keeping us impoverished," she explained. "Telling Africans they can't have electricity and economic development—except what can be produced with some wind turbines or little solar panels—is immoral. It is a crime against humanity," she added. "Hypothetical global warming a hundred years from now is worse than this?"

The World Bank estimated in 2016 that "only one in three Africans has access to electricity and for those who do, power outages can be common as cash-strapped utilities struggle to maintain steady, reliable supply because of lack of investment in their aging infrastructure."[6]

Developing nations need cheap and affordable energy so their people can escape a nasty, brutish, and short life locked in dire poverty. Not having access to reliable cheap electricity means no modern necessities like refrigerators to keep food safe, no modern dentistry, no sewage treatment to keep waste from polluting rivers and streams with deadly bacteria. Limiting development means a continuing higher infant mortality rates, shorter life expectancy, and less life-saving medical care.

No number of solar panels on top of huts made of animal dung is a long-term solution for grinding poverty. A simple fact needs to be recognized: carbon-based energy has been one of the greatest liberators of mankind.

Any so-called climate change "solutions" that attempt curtail the use of carbon-based fuels are not based on reality.

Fossil Fuels—A Power for Good

Extreme weather expert Professor Roger Pielke Jr. of the University of Colorado has accused climate activists of attempting to "promote green imperialism that helps lock in poverty." As he explained, "A recent report from the non-profit Center for Global Development estimates that $10bn invested in renewable energy projects in sub-Saharan Africa could provide electricity for 30m people. If the same amount of money went into gas-fired generation, it would supply about 90m people—three times as many."[7]

A study by development expert and former UN IPCC expert Indur Goklany found that "fossil fuels saved humanity from nature and nature from humanity." The study noted that "the cumulative contribution of various poverty-related diseases to global death and disease is 70–80 times greater than warming." Goklany, a Cato Institute scholar, found, "From 1750 to 2009, global life expectancy more than doubled, from 26 years to 69 years; global population increased 8-fold, from 760 million to 6.8 billion; and incomes increased 11-fold, from $640 to $7,300. Never before had the indicators of the success of the human species advanced as rapidly as in the past quarter millennium." The study also noted that fossil fuels are the driving force of energy for the world today.[9]

As Goklany found, "Not only have these fossil fuel–dependent technologies ensured that humanity's progress and well-being are no longer hostage to nature's whims, but they saved nature herself from being devastated by

> ## Abandon All Hope, Ye Who Enter Here
>
> "Climate policy robs the world's poor of their hopes."
> —**University of Colorado professor Roger Pielke Jr. and Daniel Sarewitz of Arizona State University**[8]

★ ★ ★

Gas *Is* Green Energy

Fossil fuels actually clean up the environment. Carbon-based energy brings modern sanitation. Water and air quality rise, bringing longer life expectancy. The wealthier nations become, the cleaner and greener their environments become and the healthier and longer the residents of these countries are able to live.

the demands of a rapidly expanding and increasingly voracious human population."[10]

If environmental activists looked at the bigger picture, they would have to admit that fossil fuels have tamed the climate for mankind. "The fossil fuel industry is not taking a safe climate and making it dangerous. They are taking a dangerous climate and making it safe," explained Alex Epstein, the president of the Center for Industrial Progress. "Anyone who contradicts me should try to go outside right now [in the brutal cold of winter] and live naturally in harmony with nature," he said. "It's not noble to use less energy, that is like saying it's noble to have less money. No. You might misuse energy, you might be inefficient—but more is always better because energy has the capacity to be productive."[11]

Statistician Caleb Rossiter is a man of the political Left who has broken with his colleagues on climate and the issue of African development. "The left wants to stop industrialization—even if the hypothesis of catastrophic, man-made global warming is false," Rossiter explained. "Western policies seem more interested in carbon-dioxide levels than in life expectancy." Rossiter, who teaches climate statistics at American University, is passionate about helping Africans gain access to modern development. "How terrible to think that so many people in the West would rather block such success stories in the name of unproved science," Rossiter said. He is angered by colleagues who use phrases such as "climate justice" or talk about fossil fuel "divestment."[12]

Rossiter pointed out, "Climate justice is a meaningless phrase. I object most strongly to 'climate justice.' Where is the justice for Africans when universities divest from energy companies and thus weaken their ability to explore for resources in Africa? Where is the justice when the U.S. discourages World Bank funding for electricity-generation projects in Africa that involve fossil fuels, and when the European Union places a 'global warming' tax on cargo flights importing perishable African goods?"[13]

As Rossiter has explained, "The average [life expectancy] in Africa is 59 years—in America it's 79. Increased access to electricity was crucial in China's growth, which raised life expectancy to 75 today from 59 in 1968." He added, "According to the World Bank, 24% of Africans have access to electricity and the typical business loses power for 56 days each year. Faced with unreliable power, businesses turn to diesel generators, which are three times as expensive as the electricity grid. Diesel also produces black soot, a respiratory health hazard."[14]

Keeping the Black Man Down

And yet President Obama told a young African leaders' town hall in South Africa in 2013 that the continent had better not even think of aspiring to U.S. standards of living. "If everybody is raising living standards to the point where everybody has got a car and everybody has got air conditioning, and everybody has got a big house, well, the planet will boil over—unless we find new ways of producing energy," Obama said.[15]

Both the American government, under Obama, and the World Bank looked at stopping the financing of coal plants in Africa. In 2013, The World Bank mulled limits to the financing of coal-fired power plants as part of the Bank's "efforts to address the impact of climate change."[16] And the Obama administration fought coal going to the world's poorest continent. According to the *Washington Post*, "In a major policy shift, Obama said he would place sharp restrictions on U.S. government financing for new coal plants overseas." The president had announced, "Today, I'm calling for an end to public financing for new coal plants overseas unless they deploy carbon-capture technologies, or there's no other viable way for the poorest countries to generate electricity." So if you're a citizen of a poor country, and you are desperately seeking a needed coal-fired electrical plant, don't count on the wealthy developed nations to help.

That was despite the fact that, as the *Washington Post* reported, "Development experts have long argued that gas- or coal-fired plants are typically the cheapest option for poorer countries, especially in dense urban areas."[17]

In 2014 Professor Roger Pielke Jr. noted that the Obama administration had "imposed a cap on emissions from energy projects of the Overseas Private Investment Corporation, a U.S. federal agency that finances international development. Other institutions of the rich world that have decided to limit support for fossil fuel energy projects include the World Bank and the European Investment Bank. Such decisions have painful consequences."[18]

This push to allow outsiders to plan African nations' economics comes at a time when Africa's development from primitive tribal life to modernization—accompanied by the fastest per capita energy consumption growth in the world—is seen by many in the environmental movement as some sort of cultural genocide.[19] Gar Smith, the former editor of the Earth Island Institute's online journal, has lamented the introduction of electricity in Africa. "I don't think a lot of electricity is a good thing. It is the fuel that powers a lot of multi-national imagery," Smith explained. And it can wreak havoc on cultures. "I have seen villages in Africa that had vibrant culture and great communities that were disrupted and destroyed by the introduction of electricity."[20]

Greenpeace founder Patrick Moore, now head of the environmental advocacy group EcoSense, believes that those who would try to limit fossil fuel energy and the electricity it provides have a "naive vision of returning to some kind of Garden of Eden, which was actually not that great because the average life span was 35." As Moore explains, "The environmentalists try to inject guilt into people for consuming, as if consuming by itself causes destruction to the environment. There is no truth to that. You have the wealthiest countries on earth with the best-looked-after environment."[21]

Climatologist Patrick Michaels has noted that "the fossil-fuel powered industrial democracies have seen a 100% increase in life expectancy since 1900. Doubling the lifetime of, say, a billion people is equivalent to saving 500 million lives."[23]

But the climate-obsessed United Nations wants to limit that life-saving technology when it comes to people of color in the developing world. UN high commissioner for human rights Mary Robinson announced bluntly, "Governments must leave fossil fuel reserves in the ground unburned and unexploited to really protect the planet. There is a global limit on a safe level of emissions."[24]

There ought to be a global limit on how much suffering the warmists can impose on the world's poor. As IRIN News reported, "A government ban on charcoal in the Chadian capital N'djamena has created what one observer called 'explosive' conditions as families desperately seek the means to cook." Chad's environment minister Ali Souleyman Dabye told the media, "Cooking is of course a fundamental necessity for every household. On the other hand... with climate change every citizen must protect his environment."[25]

The BBC noted the distress caused to poor farmers in Kenya by a UK grocery store's "carbon friendly" policies.[26] "Kenyan farmers, whose lifelong carbon emissions are negligible compared with their counterparts in the West, are fast becoming the victims of a green campaign that could threaten their livelihoods." The cause of this? "A recent bold statement by UK supermarket Tesco ushering in 'carbon friendly' measures—such as restricting

A Bad Bargain

"I'm sorry, but I am totally unwilling to trade inexpensive energy today, which is the real actual salvation of the poor today, for some imagined possible slight reduction in the temperature fifty years from now."

—climate skeptic Willis Eschenbach[22]

the imports of air freighted goods by half and the introduction of 'carbon counting' labelling—has had environmentalists dancing in the fresh produce aisles, but has left African horticulturists confused and concerned."

The Indian government is on record as having grave concerns over UN emissions reduction pacts. India told the UN in 2009, "It is morally wrong for us to reduce emissions when 40% of Indians do not have access to electricity."[27]

In the second of my primetime televised debates with Bill Nye, we discussed development versus climate policy. I asked Nye, "How is the white, wealthy Western Europe world, in Europe and the U.S., going to tell people of color, 1.3 billion in the developing world, they can't have what we have? Who is Bill Nye to tell [the developing word] they can't have carbon-based energy?"

Nye responded, "We don't want to have less. We want to do more with less. And this is where the innovations come in."

Of course, innovations can come at any time. They don't require government central planning or massive subsidies. Meanwhile dire poverty is being artificially locked in by climate change–inspired policies that keep affordable energy from the people who need it most—now. Concern for human welfare requires we reject the premise that poor nations should be limited to developing their economies only in ways that climate activists approve.

South African development activist Leon Louw slammed the warmist philosophy on development. "The third world should be doing what the first world did, which is namely to use its natural resources and build big cities and harbors on what were wetlands, harvest the timber and use it, mine the minerals, exploit the natural resources," Louw declared. "The policies that enabled the first world to become the first world—the rich to become the rich—are now denied to those who are poor."[28]

Louw critiqued the UN's "Green Climate Fund" at the 2011 UN Climate Conference in Johannesburg, South Africa. "Government to government aid is a reward for being better than anyone else at causing poverty," he explained. "It enriches the people who cause poverty.... The UN is saying to poor countries: 'Those of you who adopt more anti-prosperity, anti-jobs, and anti-growth policies—under the pretense of environmentalism—we will enrich you.'"[30]

The Associated Press described the UN climate fund as a method to "distribute tens of billions of dollars a year to poor countries to help them adapt to changing climate conditions and to move toward low-carbon economic growth." But Louw says the fund will wreak havoc on the developing world's poor. "The money goes to government and governments spend it on of course on themselves, meaning various government projects, creating bigger departments—bigger bureaucracies, it's called big bureaucratic capture. They build empires, they build conference centers, and they buy political support. They go and distribute the money to communities where they want support and votes," Louw explained.

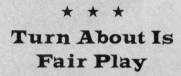

★ ★ ★

Turn About Is Fair Play

And if green groups don't like Third World development, Leon Louw doesn't care. "I can't put it any more politely. Poor countries should just say: 'Go to hell. If you don't want us to fill in our wetlands, then you bomb your big cities like Washington, a third of Holland and Rotterdam and so on, and restore them to being swamps.'"[29]

Irreplaceable

Reporting in the *Washington Post* has recognized the conflict between climate policy and the developing world's need for energy. In an article titled "In Poorer Nations, Energy Needs Trump Climate Issues," the paper quoted William Bissell, a prominent Indian entrepreneur and the author of *Making India Work*: "The United States and Europe have had the energy they needed to grow and develop. But we haven't had our 21st century yet."[31]

Even climate journalist Andrew Revkin, formerly of the *New York Times*, who now writes for ProPublica, agreed that poor nations' development is more important than climate concerns. "I'm convinced that in the world's poorest places, the transformative power of access to affordable energy—enabling everything from homework to better health to a home business using a sewing machine—trumps concerns about climate implications," Revkin wrote in 2012.[32]

Statistician Bjørn Lomborg has explained the challenges facing poor nations. "The most important threat is in fact indoor air pollution. One-third of the world's people—2.9 billion—cook and keep warm burning twigs and dung, which give off deadly fumes. This leads to strokes, heart disease and cancer, and disproportionately affects women and children. The World Health Organization estimates that it killed 4.3 million people in 2012. Add the smaller death count from outdoor pollution, and air pollution causes one in eight deaths worldwide," Lomborg wrote.

"Climate policies have a cost, and these predominantly hurt the poor," Lomborg explained. "So in choosing to spend that $10 billion on renewables, we deliberately end up choosing to leave more than 70 million people in darkness and poverty."[33]

Renewable energy such as solar and wind are the prime sources of energy that environmentalists want to mandate in order to achieve "sustainable development." Climate activist Bill McKibben makes one of the most common arguments for severe limitations on fossil fuels in the developing world when he says he would like to see the poor nations "'leapfrog the fossil-fuel age and go straight to renewables.'"[34] But that scenario is not currently realistic.

Steve Milloy countered at his Junk Science blog: "No, poor countries can't bypass fossil fuels like they bypassed land lines for cell phone." As Milloy pointed out, "Unless poor people don't need electricity at night or when the wind isn't blowing, and unless they can afford high-priced electricity sources like wind and solar (which only exist in developed countries where

★ ★ ★
"Inefficient and Unproductive"

Eco-Imperialism: Green Power, Black Death author Paul Driessen noted a 2017 American Enterprise Institute study showing how solar energy is currently "wasteful, inefficient and unproductive":

- 398,000 natural gas workers = 33.8% of all electricity generated in the United States in 2016
- 160,000 coal employees = 30.4% of total electricity
- 100,000 wind employees = 5.6% of total electricity
- 374,000 solar workers = 0.9% of total electricity

As Driessen pointed out, "It's even more glaring when you look at the amount of electricity generated per worker. Coal generated an incredible 7,745 megawatt-hours of electricity per worker; natural gas 3,812 MWH per worker; wind a measly 836 MWH for every employee; and solar an abysmal 98 MWH per worker. In other words, producing the same amount of electricity requires one coal worker, two natural gas workers—12 wind industry employees or 79 solar workers," Driessen wrote.[36]

they are significantly subsidized), then fossil fuels are not bypassable like land lines were."[35]

Green guru James Lovelock, who was once a leading promoter of climate fears, now mocks "sustainable development" as "meaningless drivel" Lovelock has explained, "We rushed into renewable energy without any thought. The schemes are largely hopelessly inefficient and unpleasant. I personally can't stand windmills at any price."[37]

He wrote in 2010, "Used sensibly, in locations where the fickle nature of wind is no drawback, it is a valuable local resource, but Europe's massive use of wind as a supplement to baseload electricity will probably be remembered as one of the great follies of the twenty-first century."[38]

And in 2013, Lovelock said, "I am an environmentalist and founder member of the Greens but I bow my head in shame at the thought that our original good intentions should have been so misunderstood and misapplied. We never intended a fundamentalist Green movement that rejected all energy sources other than renewable, nor did we expect the Greens to cast aside our priceless ecological heritage because of their failure to understand that the needs of the Earth are not separable from human needs. We need [to] take care that the spinning windmills do not become like the statues on Easter Island, monuments of a failed civilisation."[39]

CO_2, the Healthy Addiction

Climatologist John Christy of the University of Alabama in Huntsville testified to Congress on the need to eliminate energy poverty. "Oil and other carbon-based energies are simply the affordable means by which we satisfy our true addictions—long life, good health, plentiful food, internet services, freedom of mobility, comfortable homes with heating, cooling, lighting and even colossal entertainment systems, and so on. Carbon energy has made these possible," Christy said.[40]

"Rising CO_2 emissions are one indication of poverty-reduction which gives hope for those now living in a marginal existence without basic needs brought by electrification, transportation and industry. Additionally, modern, carbon-based energy reduces the need for deforestation and alleviates other environmental problems such as water and deadly indoor-air pollution. Until affordable and reliable energy is developed from non-carbon sources, the world will continue to use carbon as the main energy source," Christy added.

As Bjørn Lomborg has noted, "81 percent of the planet's energy needs are met by fossil fuels, and according to the International Energy Agency, that percentage will be almost as high in 2035 under current policies, when consumption will be much greater."[41]

Even more intriguing fossil fuels accounted for 85% of U.S. energy consumption in 1908. The U.S. Energy Information Administration (EIA) reported in 2016 that despite all the hype and rhetoric surrounding "leap frogging" to "renewable" energy, the United States was still at about the same energy mix as a century ago. "Despite the changes in fuel sources, fossil fuels have continued to make up a large percentage of U.S. energy consumption," the EIA reported.[42]

The reality is that fossil fuels are not going anywhere anytime soon. The real question: Will the developing world's poor have unfettered access to these life-saving resources?

CHAPTER 20

The Way Forward

To use the warmists' own term, 2016 was a "tipping point" in the climate debate. The U.S. presidential election changed everything.

If Hillary Clinton had been elected instead of Trump, President Obama's disastrous "climate" policies would have been continued and extended. Obama would have proven himself a transformational president, surpassing President Lyndon Johnson and rivaled only by President Franklin Delano Roosevelt in his impact on America. The Democrats had a very real chance of permanently implementing the climate agenda with domestic climate policy through the EPA and the UN Paris "commitments," but the climate change policy debate has been reset by Donald J. Trump's stunning victory.

President Obama's efforts to impose climate policy and regulations that did not even meet minimal cost-benefit analysis were halted in spectacular fashion. Trump's victory was seismic. It dramatically shifted the momentum away from policies based on the superstitious belief in catastrophic man-caused climate change and back to a rational energy policy—for now. The future is fraught with many potential roadblocks, as the Trump administration will have to navigate around or retry the 2007 Supreme Court greenhouse gas–endangerment decision (which allowed the EPA to regulate

Did you know?

★ The Obama EPA's Clean Power Plan would have only reduced temperature .023 degrees Fahrenheit by the end of the century—even on warmist assumptions

★ President Trump is keeping his promises on climate policy

★ Al Gore's embrace of state action on climate is a concession that international treaties were never necessary

CO_2 as a pollutant under the Clean Air Act[1]), other environmental lawsuits, and potentially future hostile Congresses.

The Heritage Foundation's Stephen Moore has pointed out that "Big Green" was the biggest loser in the 2016 election. "In so many ways climate change was one of the primary issues that allowed Donald Trump to crash through the blue wall of the industrial Midwest. The Democrats' preposterous opposition to building the Keystone XL Pipeline which could create as many as 10,000 high-paying construction, welding, pipefitting, electrician jobs is emblematic of how the party that is supposed to represent union workers turned their backs on their own members and their families," Moore explained.[2]

With Trump's election, climate sanity was restored to the United States. No longer do we have to hear otherwise intelligent people in charge in D.C. blather on about how UN treaties or EPA regulations will control the Earth's temperature or storminess. The election night of 2016 was one of pure enjoyment for those concerned about silly, purely symbolic, but sovereignty-threatening expensive climate policies that had been imposed on the United States without a single vote. Climate skeptics enjoyed watching the grieving faces of the mainstream media on CNN, MSNBC, NBC, ABC, and CBS, as the Trump election night shock sunk in.

Candidate Trump was the warmists' worst nightmare: the first Republican presidential nominee who ever staked out a strongly science-supported skeptical position not only on climate change claims but also on the so-called "solutions."

In a live Fox News climate debate in 2017, I expressed the feelings of my fellow climate skeptics who had been in the trenches of the climate battle for over a decade: "What EPA chief Scott Pruitt really represents—you see my smile here—this is the end of superstition in Washington. And it's actually going back to science and actual cost benefit analysis. It's very simple."[3]

★ ★ ★

Union Boss Tells It Like It Is

Labor groups came to the White House to support the Keystone Pipeline project and the Dakota Access pipelines. A vibrant energy policy brings coalitions of diverse political groups together. Labor Leader Terry O'Sullivan, head of the Laborers' International Union, had once suggested that President Obama "chose to support environmentalists over jobs" by opposing the Keystone Pipeline.[4] "Job-killers win, American workers lose," O'Sullivan said. The labor leader, whose organization had twice endorsed Obama, called the Obama administration "gutless."[5]

Dismantling a Disastrous Legacy

And in 2017 the Trump administration began in earnest to overhaul U.S. climate and energy policy, taking on the climate activists and their ill-gotten gains, which they had achieved only by bypassing democracy.

The early days of the Trump era saw rapid and unprecedented changes in climate policy. President Obama's executive orders were rescinded at breakneck pace. The EPA's Clean Power Plan climate regulations, which were strangling the coal industry, collided with President Trump's pen within the first hundred days, and the process of revoking them began.

The Keystone Pipeline was greenlit as well.

By June 2017, speculation had been rife for months on the subject of what Trump was going to do about the purely symbolic Paris climate pact. It turned out that the time for a full-scale U.S. Clexit—a U.S. exit from the UN Paris climate accord—had arrived. President Trump came down on the side of those who had argued that the UN climate treaty was nothing more than an effort to empower the UN and undermine national sovereignty while doing absolutely nothing for the climate.

"In order to fulfill my solemn duty to protect America and its citizens, the United States will withdraw from the Paris climate accord but begin

★ ★ ★

"The Only Man Standing"

When President Trump announced the U.S. withdrawal from the Paris pact, I appeared live on Al Jazeera and was asked about Trump not going along with the rest of the world.

Al Jazeera anchor: "Is Trump isolating himself? He seems to be the only man standing. Nobody seems to agree with him."

Marc Morano: "He's a leader! When Ronald Reagan was elected in 1980, he was called the reckless cowboy and many were worried. This is what leadership looks like, you buck even your allies and do what's right not only for your country but for the planet. The planet does not need nonsense about a UN climate treaty somehow saving us and now we are all doomed because we did not adhere to it. That is belief in superstition. European leaders will ultimately respect Trump more for taking a strong stand and defying all of their lobbying. This is America reborn. Trump is showing unbelievable courage.... Trump is actually standing up and willing to [take] on the religion of climate change. Again, no modern leader other than former Czech President Vaclav Klaus has shown this kind of strength."[7]

negotiations to reenter either the Paris accord or an entirely new transaction under terms that are fair to the United States," Trump declared in a White House Rose Garden ceremony on June 1, 2017. "The United States will cease all implementation of the nonbinding Paris accord," Trump said, adding that the United States would stop funding the United Nations' Green Climate Fund.[6]

"As someone who cares deeply about our environment, I cannot in good conscience support a deal which punishes the United States," the president said. "The Paris accord is very unfair at the highest level to the United States."

Of course, the planet will not care one way or the other about the fate of the UN Paris pact.

"The UN treaty is nothing but a paper tiger: Its only legal underpinning is that all nations submitted promises—but those promises do not need to be kept," Danish statistician Bjørn Lomborg wrote.[8] He also pointed out that the EPA's Clean Power Plan—"the primary measure America offered to achieve the promised [UN emission] cuts"—"would have achieved just a third of the U.S. promises under the Paris agreement. If it had remained in effect for the entire century, my peer-reviewed research using UN climate models found that it would have reduced temperature rises by an absolutely trivial 0.023 Fahrenheit at the end of this century." Lomborg explained that the Paris pact will most likely fail due to the focus on "inefficient solar and wind and Teslas that feel good but actually don't do much" to reduce CO_2 emissions.[9]

An August 2017 analysis in the journal *Nature* backed up Lomborg's views on the UN Paris climate pact. "No major advanced industrialized country is on track to meet its pledges to control the greenhouse-gas emissions that cause climate change. Wishful thinking and bravado are eclipsing reality," the analysis found. "It is easy for politicians to make promises to impatient voters and opposition parties. But it is hard to impose high costs on powerful, well-organized groups. No system for international governance can erase these basic political facts. Yet the Paris agreement has unwittingly fanned the flames by letting governments set such vague and unaccountable pledges."[10]

Sound Policy

Back on January 14, 2016, I had authored an editorial on what the next president needed to do to dismantle the symbolic but costly climate agenda:

> Republicans need to get their act together quickly in order to prevent Obama's climate legacy from being cemented.... The GOP nominee for president in 2016 must present a basic plan to

roll back Obama's climate regulations. Here is a simple break-down of what is needed:

1. Repeal all EPA climate regulations;
2. Withdraw the U.S. from any Paris agreement (nonbinding) "commitments";
3. Withdraw the U.S. from the UN climate treaty process entirely;
4. The U.S. should defund the UN IPCC climate panel;
5. Start praising carbon based energy as one of the greatest lib-erators of mankind and the best hope for the developing world's poor.

Anything short of this clear and comprehensive approach will lead to failure and guarantee Obama's climate policies will become permanent in the U.S. The Republicans need to get a coherent plan and articulate their course of action.[11]

To my delight, five months later, on May 26, 2016, candidate Trump released his "100-day action plan" on energy and climate during a speech in Bismarck, North Dakota. The plan echoed my analysis of what the next president needed to accomplish:

We're going to rescind all the job-destroying Obama executive actions including the Climate Action Plan and the Waters of the U.S. rule.

We're going to save the coal industry and other industries threatened by Hillary Clinton's extremist agenda.

I'm going to ask Trans Canada to renew its permit application for the Keystone Pipeline.

★ ★ ★

Triggered by Trump

On November 16, 2016, at the UN climate summit in Marrakesh, Morocco, I found out firsthand just how much the climate crowd fears President Trump and his policies. When I unfurled a life-sized image of Trump, all hell broke loose. UN climate delegates reacted like vampires to a cross.

The Associated Press reported on my day: "An American climate change skeptic has shredded a copy of the Paris agreement on global warming at the U.N. climate conference. Marc Morano, who runs a climate skeptic website, was led away by security guards after the stunt outside the media center in Marrakech. Morano put the document in a paper shredder and said that's what will happen to the Paris deal once Trump takes office. Wearing a red Trump hat, he said 'the delegates here seem to be in deep denial about President-elect Trump's policies.' As security guards led him away, he said 'we will not be silenced.'"[12]

Armed UN "climate cops" swarmed me and removed me from the UN conference. My CFACT colleague Craig Rucker and I were literally pushed out into the desolate desert outside the perimeter of the Marrakesh conference center. Moments later, armed UN security commandeered my briefcase and confiscated my papers inside, which were never to be returned.

We're going to lift moratoriums on energy production in federal areas.

We're going to revoke policies that impose unwarranted restrictions on new drilling technologies. These technologies create millions of jobs with a smaller footprint than ever before.

We're going to cancel the Paris Climate Agreement and stop all payments of U.S. tax dollars to U.N. global warming programs.

Any regulation that is outdated, unnecessary, bad for workers, or contrary to the national interest will be scrapped. We will also eliminate duplication, provide regulatory certainty, and trust local officials and local residents.

> ★ ★ ★
> ## We're Not out of the Woods Yet
> President Trump's withdrawal from the UN Paris pact will not take full effect until November 4, 2020—the day after the next presidential election. The process to extricate the United States is long, and it could be reversed by the next president.[14]

Any future regulation will go through a simple test: is this regulation good for the American worker? If it doesn't pass this test, the rule will not be approved.[13]

Even more astonishing, after President Trump was sworn in as president, he actually stuck very close to his "100-day action plan" on climate and energy, keeping the pledges that he had made during the campaign.

In a big win for science, cost-benefit analysis, our economy, and the sovereignty of the United States, the climate agenda has taken severe blows. But it remains to be seen if it is defeated or just suffering a temporary setback.

So far, the signs are promising. The climate campaigners and their allies in the media have plenty of valid reason to be afraid. An October 2016 *Washington Post* editorial warned that "a President Trump could wreck progress on global warming."[15] Of course, since the polices do not even impact CO_2 levels, they are merely symbolic. "Wreck progress on alleged climate 'solutions'" would be a more apt description—of a course of action that's long overdue.

Even Al Gore has admitted, during a 2017 Fox News interview with Chris Wallace, that the UN Paris pact is symbolic:

> WALLACE: "You would agree that even if all 195 nations, now 194, met their targets, it still wouldn't solve the problem."
>
> GORE: "That is correct. However, it sends a very powerful signal to business and industry and civil society, and countries around the world."[16]

But the *Post* was right about Trump taking a wrecking ball to the climate change agenda.

Former Obama EPA Chief Gina McCarthy was so distressed, she admitted she has turned to the bottle to cope as Trump reveres climate policy. *Politico* reported, "McCarthy, who returned to her native Boston after the White House handover, admitted that she has turned to one of her city's tried-and-true methods of coping with frustration." As McCarthy explained, "We drink a lot of coffee during the day and other things at night. And night comes earlier and earlier." Hiccup.[17]

★ ★ ★

Think Globally, Act Locally

Gore sings the praises of governors and mayors who have taken it upon themselves to "compensate" for Trump's Clexit from the UN pact.[20] Ironically, in praising this decentralized approach, Gore is inadvertently conceding that a massive international treaty was never necessary in the first place.

An Inconvenient Sequel: Truth to Power, Al Gore's sequel to his 2006 *An Inconvenient Truth*, lambasted the dismantling of the climate agenda. Gore was initially hopeful that he could persuade Trump to keep Obama's climate policies intact. By July 2017, though, the former vice president was lamenting, "I thought Trump would come to his senses, but I was wrong." But he added, "We're going to win this."[18]

One of my personal favorite "solutions" proposed to address supposed man-made global warming came from Lord Christopher Monckton, the former Thatcher advisor who was my fellow travel companion at many UN summits around the world from Durban, South Africa, to Rio De Janeiro. "The right response to the non-problem of global warming is to have the courage to do nothing," Monckton told the U.S. Congress in 2009.[19]

What America needs right now is an articulate voice of climate change skepticism to challenge the Democrats, the media, and activists within the government. My one critique so far of Trump on climate during his first year in office is that there is a huge vacuum when it comes to administration officials actually challenging climate change claims, and this silence may result in only half-hearted measures to reverse Obama's climate policies as the Trump administration progresses. In 2006, when I was working at the U.S. Senate Environment and Public Works Committee, the public was hungry for a scientific smackdown of Gore and the UN's claims of settled science. When committee chair Senator James Inhofe, gave an hour-long speech presenting climate science rebuttals, the public ignited. I vividly recall my phone ringing off the hook when Senator Inhofe's speech was linked on the Drudge Report. When majority staff director Andrew Wheeler came into my office, I randomly picked up calls on speakerphone so he could hear the public thanking us for not being intimidated into silence when it came to climate science.

The 2017 Trump administration's proposed climate reform strategy is fine—as far as it goes. But the best way forward on climate change would be for President Trump to fill the science czar position vacated by John Holdren with a strong well credentialed scientist like Princeton physicist Will Happer. Choosing Happer or someone similar as science czar would enable the Trump administration to start pushing back strongly on the overwrought and oftentimes absurd climate claims. A strong climate skeptic science czar would have an unequaled platform from which to steer climate policy and influence public opinion.

The Fight Goes On

The battle over climate change will continue. The climate movement is better financed, organized, and media-hyped than all the previous environmental scares combined. The movement also has key institutional support from the United Nations and academia. The green movement essentially put all of their chips onto the climate scare, and it is unlikely that they will back down anytime soon.

So let's review. By virtually every measure—from global temperatures and climate model predictions to polar bears, sea level rise, and extreme weather events—the scientific claims of the global warming movement are falling short, or the science is actually going in the opposite direction. The fears that rising levels of carbon dioxide are a major threat facing humanity do not hold up to scientific scrutiny. The geologic history of the Earth demonstrates that carbon dioxide is drowned out by many, many other variables in our earth's climate system. Carbon dioxide is not the control knob of the climate.

Throughout human history, concerns about the weather have been used to generate fear and manipulate the public. The history of climate change alarm includes swings from cooling scares to warming scares.

And the alleged scientific "consensus" is merely an illusion carefully crafted by a partisan campaign to promote global warming fears and their so-called solutions. The Climategate scandal revealed that the upper echelons of UN scientists were colluding to craft a man-made global warming narrative, suppress scientists and studies, and threaten journal editors that did not support the party line of the climate movement. An increasing number of prestigious scientists from around the world are re-examining the evidence, publicly declaring themselves skeptical, and rejecting the claims being put forth by the United Nations, Al Gore, and climate activists and hyped by the mainstream media.

The UN has publicly stated that climate policy will redistribute wealth. As far as any effect on the climate, their so-called solutions are only symbolic. But the effects on the developing world's poor will be devastating.

It bears repeating: if we actually faced a man-made climate crisis and we had to rely on the UN or the EPA or Congress to save us, we would all be doomed! But more important, if we actually did face a catastrophic global warming, the last "solution" we would want to seek would be one that saddles us with sovereignty-threatening, central-planning, wealth-redistributing, economy-crippling regulations and the most expensive treaty in world history. If we did face a man-made climate change crisis, we would want to unleash the free market and entrepreneurship to come up with new technologies and make them viable and affordable—without banning or regulating current fossil fuel energy out of existence until we had replacements. If Al Gore is correct in his assertions that there are financial fortunes to be made for young entrepreneurs and inventors in developing new forms of energy—and Al Gore himself has already made his climate fortune many times over—then all that is really needed is advancing technology.

The day Americans, or anyone on planet Earth, can go to their local Walmart and buy a solar panel and install it on their roof and get off the grid is the day climate "solution" debate ends. There is no need for central planning, or banning energy that is cheap and abundant in favor of energy that needs massive subsidies and is not yet ready for prime time.

But the climate change activists have betrayed again and again—by their manipulated science, by the trickery that props up the "consensus" in favor of their panic, by their admitted ambitions to control the world's economy and redistribute its wealth, and finally by their gross hypocrisy in their own use of energy—that their real motivation is not saving the planet.

It's up to the rest of us to save the planet from them.

UPDATE: Marc Morano Destroys the Green New Deal

Submitted Written Testimony of Marc Morano, publisher of Climate Depot and author of the bestselling *Politically Incorrect Guide to Climate Change* and former staff of U.S. Senate Environment & Public Works Committee

Presented at the Congressional Western Caucus Policy Forum and Press Conference Examining the Green New Deal Bill introduced by Alexandria Ocasio-Cortez (NY-14) at 1300 Longworth House Office Building on Wednesday, February 27, 2018, and edited for publication

I want to thank Congressional Western Caucus for hosting this hearing on the Green New Deal. My background is in political science, which happens to be an ideal background for examining the Green New Deal and man-made global warming claims they are based upon. I am the author of the best-selling 2018 book, *The Politically Incorrect Guide to Climate Change.*

I have been passionate about environmental issues since I began my career in 1991, having produced a documentary on the myths surrounding the Amazon Rainforest in 2000,[1] and I was a fully credentialed investigative journalist who reported extensively on environmental and energy issues such as deforestation, endangered species, pollution, and climate change. In 2016, I wrote and starred in the film *Climate Hustle*, which debuted in over four hundred theaters in the U.S. and Canada.

In my capacity as communications director for the Senate Environment and Public Works Committee under Senator James Inhofe, I was speechwriter and hosted the award-winning U.S. Senate Committee on Environment and Public Works blog.[2] I released the first ever U.S. government "Skeptic's Guide to Debunking Global Warming Alarmism"[3] in 2006. I also authored the 255-page Senate report of over seven hundred dissenting scientists, originally published in 2007 and updated in 2008 [and] 2009. In 2010, the number of dissenting international scientists exceeded one thousand.[4] I am now the publisher of the award-winning Climate Depot[5] and work daily with scientists who examine the latest peer-reviewed studies and data on the climate as well as the feasibility of the alleged solutions.

The Green New Deal is neither "green" or "new" and it is a raw deal. It is one big bowl of crazy.

A few key points:

- "Global warming" is merely the latest environmental scare with the same solutions of wealth redistribution and central planning. "Global warming" is merely the latest environmental scare with the same big government solution.
- The Green New Deal has very little to do with the environment or climate.
- The deal claims free college or trade schools for every citizen.
- The government will ensure "healthy food" to all, "safe, affordable, adequate housing," incomes for all who are "unable or unwilling" to work.
- It seeks to go after meat eating and "farting cows."
- It will end all traditional forms of energy in the next ten years. The Green New Deal is "a ten-year plan to mobilize every aspect of American society at a scale not seen since World War II to achieve net-zero greenhouse gas emissions."

The cost of the Green New Deal is not cheap. As Bloomberg News has reported, the "Green New Deal Could Cost $93 Trillion," according to the American Action Forum:

"The so-called Green New Deal may tally between $51 trillion and $93 trillion over 10-years, concludes American Action Forum, which is run by Douglas Holtz-Eakin, who directed the non-partisan CBO from 2003 to 2005. That includes between $8.3 trillion and $12.3 trillion to meet the plan's call to eliminate carbon emissions from the power and transportation sectors and between $42.8 trillion and $80.6 trillion for its economic agenda including providing jobs and health care for all."[6]

Recycling the Same Solutions

Representative Alexandria Ocasio-Cortez of New York has acknowledged that her "deal" will require "massive government intervention."[7] The "Green New Deal seeks a massive transformation of our society."[8]

Ocasio-Cortez is echoing the rhetoric of former UN Climate Chief Christiana Figureres, who sought "centralized transformation" that she said was "going to make the life of everyone on the planet very different" in order to fight "global warming."[9]

But as I said earlier, the "Green New Deal" is neither "green" or "new." The environmental Left has been using green scares to push for the same solutions we see today—wealth redistribution, central planning, sovereignty limiting treating—since the overpopulation scares of the 1960s and 1970s.

A former Ocasio-Cortez campaign manager called the "Green New Deal" a plan to "redistribute wealth and power" from the rich to the poor.[10] In this the Green New Deal borrows from previous proposed solutions. In 2010 a UN IPCC office admitted that the UN was seeking to "distribute de facto the world's wealth by climate policy" and said, "This has almost nothing to do with environmental policy anymore."[11]

Different Environmental Scare, Same Solution

In 1974, future Obama science czar John Holdren proposed "redistribution of wealth" to battle environmental degradation. Holdren testified to the U.S. Senate Commerce Committee, "The neo-Malthusian view proposes conscious accommodation to the perceived limits to growth via population limitation and redistribution of wealth in order to prevent the 'overshoot' phenomenon. My own sympathies are no doubt rather clear by this point. I find myself firmly in the neo-Malthusian camp."

In the 2019 Green New Deal solution, as the *New York Post* reported, "AOC Explains Why 'Farting Cows' Were Considered in the Green New Deal": "Maybe we shouldn't be eating a hamburger for breakfast, lunch, and dinner."[12]

She is borrowing from the UN: "Christiana Figueres, the former United Nations official responsible for the 2015 Paris climate agreement, has a startling vision for restaurants of the future: Anyone who wants a steak should be banished. 'How about restaurants in ten to fifteen years start treating carnivores the same way that smokers are treated?' Figueres suggested during a recent conference. 'If they want to eat meat, they can do it outside the restaurant,'" according to Bjørn Lomborg.[13]

In the 2019 Green New Deal solution, the "entire economy would operate under the Green New Deal" and "the government would have 'appropriate ownership stakes' in ALL Green New Deal businesses," according to Steve Milloy.[14]

Flashback to the 1970 proposed solution to overpopulation: Amherst College professor Leo Marx warned in 1970 about the "global rate of human population growth. All of this is only to say that, on ecological grounds, the case for world government is beyond argument."[15]

There is nothing new about the Green New Deal. "Global warming" is merely the latest alleged environmental scare that is being substituted to push the same solutions. Instead of arguing the merits of the economic and political changes of the Green New Deal, they are using—in the words of Al Gore—a "torqued up"[16] climate change scare to urge quick imposition of the policies to protect us from a climate emergency.

In my book *The Politically Incorrect Guide to Climate Change* I showcase how they use the same hysteria for the different environmental scares in the 1970s, whether it's resource scarcity, over-population, rainforest clearing, et cetera.

They will say, *we need a global solution; we need global governance; we need wealth redistribution; we need sovereignty threatening treaty, or some kind of economic activity limiting.* No matter what environmental scare in the past that they tried to scare people with, it was the same solutions they're proposing now.

In the book, I go back and show over and over that global warming is merely the latest scare they're using to get their agenda. I feature climate activist Naomi Klein, who's an adviser to Pope Francis and wrote the book *Capitalism vs. the Climate.* Klein actually says that they would be seeking the same solutions even if there was no global warming and that essentially, capitalism is incompatible with a livable climate. She urges people that they need to jump on this because solving global warming will solve what we've been trying to achieve all along.[17]

Even the *New York Times* recognizes the Green New Deal as a cover for other non-environmental issues. The *New York Times* editorial board: "Is the Green New Deal aimed at addressing the climate crisis? Or is addressing the climate crisis merely a cover for a wish-list of progressive policies and a not-so-subtle effort to move the Democratic Party to the left? At least some candidates—Amy Klobuchar of Minnesota among them—seem to think so. Read literally, the resolution wants not only to achieve a carbon-neutral energy system but also to transform the economy itself."[18]

Even the *Washington Post* is souring on it. "They should not muddle this aspiration with other social policy, such as creating a federal jobs guarantee, no matter how desirable that policy might be," the *Washington Post* editorial board wrote. The *Post* also called the Green New Deal's goal of reaching "net-zero" greenhouse gas emissions within ten years "impossible" and criticized the resolution's "promise to invest in known fiascos such as high-speed rail."[19]

The climate activists are openly using climate scare tactics to achieve their ends. And in order to get those ends achieved, they have to hype and scare. It's been a very effective strategy because they've bullied Republican politicians, who should know better, into at least submissiveness and silence.

AOC and Senator Markey have bungled the release of the Green New Deal. They had to pull parts of it from their website. There's a whole dispute over what they meant on nuclear power. They haven't even gotten this straightened out.

But what they do have straightened out: this is the litmus test for the 2020 Democratic contenders. And in a way they have given anyone who cares about free markets, liberty, and science a grand opening to expose anyone who signs on to this plan.

Another aspect that is remarkable is that the Green New Deal is not sitting well with many environmental activists and other factions of the Democratic Party base.

Prominent environmentalist Michael Shellenberger, president of Environmental Progress and an activist *Time* magazine called a "hero of the environment,"[20] wrote, "I am calling bullshit not just on @AOC but on her progressive enablers in the news media who are giving her a pass on the most crucial test of moral and political leadership of our time when it comes to climate change: a person's stance on nuclear power.... I am calling bullshit on climate fakery. Anyone who is calling for phasing out nuclear is a climate fraud perpetuating precisely the gigantic 'hoax' that [Oklahoma] Sen. James Inhofe famously accused environmentalists of perpetuating.... if you want to be a self- respecting progressive or journalist who is fairly considering or covering the climate issue, please stop giving Ocasio-Cortex and other supposedly climate-concerned greens a pass. THEY ARE INCREASING EMISSIONS."[21]

Labor leader Terry O'Sullivan, general president of the Laborers' International Union of North America (LIUNA), which endorsed Barack Obama and Hillary Clinton for president,[22] came out swinging against the Green New Deal from New York Democrat Representative Alexandria Ocasio-Cortez: "It is exactly how not to win support for critical measures to curb climate change.... It is difficult to take this unrealistic manifesto seriously, but the economic and social devastation it would cause if it moves forward is serious and rea.... threatens to destroy workers' livelihoods, increase divisions and inequality, and undermine the very goals it seeks to reach. In short, it is a bad deal."[23]

In summary, the Green New Deal has to be opposed, exposed, and defeated. We must challenge the economics, ideology, and science claims of this deal. I thank the Western Caucus for this opportunity and look forward to them leading the battle.

Thank you.

Acknowledgments

This book would not have been possible without the tireless efforts of a band of climate-skeptical bloggers and vocal scientists who have devoted themselves to revealing the truth behind the climate claims put forth by the UN, Al Gore, the media, and academia. It is not possible to thank or even list them all, but here is a partial honor roll of scientists and bloggers whose work was invaluable in writing this book:

Tom Nelson	James Taylor
Anthony Watts	Steve Milloy
Tony Heller	Will Happer
Pierre Gosselin	Paul Homewood
Joanne Nova	Benny Peiser
Steve McIntyre	David Legates
Tim Ball	Michael Bastasch
John L. Daly	Bob Carter
Tom Harris	Willie Soon
Caleb Rossiter	John Stossel
Joe Bast	Joe Bastardi

Ryan Maue

Patrick Moore

Robert Giegengack

John Coleman

Judith Curry

Lubos Motl

Roger Pielke Sr.

Alex Epstein

James Delingpole

Christopher Monckton

Joseph D'Aleo

Roger Pielke Jr.

Roy Spencer

Philip Stott

Notes

Chapter 1:
The Education of a Climate Denier

1. Marc Morano, "Amazon Rainforest: Clear-Cutting the Myths New Television Special Refutes Myths & Lies of Environmentalists," American Investigator TV, June 6, 2000, http://web.archive.org/web/20010201065600/http://ai-tv.com:80/.

2. Marc Morano and Mick Curran, *Climate Hustle*, 2016, http://www.climatehustle.com/.

3. Brian Merchant, "Rolling Stone Calls Out the 17 Worst 'Climate Killers' in the US," TreeHugger, January 11, 2010, https://www.treehugger.com/corporate-responsibility/rolling-stone-calls-out-the-17-worst-climate-killers-in-the-us.html.

4. Katie Connolly, "King of the Skeptics," *Newsweek*, December 17, 2009, http://www.cfact.org/2009/12/17/cfacts-morano-king-of-the-skeptics/.

5. *Newsnight*, BBC TV, December 4, 2009, https://youtu.be/DnBawFsxb-M.

6. John H. Richardson, "This Man Wants to Convince You Global Warming Is a Hoax," *Esquire*, March 20, 2010, http://www.esquire.com/news-politics/a7078/marc-morano-0410/.

7. Juliet Eilperin, "Marc Morano's 15 Minutes of Fame," *Washington Post*, March 30, 2010, http://views.washingtonpost.com/climate-change/post-carbon/2010/03/marc_moranos_15_minutes_of_fame.html.

8. Al Gore, "Climate 'Misinformer of the Year,'" Al Gore's Blog, December 30, 2012, http://blog.algore.com/2012/12/climate_misinformer_of_the_yea_1.html.

9. Justin Chang, "Telluride Film Review: 'Merchants of Doubt,'" *Variety*, September 4, 2014, http://variety.com/2014/film/reviews/telluride-film-review-merchants-of-doubt-1201297810/.

10. Gautami Sharma, "The Truth Revealed in 'Merchants of Doubt,'" *Daily Californian*, March 12, 2015, http://www.dailycal.org/2015/03/12/truth-revealed-merchants-doubt/.

11. Lindsay Abrams, "'Merchants of Doubt': Meet the Sleazy Spin Doctors Who Will Stop at Nothing to Obscure the Truth," Salon, March 6, 2015, 05:28http://www.salon.com/2015/03/06/merchants_of_doubt_meet_the_sleazy_spin_doctors_who_will_stop_at_nothing_to_obscure_the_truth/.

12. Alan Scherstuhl, "Merchants of Doubt Shows How Corporate Lies Infect Our Minds, From Tobacco to Climate Change," *LA Weekly*, March 4, 2015, http://www.laweekly.com/film/merchants-of-doubt-shows-how-corporate-lies-infect-our-minds-from-tobacco-to-climate-change-5416545.

13. David Lewis, "'Merchants of Doubt': Exposing Climate Change 'Experts,'" San Francisco Chronicle, March 12, 2015, http://www.sfgate.com/movies/article/Merchants-of-Doubt-Exposing-climate-change-6127813.php.

14. Andrew O'Hehir, "Climate Deniers and Other Pimped-Out Professional Skeptics: The Paranoid Legacy of Nietzsche's 'Problem of Science,'" Salon.com, March 7, 2015, http://www.salon.com/2015/03/07/climate_denial_the_ghosts_of_the_cold_war_and_the_legacy_of_nietzsche/.

15. Lindsay Abrams, "'Merchants of Doubt': Meet the Sleazy Spin Doctors Who Will Stop at Nothing to Obscure the Truth," Salon.com, March 6, 2015, 05:28http://www.salon.com/2015/03/06/merchants_of_doubt_meet_the_sleazy_spin_doctors_who_will_stop_at_nothing_to_obscure_the_truth/.

16. Jeremy Martin, "The Hazy Science of Climate Change Deniers," The Daily Good, March 13, 2015, https://www.good.is/articles/merchants-of-doubt.

17. Steven Searls, "Evil Personified: Marc Morano, Climate Denier Exposed in 'Merchants of Doubt,'" Daily Kos, March 20, 2015, https://www.dailykos.com/story/2015/3/20/1372121/-Evil-Personified-Marc-Morano-Climate-Denier-Exposed-as-Bully-in-Merchants-of-Doubt.

18. Marc Morano, "Amazon Rainforest: Clear-Cutting the Myths: New Television Special Refutes Myths & Lies of Environmentalists," American Investigator TV, June 6, 2000, http://web.archive.org/web/20010201065600/http://ai-tv.com:80/.

19. Ben Pappas, "Stars Confront New Attack on Rain Forest," *Us Weekly* magazine, June 26, 2000, http://web.archive.org/web/20010201065600/http://ai-tv.com:80/; Extra TV, "Rumble in the Jungle," June 26, 2000, http://web.archive.org/web/20010201065600/http://ai-tv.com:80/.

20. Elisabeth Rosenthal, "New Jungles Prompt a Debate on Rain Forests," *New York Times*, January 29, 2009, http://www.nytimes.com/2009/01/30/science/earth/30forest.html?_r=2&hp&mtrref=undefined&gwh=082C1E03D612476DF3CB50766B2D89A3&gwt=pay.

21. Marc Morano, "Save the Trees, Use More Wood," CNS News, May 1, 2002, http://www.cnsnews.com/public/Content/Article.aspx?rsrcid=5020.

22. Brendan Demelle, "Top 10 Climate Deniers," "Before the Flood," November 7, 2016, https://www.beforetheflood.com/explore/the-deniers/top-10-climate-deniers/.

23. Marc Morano and Matt Dempsey, ,"A Skeptic's Guide to Debunking Global Warming Alarmism," U.S. Senate Environment & Public Works Committee, December 2006, https://www.epw.senate.gov/public/_cache/files/5/6/56dd129d-e40a-4bad-abd9-68c808e8809e/01AFD79733D77F24A71FEF9DAFCCB056.skeptics-guide.pdf.

24. Emily Heil, "Drudge, Global Warming Shut Down Senate Site," *The Hill*, January 23, 2007, https://www.epw.senate.gov/public/index.cfm/press-releases-all?ID=4fcf378e-802a-23ad-470b-321ac83dd4de.

25. Matthew Dempsey, "Inhofe EPW Website Wins Coveted Gold Mouse Award," Senate Environment & Public Works Committee, January 14, 2008, https://webcache.googleusercontent.com/search?q=cache:huStxXle1ZAJ:https://www.epw.senate.gov/public/index.cfm/in-the-news%3FID%3D79541C08-802A-23AD-4789-56296E2D061B+&cd=1&hl=en&ct=clnk&gl=us.

26. Marc Morano, "More Than 700 International Scientists Dissent Over Man-Made Global

Warming Claims," U.S. Senate Environment & Public Works Committee, March 16, 2009, https://www.epw.senate.gov/public/index.cfm/press-releases-all?ID=10fe77b0-802a-23ad-4df1-fc38ed4f85e3.

27. Marc Morano, "Special Report: More Than 1000 International Scientists Dissent Over Man-Made Global Warming Claims—Challenge UN IPCC & Gore," Climate Depot, December 8, 2010, http://www.climatedepot.com/2010/12/08/special-report-more-than-1000-international-scientists-dissent-over-manmade-global-warming-claims-challenge-un-ipcc-gore-2/,

28. Riley E. Dunlap, "Climate-Change Views: Republican-Democratic Gaps Expand," Gallup, January 23, 2007, http://www.gallup.com/poll/107569/ClimateChange-Views-RepublicanDemocratic-Gaps-Expand.aspx.

29. Marc Morano, "Climate Depot Wins Another Award! Daily Caller: Climate Depot Shares '2010 Award for Political Incorrectness' with Sen. Inhofe!" Climate Depot, December 29, 2010, http://www.climatedepot.com/2010/12/29/climate-depot-wins-another-award-daily-caller-climate-depot-shares-2010-award-for-political-incorrectness-with-sen-inhofe/.

30. Rush Limbaugh, "ClimateDepot.com: Substantial Fraud in Climate Change Hoax?" *The Rush Limbaugh Show*, Nov 20, 2009, https://www.rushlimbaugh.com/daily/2009/11/20/climatedepot_com_substantial_fraud_in_climate_change_hoax/.

31. Neil Cavuto, "Inconvenient Truth About Green Agenda," Fox News Channel, Nov. 26, 2010, http://video.foxnews.com/v/4437320/?#sp=show-clips.

32. Donna Laframboise, "Rajendra Pachauri's Resignation Letter," BigPicNews.com, February 24, 2015, https://nofrakkingconsensus.com/2015/02/24/rajendra-pachauris-resignation-letter/,

33. Suzanne Goldenberg, "IPCC chairman dismisses climate report spoiler campaign," UK *Guardian*, September 2013, https://www.theguardian.com/environment/2013/sep/19/ipcc-chairman-climate-report?CMP=twt_fd.

34. Suzanne Goldenberg, "IPCC chairman dismisses climate report spoiler campaign," UK *Guardian*, September 2013, https://www.theguardian.com/environment/2013/sep/19/ipcc-chairman-climate-report?CMP=twt_fd.

35. Dr. Richard Tol testimony, Full Committee Hearing—Examining the UN Intergovernmental Panel on Climate Change Process, U.S. House Committee on Science, Space and Technology, May 29, 2014, http://science.house.gov/hearing/full-committee-hearing-examining-un-intergovernmental-panel-climate-change-process.

36. Elizabeth Kolbert, "Global Warming Talks Progress Is 'Slow But Steady'—UN Climate Chief," November 21, 2012, https://www.theguardian.com/environment/2012/nov/21/global-warming-talks-progress-un-climate-chief.

37. Former EPA chief Gina McCarthy testimony, "Full Committee Hearing—Examining EPA's Regulatory Overreach," U.S. House Committee on Science, Space and Technology, July 9, 2015, https://science.house.gov/legislation/hearings/examining-epa-s-regulatory-overreach.

38. UK Professor Emeritus of Biogeography Philip Stott of the University of London, "It's All Unravelling," "Global Warming Politics," May 2, 2008, http://web.mac.com/sinfonia1/Global_Warming_Politics/A_Hot_Topic_Blog/Entries/2008/5/2_It%E2%80%99s_All_Unravelling.html.

39. Steve Milloy, "Climategate 2.0: Jones Says 2-Degree C Limit 'Plucked out of Thin Air,'" JunkScience.com, November 23, 2011, http://junkscience.com/2011/11/climategate-2-0-jones-says-2o-limit-plucked-out-of-thin-air/.

Chapter 2:
Climate Change Déjà Vu

1. Brian Williams, "March 31: Nightly News Monday Broadcast," *NBC Nightly News*, March 31, 2014, http://www.nbcnews.com/video/nightly-news-netcast/54832241.

2. Tom Brokaw, "Global Warming: What You Need to Know," Discovery Channel, July 16, 2006, https://www.youtube.com/watch?v=rJBOT5A9syQ.

3. Larry Bell, "Rising Tides of Terror: Will Melting Glaciers Flood Al Gore's Coastal Home?," *Forbes*, June 26, 2012, https://www.forbes.com/sites/larrybell/2012/06/26/rising-tides-of-terror-will-melting-glaciers-flood-al-gores-coastal-home/#74d38f574ee8.

4. Anne Thompson, "March 31: Nightly News Monday Broadcast," *NBC Nightly News*, March 31, 2014, http://www.nbcnews.com/video/nightly-news-netcast/54832241.

5. Tom Brokaw, "Global Warming: What You Need to Know," Discovery Channel, July 16, 2006, https://www.youtube.com/watch?v=rJBOT5A9syQ.

6. Thomas Jefferson, "Notes on the State of Virginia," 1799, http://xroads.virginia.edu/~hyper/jefferson/ch07.html.

7. Lawrence Solomon, "History Trumps Climate Scientists," *Financial Post*, April 18, 2013, http://business.financialpost.com/opinion/lawrence-solomon.

8. "Climate Changes," *Mercury*, June 6, 1914, http://trove.nla.gov.au/newspaper/article/10373644.

9. Walter Sullivan, "The Changing Face of the Arctic; The Changing Face of the Arctic," *New York Times*, October 19, 1958, http://www.nytimes.com/1958/10/19/archives/the-changing-face-of-the-arctic-the-changing-face-of-the-arctic.html.

10. "Imaginary Changes of Climate," *Brisbane Courier*, January 10, 1871, http://trove.nla.gov.au/newspaper/article/1298497.

11. "Climate Never Changes," *Adelaide News*, February 1, 1933, http://trove.nla.gov.au/newspaper/article/133062563?searchTerm=climate%20changing&searchLimits=.

12. Walter Orr Roberts, "Climate Change and Its Effect on the World," National Center for Atmospheric Research, Boulder, 1974, http://www.iaea.org/Publications/Magazines/Bulletin/Bull165/16505796265.pdf.

13. Peter Gwynne, "The Cooling World," Newsweek, April 28, 1975, http://www.denisdutton.com/cooling_world.htm.

14. Ibid.

15. Harold M. Schmeck Jr., "Climate Changes Endanger World's Food Output," *New York Times*, August 8, 1974, http://www.nytimes.com/1974/08/08/archives/climate-changes-endanger-worlds-food-output-scientists-view-global.html.

16. Thomas C. Peterson, William M. Connolley, and John Fleck, "The Myth of the 1970s Global Cooling Scientific Consensus," American Meteorological Society, September 2008, http://journals.ametsoc.org/doi/pdf/10.1175/2008BAMS2370.1.

17. James Delingpole, "Massive Cover-Up Exposed: Lying Alarmists Rebranded 70s Global Cooling Scare as a Myth," Breitbart, September 14, 2016, http://www.breitbart.com/london/2016/09/14/massive-cover-exposed-lying-alarmists-rebranded-70s-global-cooling-scare-myth/.

18. Steven Goddard, "Forty Years of Climate Insomnia," Real Science, April 26, 2014, https://stevengoddard.wordpress.com/2014/04/26/forty-years-of-climate-insomnia/, quoting "Climate and the President's Science Advisor," Living on Earth: PRI's Environmental News Magazine, April 25, 2014, http://www.loe.org/shows/segments.html?programID=14-P13-00017&segmentID=3#.U1uo5R9cg58.twitter. See also Marc Morano, "John Holdren in 1971: 'New Ice Age' Likely," Climate Depot, January 8, 2014, http://www.climatedepot.com/2014/01/08/flashback-john-holdren-in-1971-new-ice-age-likely/.

19. *Ukiah Daily Journal*, November 20, 1974, http://www.climatedepot.com/2015/12/17/flashback-1974-60-theories-have-been-advanced-to-explain-the-global-cooling/.

20. Leonard Nimoy, "The Coming Ice Age," *In Search Of…*, 1978, https://www.youtube.com/watch?v=L_861us8D9M.

21. Myles Allen, "Stephen Schneider Obituary," *Guardian*, July 21, 2010, http://www.guardian.co.uk/science/2010/jul/21/stephen-schneider-obituary.

22. Walter Sullivan, "Experts Fear Great Peril If SST Fumes Cool Earth, *New York Times*, December 21, 1975, http://www.nytimes.com/1975/12/21/archives/experts-fear-great-peril-if-sst-fumes-cool-earth.html.

23. Steven Goddard, "1970s Global Cooling Scare," Real Science, https://stevengoddard.wordpress.com/1970s-ice-age-scare/.

24. "A Study of Climatological Research as It Pertains to Intelligence Problems," Office of Research and Development of the Central Intelligence Agency, August 1974, http://www.climatemonitor.it/wp-content/uploads/2009/12/1974.pdf.

25. David Archibald, "The CIA Report and the Warning from Wisconsin," Watts Up with That, May 25, 2012, http://wattsupwiththat.com/2012/05/25/the-cia-documents-the-global-cooling-research-of-the-1970s/.

26. "C.I.A. Warning: Changes to Climate to Bring Upheaveal," *Canberra Times*, July 21, 1976, http://trove.nla.gov.au/newspaper/article/110818238.

27. Harry Reasoner and Howard K. Smith, *ABC Evening News*, ABC News, January 18, 1977.

28. Associated Press, "There's a New Ice Age Coming!" *Windsor Star*, September 9, 1972, https://news.google.com/newspapers?id=lzI_AAAAIBAJ&sjid=PlEMAAAAIBAJ&dq=climate%20expert%20new%20ice%20age%20coming%20hubert%20lamb&pg=4365%2C2786655.

29. "Walter Sullivan, International Team of Specialists Finds No End in Sight to 30-Year Cooling Trend in Northern Hemisphere," *New York Times*, January 5, 1978, http://www.nytimes.com/1978/01/05/archives/international-team-of-specialists-finds-no-end-in-sight-to-30year.html.

30. United Press International, "Air Pollution May Trigger Ice Age, Scientists Feel," *Telegraph*, December 5, 1974, https://news.google.com/newspapers?id=ZaQrAAAAIBAJ&sjid=uPwFAAAAIBAJ&pg=5191,1022557.

31. Harold M. Schmeck Jr., "Climate Changes Called Ominous; Scientists Warn Predictions Must Be Made Precise to Avoid Catastrophe," *New York Times*, January 19, 1975, http://query.nytimes.com/gst/abstract.html?res=9505E0D9143AE034BC4152DFB766838E669EDE&legacy=true.

32. Rick Van Sant, "Geologist Says Winters Getting Colder," *Middlesboro Daily News*, January 16, 1978, https://news.google.com/newspapers?id=mylCAAAAIBAJ&sjid=xaoMAAAAIBAJ&pg=6147,666578.

33. Walter Sullivan, "Science; Worrying About a New Ice Age," *New York Times*, February 23, 1969, http://query.nytimes.com/gst/abstract.html?res=9F06E0DC113FE63ABC4B51DFB4668382679EDE&legacy=true.

34. David R. Boldt, "Colder Winters Held Dawn of New Ice Age," *Washington Post*, January 11, 1970, http://pqasb.pqarchiver.com/washingtonpost_historical/doc/147902052.html?FMT=ABS&FMTS=ABS:FT&type=historic&date=JAN%2011,%201970&author=Washington%20Post%20Staff%20WriterBy%20David%20R.%20Boldt&pub=The%20Washington%20Post&edition=&startpage=&desc=Colder%20Winters%20Held%20Dawn%20of%20New%20Ice%20Age.

35. John H. Douglass, "Climate Change: Chilling Possibilities," *Science News*, March 1, 1975, https://www.sciencenews.org/sites/default/files/8983.

36. Walter Sullivan, "Scientists Agree World Is Colder," *New York Times*, January 30, 1961, http://query.nytimes.com/gst/abstract.html?res=

9F01E0DA133FE13ABC4850DFB766838A679EDE
&legacy=true.

37. "British Climate Scientist Predicts New Ice Age," *Christian Science Monitor*, September 23, 1972, http://pqasb.pqarchiver.com/csmonitor_historic/doc/511356590.html?FMT=CITE&FMTS=CITE:AI&type=historic&date=Sep%2023,%201972&author=&pub=Christian%20Science%20Monitor&edition=&startpage=&desc=British%20climate%20expert%20predicts%20new%20Ice%20Age.

38. George Getze, "New Ice Age Coming—It's Already Getting Colder," *Los Angeles Times*, October 24, 1971, http://pqasb.pqarchiver.com/latimes/doc/156732829.html?FMT=ABS&FMTS=ABS:AI&type=historic&date=Oct%2024,%201971&author=GEORGE%20GETZE&pub=Los%20Angeles%20Times&edition=&startpage=&desc=New%20Ice%20Age%20Coming—-It%27s%20Already%20Getting%20Colder.

39. James D. Hays, "The Ice Age Cometh," *Canberra Times*, May 31, 1975, reproduced at https://stevengoddard.wordpress.com/1970s-ice-age-scare/.

40. Steven Goddard, "Time Magazine Goes Both Ways on the Polar Vortex," Real Science, January 7, 2014, https://stevengoddard.wordpress.com/2014/01/07/time-magazine-goes-both-ways-on-the-polar-vortex/.

41. James P. Sterba, "Climatologists Forecast Stormy Economic Future," *New York Times*, July 12, 1976, http://www.nytimes.com/1976/07/12/archives/climatologists-forecast-stormy-economic-future-climatologists.html.

42. James P. Sterba, "Problems from Climate Changes Foreseen in a 1974 C.I.A Report," *New York Times*, February 02, 1977, http://query.nytimes.com/gst/abstract.html?res=9A0CE2DD103BE334BC4A53DFB466838C669EDE&legacy=true.

43. Vincent Gray, "The Perversion of Science," NZClimate Truth Newsletter, July 8, 2010, http://climaterealists.com/index.php?id=5969.

44. David Kear, "Global Warming Alias Climate Change—the Non-Existent, Incredibly Expensive Threat to Us All, including to Our Grandchildren," New Zealand Center for Political Research, April 6, 2014, http://www.nzcpr.com/global-warming-alias-climate-change/.

45. Donna Laframboise, "The WWF Activist in Charge at the IPCC," PicBigNews, March 30, 2014, https://nofrakkingconsensus.com/2014/03/30/the-wwf-activist-in-charge-at-the-ipcc/.

46. Donna Laframboise, "IPCC Invites in the Activists," PicBigNews, March 6, 2013, https://nofrakkingconsensus.com/2013/03/06/ipcc-invites-in-the-activists/.

47. Richard Tol, "Examining the UN Intergovernmental Panel on Climate Change Process," U.S. House of Representatives Full Science Committee Hearing, May 29, 2014, http://science.house.gov/sites/republicans.science.house.gov/files/documents/HHRG-113-SY-WState-RTol-20140529_0.pdf.

48. Philip Shabecoff, "Global Warming Has Begun, Expert Tells Senate," *New York Times*, June 24, 1988, http://www.nytimes.com/1988/06/24/us/global-warming-has-begun-expert-tells-senate.html?pagewanted=all.

49. Marc Morano, "James Hansen's Former NASA Supervisor Declares Himself a Skeptic," U.S. Senate Environment & Public Works Committee, January 27, 2009, https://webcache.googleusercontent.com/search?q=cache:h9dVPoJyl2AJ:https://www.epw.senate.gov/public/index.cfm/press-releases-all%3FID%3D1a5e6e32-802a-23ad-40ed-ecd53cd3d320+&cd=2&hl=en&ct=clnk&gl=us.

50. Deborah Amos, Frontline, PBS, Apr 24, 2007, http://www.pbs.org/wgbh/pages/frontline/hotpolitics/etc/script.html.

51. Victor Cohn, "U.S. Scientist Sees New Ice Age," *Washington Post*, July 9, 1971, http://hockeyschtick.blogspot.com/2014/05/flashback-hansens-climate-model-says.html.

52. Glenn Kessler, "Setting the Record Straight: The Real Story of a Pivotal Climate-Change Hearing," *Washington Post*, March 30, 2015, https://www.washingtonpost.com/news/fact-checker/wp/2015/03/30/setting-the-record-straight-the-real-story-of-a-pivotal-climate-change-hearing/?utm_term=.4c3fa42ca7a4.

53. *20/20*, ABC News, August 30, 2006.

54. Howard K. Smith, *ABC Evening News*, ABC News, January 18, 1977.

55. "Cold Comfort: Scientists Anticipate Conclusion of Current Warm Cycle," *New Castle News*, April 23, 1973, https://realclimatescience.com/2016/09/exxon-didnt-know-and-neither-did-anyone-else-3/.

56. Douglas Fischer, "Armadillo Moves North across a Warmer North America," *Scientific American*, June 20, 2011, https://www.scientificamerican.com/article/armadillo-moves-north-across-warmer-north-america/.

Chapter 3: "Pulled from Thin Air": The 97 Percent "Consensus"

1. Jay Richards, "When to Doubt a Scientific 'Consensus,'" American Enterprise Institute, March 16, 2010, http://www.aei.org/publication/when-to-doubt-a-scientific-consensus/.

2. Joseph Bast and Roy Spencer, "The Myth of the Climate Change '97%,'" *Wall Street Journal*, May 26, 2014, https://www.wsj.com/articles/joseph-bast-and-roy-spencer-the-myth-of-the-climate-change-97-1401145980?mod=rss_opinion_main&tesla=y.

3. Valerie Richardson, "Obama's 97 Percent Climate Change Consensus Includes 'Deniers,'" *Washington Times*, April 25, 2016, http://www.washingtontimes.com/news/2016/apr/25/obamas-97-percent-climate-change-consensus-include/.

4. Brian Stelter, "Climate Change Is Not Debateable (sic)" *Reliable Sources*, CNN, February 23, 2014, http://cnnpressroom.blogs.cnn.com/2014/02/23/climate-change-is-not-debateable/.

5. Tom Friedman, *Face the Nation*, CBS News, April 6, 2014, http://ijr.com/2014/04/127350-nyt-reporter-compares-climate-change-skeptics-radical-marxists/.

6. Interview of Robert Giegengack, Marc Morano and Mick Curran, *Climate Hustle*, 2016, www.ClimateHustle.com.

7. Interview of Caleb Rossiter, Morano and Curran, *Climate Hustle*.

8. Interview of Judith Curry, Morano and Curran, *Climate Hustle*.

9. Interview of Leighton Steward, Morano and Curran, *Climate Hustle*, 2016.

10. ITN News interview of Philip Stott, Morano and Curran, *Climate Hustle*.

11. ITN News interview of Roy Spencer, Morano and Curran, *Climate Hustle*.

12. Marc Morano, "MIT Climate Scientist Dr. Richard Lindzen Mocks 97% Consensus: 'It Is Propaganda,'" Climate Depot, February 15, 2016, http://www.climatedepot.com/2016/02/15/mit-climate-scientist-dr-richard-lindzen-mocks-97-consensus-it-is-propaganda/.

13. Interview of Robert Giegengack, Morano and Curran, *Climate Hustle*.

14. Pamela Leavey, "Kerry and Gingrich Face Off on Climate Change," *Democratic Daily*, April 10, 2007, http://blog.thedemocraticdaily.com/?p=5605.

15. Marc Morano, "Climate Depot's TV Debate in Copenhagen: UK Warming Prof. Falsely Claims '5000 Leading Climate Scientists' in UN IPCC—Morano Counters: 'You Need to Apologize and Retract That Immediately,'" Climate Depot, December 12, 2009, http://www.climatedepot.com/2009/12/12/climate-depots-tv-debate-in-copenhagen-uk-warming-prof-falsely-claims-5000-leading-climate-scientists-in-un-ipcc-morano-counters-you-need-to-apologize-and-retract-that-immediately/.

16. Jean Goodwin, "Morano Analysis #8: Repeating Oneself All Over Again," Between Scientists & Citizens, March 22, 2010, https://scientistscitizens.wordpress.com/2010/03/22/morano-analysis-8/.

17. Marc Morano, "Special Report: More Than 1000 International Scientists Dissent over Man-Made Global Warming Claims—Challenge UN IPCC & Gore," Climate Depot, December 8, 2010, http://www.climatedepot.com/2010/12/08/special-report-more-than-1000-international-scientists-dissent-over-manmade-global-warming-claims-challenge-un-ipcc-gore-2/.

18. Suzanne Goldenberg, "IPCC Report: Climate Change Felt 'on All Continents and across the Oceans,'" *Guardian*, March 28, 2014, https://www.theguardian.com/environment/2014/mar/28/ipcc-report-climate-change-report-human-natural-systems?CMP=twt_gu.

19. Ibid.

20. Robert Stavins, "Is the IPCC Government Approval Process Broken?" An Economic View of the Environment, April 25, 2014, http://www.robertstavinsblog.org/2014/04/25/is-the-ipcc-government-approval-process-broken-2/.

21. David Rose, "Top Climate Expert's Sensational Claim of Government Meddling in Crucial UN Report," *Daily Mail*, April 26, 2014, http://www.dailymail.co.uk/news/article-2614097/Top-climate-experts-sensational-claim-government-meddling-crucial-UN-report.html.

22. Mike Hulme and Martin Mahony, "Climate Change: What Do We Know about the IPCC?" *Progress in Physical Geography*, April 12, 2010, http://mikehulme.org/wp-content/uploads/2010/01/Hulme-Mahony-PiPG.pdf.

23. Paul Chesser, "Chesser/Schlesinger Debate, Part II," Global Warming, February 16, 2009, http://www.globalwarming.org/2009/02/16/christyschlesinger-debate-part-ii/.

24. Interview of Richard Tol, Morano and Curran, *Climate Hustle*, 2016.

25. Morano, "Special Report: More Than 1000 International Scientists."

26. Bill Murray, "AMS Survey of Weathercasters on Climate Change," AlabamaWx, November 15, 2009, http://www.alabamawx.com/?p=24574; Marco Morano, "American Meteorological Society Members Reject Man-Made Climate Claims: 75% Do Not Agree with UN IPCC Claims—29% Agree 'Global Warming Is a Scam,'" Climate Depot, November 16, 2009, http://www.climatedepot.com/2009/11/16/american-meteorological-society-members-reject-manmade-climate-claims-75-do-not-agree-with-un-ipcc-claims-29-agree-global-warming-is-a-scam/.

27. Interview of Tom Harris, Morano and Curran, *Climate Hustle*.

28. Morano, "Special Report: More Than 1000 International Scientists."

29. "Editorial: Tax Dollars Perpetuate Global-Warming Fiction," *Washington Times*, May 25, 2010, http://www.washingtontimes.com/news/2010/may/25/tax-dollars-perpetuate-global-warming-fiction/.

30. Marc Morano, "NAS Urges Carbon Tax, Becomes Advocacy Group," May 20, 2010, Climate Deport, http://www.climatedepot.com/2010/05/27/cicerones-shame-nas-urges-carbon-tax-becomes-advocacy-group-political-appointees-heading-politicized-scientific-institutions-that-are-virtually-100-dependent-on-govt-funding-2/.

31. "DEPS FAQ: What Do All These Acronyms Mean?" The National Academy of Sciences Engineering Medicine, 2017, http://sites.nationalacademies.org/DEPS/DEPS_037300.

32. Richard S. Lindzen, "Global Warming: How to Approach the Science," Testimony before the House Subcommittee on Science and Technology, November 17, 2010, https://wattsupwiththat.files.wordpress.com/2010/11/lindzen_testimony_11-17-2010.pdf.

33. Marc Morano, "MIT Climate Scientist Dr. Richard Lindzen: Believing CO$_2$ Controls the Climate 'Is Pretty Close to Believing in Magic,'" May 1, 2017, http://www.climatedepot.com/2017/05/01/mit-climate-scientist-dr-richard-lindzen-believing-co2-controls-the-climate-is-pretty-close-to-believing-in-magic/.

34. William Happer, "Quote of the Week," CO$_2$ Coalition, June 3, 2017, http://co2coalition.org/2017/06/03/the-week-that-was-quote-of-the-week/.

35. John Cook, et al., "Quantifying the Consensus on Anthropogenic Global Warming in the Scientific Literature," *Environmental Research Letters* 8:2 (May 2013), http://iopscience.iop.org/article/10.1088/1748-9326/8/2/024024/meta;jsessionid=DDD4F9D426CA8AD1FD6336942A27202F.c3.iopscience.cld.iop.org.

36. Richard S. J. Tol, "Quantifying the Consensus on Anthropogenic Global Warming in the Literature: A Re-Analysis," *Energy Policy* 73 (October 2014), 709.

37. Anthony Watts, "Busted: Tol Takes on Cook's '97% Consensus' Claim with a Re-Analysis, Showing the Claim Is 'Unfounded,'" Watts Up with That, June 4, 2014, https://wattsupwiththat.com/2014/06/04/tol-takes-on-cooks-97-consensus-claim-with-a-re-analysis-showing-the-claim-is-unfounded/.

38. Andrew Montford, "Consensus? What Consensus?" Global Warming Policy Foundation, http://www.thegwpf.org/content/uploads/2013/09/Montford-Consensus.pdf.

39. Joseph Bast and Roy Spencer, "The Myth of the Climate Change '97%,'" *Wall Street Journal*, May 26, 2014, https://www.wsj.com/articles/joseph-bast-and-roy-spencer-the-myth-of-the-climate-change-97-1401145980?mod=rss_opinion_main&tesla=y.

40. Neela Banerjee, "The Most Hated Climate Scientist in the US Fights Back," *Yale Alumni Magazine*, March/April 2013, https://yalealumnimagazine.com/articles/3648; Marc Morano, "Yale Alumni Magazine Profiles Michael Mann," Climate Depot, March 6, 2013, http://www.climatedepot.com/2013/03/06/yale-alumni-magazine-profiles-michael-mann-features-climate-depots-morano-saying-mann-is-the-embodiment-of-everything-that-is-wrong-with-climate-science-today-he-is-a-hardcore-political-activist/.

41. Christopher Monckton, "Honey, I Shrunk the Consensus," September 24, 2013, http://joannenova.com.au/2013/09/monckton-honey-i-shrunk-the-consensus/.

42. Interview of Christopher Monckton, Morano and Curran, *Climate Hustle*.

43. Valerie Richardson, "Obama's 97 Percent Climate Change Consensus Includes 'Deniers,'" *Washington Times*, April 25, 2016, http://www.washingtontimes.com/news/2016/apr/25/obamas-97-percent-climate-change-consensus-include/.

44. Marc Morano, "'It's All Wrong': UN IPCC Lead Author Dr. Richard Tol Slams media for False Claims about Alleged 97% consensus," Climate Depot, September 3, 2015, http://www.climatedepot.com/2015/09/03/its-all-wrong-un-convening-lead-author-dr-richard-tol-slams-media-for-false-claims-about-alleged-97-consensus/.

45. Lawrence Solomon, "Lawrence Solomon: 75 Climate Scientists Think Humans Contribute to Global Warming," *National Post*, December 30, 2010, http://nationalpost.com/full-comment/lawrence-solomon-75-climate-scientists-think-humans-contribute-to-global-warming/.

46. Jeff Dunetz, "Pew Research: Americans Don't Believe The '97% Consensus of Climate Scientists' Claim," The Lid, Dec 8, 2016, http://lidblog.com/bogus-97-study/.

47. Interview of Judith Curry, Morano and Curran, *Climate Hustle*.

48. Interview of Caleb Rossiter, Morano and Curran, *Climate Hustle*.

Chapter 4:
The Tail Does Not Wag the Dog

1. "A Jacket for the Planet," NASA website, November 20, 2015, https://spaceplace.nasa.gov/atmosphere/en/.

2. Tom Brokaw, "Global Warming: What You Need to Know," Discovery Channel, July 16, 2006, https://www.youtube.com/watch?v=rJBOT5A9syQ.

3. Michael Mann, "Climate Danger Threshold Approaching," RT TV, March 25, 2014, https://www.youtube.com/watch?v=daIYW2Tk6uw.

4. Marc Morano and Mick Curran, *Climate Hustle*, 2016, www.ClimateHustle.com.

5. NASA Earth Observatory, Svante Arrhenius (1859-1927), https://earthobservatory.nasa.gov/Features/Arrhenius/arrhenius_3.php.

6. Marelene Cimons, "A Graphical Look at Presidents' Environmental Records," Think Progress, February 15, 2016, https://thinkprogress.org/a-graphical-look-at-presidents-environmental-records-f232f07005d0/.

7. "How the Burning of Coal Vitiates the Atmosphere," *North Otago Times*, April 16, 1910, https://paperspast.natlib.govt.nz/newspapers/NOT19100416.2.32.24.

8. Herbert Quick, "The Coming Exodus to Siberia and Canada," *Pueblo Leader*, January 28, 1913, https://news.google.com/newspapers?id=0jlcAAAAIBAJ&sjid=-1UNAAAAIBAJ&pg=2049,1086624&dq=greenland+farming&hl=en.

9. "Electricity and the Growth of Children," *Rodney and Otamatea Times*, October 25, 1911, https://paperspast.natlib.govt.nz/newspapers/ROTWKG19111025.2.56.

10. John Coleman, "The Amazing Story behind the Global Warming Scam," KUSI News, February 11, 2009, http://www.kusi.com/story/13167480/the-amazing-story-behind-the-global-warming-scam.

11. Ibid.

12. Al Gore, @algore, Twitter, May 10, 2013, https://twitter.com/algore/status/332865247897600000.

13. Justin Gillis, "Heat-Trapping Gas Passes Milestone, Raising Fears," *New York Times*, May 10, 2013, http://www.nytimes.com/2013/05/11/science/earth/carbon-dioxide-level-passes-long-feared-milestone.html?smid=tw-share&pagewanted=all.

14. Nicholas Thompson, "Terrible News about Carbon and Climate Change," *New Yorker*, May 12, 2013, http://www.newyorker.com/online/blogs/elements/2013/05/terrible-news-about-carbon-and-climate-change.html.

15. Editorial, "Climate Change: Swift Political Action Can Avert a Carbon Dioxide Crisis," *Guardian*, May 11, 2013, https://www.theguardian.com/commentisfree/2013/may/12/observer-editorial-carbon-dioxide-levels-high?CMP=twt_fd.

16. Paul Mulshine, "Climatologists are no Einsteins, Says His Successor," *Star Ledger*, April 3, 2013, http://blog.nj.com/njv_paul_mulshine/2013/04/climatologists_are_no_einstein.html.

17. "Famed Physicist Freeman Dyson Predicts the Future," IEEE Spectrum Interview, June 30, 2014, http://spectrum.ieee.org/video/aerospace/space-flight/famed-physicist-freeman-dyson-predicts-the-future; http://www.climatedepot.com/2015/02/19/renowned-princeton-physicist-freeman-dyson-i-like-carbon-dioxide-warns-climate-excerpts-have-set-themselves-up-as-being-the-guardians-of-the-truth-that-is-a-dangerous-situation/.

18. Will Happer, "Update on the Latest Global Warming Science," U.S. Senate Environment & Public Works Committee testimony, February 25, 2009, http://www.climatedepot.com/2017/01/14/

flashback-2009-prominent-scientist-tells-congress-earth-in-co2-famine/.

19. "Geologists Reconstruct the Earth's Climate Belts between 460 and 445 Million Years Ago," *Proceedings of the National Academy of Sciences of the USA*, August 9, 2010, https://www.eurekalert.org/pub_releases/2010-08/uol-aae080910.php.

20. "Stomatal Proxy Record of CO_2 Concentrations from the Last Termination Suggests an Important Role for CO_2 at Climate Change Transitions," *Quaternary Science Reviews*, May 15, 2013, http://hockeyschtick.blogspot.com/2013/03/new-paper-finds-co2-levels-were-higher.html.

21. Thom Hartmann, *Carbon*, August 20, 2014, http://www.climatedepot.com/2014/08/21/leonardo-dicaprio-cimate-change-video-declares-carbon-dioxide-to-be-poissen/.

22. Happer, "Update."

23. Harrison H. Schmitt and William Happer, "In Defense of Carbon Dioxide," *Wall Street Journal*, May 8, 2013, http://online.wsj.com/article/SB10001424127887323352840457845248365606 7190.html?mod=rss_opinion_main.

24. Richard Lindzen, "Thoughts on the Public Discourse over Climate Change," Merion West, April 25, 2017, http://merionwest.com/2017/04/25/richard-lindzen-thoughts-on-the-public-discourse-over-climate-change/.

25. Schmitt and Happer, "In Defense of Carbon Dioxide."

26. "Geologists Reconstruct the Earth's Climate Belts."

27. Reid Bryson, "The Faithful Heretic," *Wisconsin Energy Cooperative News*, May 2007, http://wecnmagazine.com/2007issues/may/may07.html.

28. Marc Morano "More Than 1000 International Scientists Dissent Over Man-Made Global Warming Claims," Climate Depot, December 8, 2010, http://www.cfact.org/pdf/2010_Senate_Minority_Report.pdf.

29. UK Professor Emeritus of Biogeography Philip Stott of the University of London, "It's All Unravelling," Global Warming Politics," May 2, 2008, http://web.mac.com/sinfonia1/Global_Warming_Politics/A_Hot_Topic_Blog/Entries/2008/5/2_It%E2%80%99s_All_Unravelling.html.

30. Lindzen, "Thoughts on the Public Discourse."

31. Morano "More Than 1000 International Scientists Dissent."

32. Nir Shaviv, "The IPCC AR5—First impressions," Science Bits, October 2, 2013, http://sciencebits.com/AR5-FirstImpressions.

33. Interview of Patrick Moore in Morano and Curran, *Climate Hustle*.

34. Henrik Svensmark, "Evidence of Nearby Supernovae Affecting Life on Earth," Royal Astronomical Society in London, April 24, 2012, https://calderup.wordpress.com/2012/04/24/a-stellar-revision-of-the-story-of-life/.

35. Patrick Moore, "Natural Resource Adaptation: Protecting Ecosystems and Economies," testimony before the U.S. Senate Environment & Public Works Committee, February 25, 2014, https://www.epw.senate.gov/public/_cache/files/415b9cde-e664-4628-8fb5-ae3951197d03/22514hearingwitnesstestimonymoore.pdf.

36. Spencer R. Weart, "Simple Question, Simple Answer…Not," Real Climate, September 8, 2008, http://www.realclimate.org/index.php/archives/2008/09/simple-question-simple-answer-no/#more-595.

37. Morano "More Than 1000 International Scientists Dissent."

38. Lindzen, "Thoughts on the Public Discourse."

39. Morano, "More Than 1000 International Scientists Dissent."

40. Lubos Motl, "Why We Should Work Hard to Raise the CO_2 Concentration," Reference Frame, May 10, 2013, https://motls.blogspot.com/2013/05/why-we-should-work-hard-to-raise-co2.html.

41. Robert M. Carter interviewed in Morano and Curran, *Climate Hustle*, 2016, www.ClimateHustle.com.

42. Moore, "Natural Resource Adaptation."

43. Interview of Leighton Steward in Morano and Curran, *Climate Hustle*.

44. Anders Bolling, "We Create a Lot of Anxiety without Being Justified," *Dagens Nyheter*, February 3, 2013, http://translate.google.com/translate?sl=auto&tl=en&u=http%3A//www.dn.se/nyheter/vetenskap/vi-skapar-en-valdig-angslan-utan-att-det-ar-befogat.

45. Bolling, "We Create a Lot of Anxiety."

46. Marc Morano, "Climate Scientist Dr. Chris de Freitas: 'Current Warm Phase...Is Not Unprecedented,'" Climate Depot, http://www.climatedepot.com/2013/05/28/climate-scientist-dr-chris-de-freitas-current-warm-phase-is-not-unprecedented-from-the-results-of-research-to-date-it-appears-the-influence-of-increasing-co2-on-global-warming-is-almost/.

47. Tom Wysmuller, "The Recent Temperature and CO_2 Disconnect," The Colder Side, 2012, http://www.colderside.com/Colderside/Temp_%26_CO2.html.

48. Ian Plimer, "ETS Forum—It's All in the Rocks," Quadrant Online, August 8, 2009, http://quadrant.org.au/opinion/doomed-planet/2009/08/ian-plimer/.

49. Andrew Bolt, "Sceptics Create a Best-Seller," *Herald Sun*, August 5, 2009, http://www.heraldsun.com.au/blogs/andrew-bolt/sceptics-create-a-bestseller/news-story/eb84217e45efdbd14ec305917d3ec2e2.

50. John Marchese, "SCIENCE: Al Gore Is a Greenhouse Gasbag," *Philadelphia*, February 2, 2007, http://www.phillymag.com/articles/science-al-gore-is-a-greenhouse-gasbag/.

51. Interview of Robert Giegengack in Morano and Curran, *Climate Hustle*.

52. Robert Giegengack, "An Inaccurate Truth?" *Pennsylvania Gazette*, May/June 2007, http://www.upenn.edu/gazette/0507/gaz01.html.

53. Interview of Robert Giegengack in Morano and Curran, *Climate Hustle*.

54. Interview of Patrick Moore in Morano and Curran, *Climate Hustle*.

55. Morano "More Than 1000 International Scientists Dissent.

56. Laurie David and Cambria Gordon, *The Down-to-Earth Guide to Global Warming* (Orchard Books, 2007).

57. "A Fundamental Scientific Error in 'Global Warming' Book for Children," Science and Public Policy Institute, September 2007, http://scienceandpublicpolicy.org/images/stories/papers/other/david_book.pdf.

58. Laurie David, "The Children's Book That Has Global Warming Deniers Up in Arms," Huffington Post, September 19, 2007, http://www.huffingtonpost.com/laurie-david/the-childrens-book-that-h_b_64998.html.

59. Roy Spencer, *An Inconvenient Deception: How Al Gore Distorts Climate Science and Energy Policy* (Amazon Digital Services, 2017).

60. Ibid.

61. Anthony Watts, "Video Analysis and Scene Replication Suggests That Al Gore's Climate Reality Project Fabricated Their Climate 101 Video 'Simple Experiment,'" Watts Up with That, September 28, 2011, https://wattsupwiththat.com/2011/09/28/video-analysis-and-scene-replication-suggests-that-al-gores-climate-reality-project-fabricated-their-climate-101-video-simple-experiment/.

62. Interview of Robert Giegengack, Morano and Curran, *Climate Hustle*.

Chapter 5:
The Ice Caps Are Melting!

1. Geoffrey Lean, "Why Antarctica Will Soon Be the Only Place to Live—Literally," *Independent*,

May 1, 2004, http://www.independent.co.uk/environment/why-antarctica-will-soon-be-the-only-place-to-live-literally-58574.html.

2. "Antarctic Sea Ice Reaches New Record Maximum," NASA, October 7, 2014, https://www.nasa.gov/content/goddard/antarctic-sea-ice-reaches-new-record-maximumfds.

3. "Inconvenient Truth: Antarctica Sea Ice Extent Growing 1.43% per Year," The Hockey Schtick, September 9, 2011, http://hockeyschtick.blogspot.com/2011/09/inconvenient-truth-antarctica-sea-ice.html; "Antarctic Sea Ice Reaches New Record."

4. John Turner, et al., "An Initial Assessment of Antarctic Sea Ice Extent in the CMIP5 Models," American Meteorological Society, February 27, 2013, http://journals.ametsoc.org/doi/abs/10.1175/JCLI-D-12-00068.1.

5. Marc Morano, "Clitantic Scientists Trapped in Antarctic Ice Claim Expanding Sea Ice Caused by 'Global Warming'—but Data and Studies Refute Claims," Climate Depot, January 1, 2014, http://www.climatedepot.com/2014/01/01/clitantic-ship-scientists-trapped-in-ice-claim-expanding-sea-ice-caused-by-global-warming-but-data-and-studies-refute-claims/.

6. Maria-José Viñas, "NASA Study: Mass Gains of Antarctic Ice Sheet Greater than Losses," NASA, October 30, 2015, https://www.nasa.gov/feature/goddard/nasa-study-mass-gains-of-antarctic-ice-sheet-greater-than-losses/.

7. "Antarctic Temperatures Disagree with Climate Model Predictions," Ohio State University, February 15, 2007, https://www.eurekalert.org/pub_releases/2007-02/osu-atd021207.php.

8. "New Paper Finds the Majority of East Antarctic Glaciers Have Advanced in Size since 1990," The Hockey Schtick, August 2013, http://hockeyschtick.blogspot.com/2013/08/new-paper-finds-majority-of-east.html.

9. P. Gosselin, "Alfred Wegener Institute Deputy Director, Veteran Polar Scientist: 'We Have a Light Cooling Trend,'" No Tricks Zone, December 4, 2011, http://notrickszone.com/2011/12/04/antarctic-neumayer-iii-deputy-director-veteran-polar-scientist-we-have-a-light-cooling-trend/.

10. N. A. N. Bertler, et al., "Cold Conditions in Antarctica during the Little Ice Age—Implications for Abrupt Climate Change Mechanisms," Earth and Planetary Science Letters 308 (2011): 41–51, http://www.co2science.org/data/mwp/studies/l2_victoriaglacier.php.

11. Bruce Leshan, "Sea Level Rise Could Lap at DC Memorials," WUSA, May 14, 2014, http://www.wusa9.com/news/local/dc/sea-level-rise-could-lap-at-dc-memorials/285179702.

12. Marc Morano, "Update: AP's Seth Borenstein at It Again Hyping Antarctic Melt Fears—Recycles Same Claims from 2014, 1990, 1979, 1922 & 1901!—Climate Depot's Point-By-Point Rebuttal," Climate Depot, February 28, 2015, http://www.climatedepot.com/2015/02/28/aps-seth-borenstein-at-it-again-hyping-antarctic-melt-fears-climate-depots-point-by-point-rebuttal/.

13. Ibid.

14. Ibid.

15. Justin Haskins, "New Report about Antarctica Is Horrible News for Global Warming Alarmists," The Blaze, April 30, 2017, http://www.theblaze.com/news/2017/04/30/new-report-about-antarctica-is-horrible-news-for-global-warming-alarmists/.

16. "Media Hype on 'Melting' Antarctic Ignores Record Ice Growth," U.S. Senate Environment and Public Works Committee, March 27, 2008, https://www.epw.senate.gov/public/index.cfm/press-releases-all?ID=F1F2F75F-802A-23AD-4701-A92B4EBBCCBF.

17. "Variable Crustal Thickness beneath Thwaites Glacier Revealed from Airborne Gravimetry, Possible Implications for Geothermal Heat Flux in West Antarctica," Earth and Planetary Science Letters 407 (December 1, 2014), http://www.sciencedirect.com/science/article/pii/

S0012821X1400578; "New Paper Finds West Antarctic Glacier Likely Melting from Geothermal Heat Below," The Hockey Schtick, October 11, 2014, http://hockeyschtick.blogspot.com/2014/10/new-paper-finds-west-antarctic-glacier.html.

18. "Scientists Find 91 Volcanoes under Antarctic Ice," Sky News, August 13, 2017, http://www.msn.com/en-gb/news/world/scientists-find-91-volcanoes-under-antarctic-ice/ar-AApYDuf?li=AAmiR2Z&ocid=spartandhp.

19. Marc Morano, "Flashback 1977 Shocker: Scientists Knew Man Was Not Responsible! West Antarctic Ice Sheet Collapse 'Has Nothing to Do with Climate, Just the Dynamics of Unstable Ice,'" Antarctic Ice Collapse Perils World's Coastal Cities," Independent Press-Telegram, January 23, 1977, http://www.climatedepot.com/2015/11/03/flashback-1977-shocker-scientists-knew-man-was-not-responsible-west-antarctic-ice-sheet-collapse-has-nothing-to-do-with-climate-just-the-dynamics-of-unstable-ice/.

20. James Watts, "The Guardian's Suzanne Goldenberg Jumps the Shark Again—Gets Called Out by NYT," Watts Up with That? May 12, 2014, https://wattsupwiththat.com/2014/05/12/the-guardians-suzanne-goldenberg-jumps-the-shark-again-gets-called-out-by-nyt/.

21. Sean Green, "After Antarctica Sheds a Trillion-Ton Block of Ice, the World Asks: Now What?" *Los Angeles Times*, July 13, 2017, http://www.latimes.com/science/sciencenow/la-sci-sn-larson-ice-sheet-20170713-htmlstory.html.

22. Interview of Don Easterbrook, Marc Morano and Curran, *Climate Hustle*, 2016, www.ClimateHustle.com.

23. "Recent Regional Climate Cooling on the Antarctic Peninsula and Associated Impacts on the Cryosphere," *Science of the Total Environment*, February 15, 2017, http://www.sciencedirect.com/science/article/pii/S0048969716327152.

24. "Antarctic Ice More Stable than Thought: Antarctic Ice Sheet Has Been Stable for Millions Of Years: Study," Universities of Edinburgh and Northumbria, May 8, 2017, http://mailchi.mp/thegwpf/antarctic-ice-more-stable-than-thought?e=f4e33fdd1e.

25. Anthony Watts, "You Ask, I Provide. November 2nd, 1922. Arctic Ocean Getting Warm; Seals Vanish and Icebergs Melt," Watts Up with That?, March 16, 2008, https://wattsupwiththat.com/2008/03/16/you-ask-i-provide-november-2nd-1922-arctic-ocean-getting-warm-seals-vanish-and-icebergs-melt/.

26. Marc Morano, "Don't Panic! Arctic Ice Hits 'Record' Low!? Climate Depot Explains Arctic melting hype," Climate Depot, August 27, 2012, http://www.climatedepot.com/2012/08/27/dont-panic-arctic-ice-hits-record-low-climate-depot-explains-arctic-melting-hype/.

27. Marc Morano, "120 Years of Climate Scares—70s Ice Age Scare," Climate Depot, May 23, 2017, http://www.climatedepot.com/2017/05/23/120-years-of-climate-scares-70s-ice-age-scare/.

28. Marc Morano, "NASA Finally Admits It Arctic Cyclone in August 'Broke Up' and 'Wreaked Havoc' on Sea Ice—Reuters Reports Arctic Storm Played 'Key Role' in Ice Reduction," Climate Depot, September 24, 2012, http://www.climatedepot.com/2012/09/24/nasa-finally-admits-it-arctic-cyclone-in-august-broke-up-and-wreaked-havoc-on-sea-ice-reuters-reports-arctic-storm-played-key-role-in-ice-reduction/.

29. Morano, "Don't Panic!"

30. Paul Watson, "Arctic's 'Canary in a Coal Mine,'" *Star*, July 28, 2009, https://www.thestar.com/news/canada/2009/07/28/arctics_canary_in_a_coal_mine.html.

31. Paul Homewood, "Arctic Ice Growing Rapidly," Not a Lot of People Know That, September 15, 2016, https://notalotofpeopleknowthat.wordpress.com/2016/09/15/arctic-ice-growing-rapidly/.

32. David Whitehouse, "A Ten-Year Hiatus in Arctic Ice Decline?" Global Warming Policy Foundation, September 22, 2016, https://www.thegwpf.com/a-ten-year-hiatus-in-arctic-ice-decline/.

33. Maria-José Viñas, "End-of-Summer Arctic Sea Ice Extent Is Eighth Lowest on Record," Phys. Org, September 19, 2017, https://phys.org/news/2017-09-end-of-summer-arctic-sea-ice-extent.html.

34. Kenneth Richard, "Since 2005, Arctic Sea Ice Has Pivoted to A Slightly Increasing Trend, with No Detectable Arctic Warming," No Tricks Zone, June 26, 2017, http://notrickszone.com/2017/06/26/since-2005-arctic-sea-ice-has-pivoted-to-a-slightly-increasing-trend-with-no-detectable-arctic-warming/.

35. Ron Clutz, "Overachieving September Arctic Ice," Science Matters, October 1, 2017, https://rclutz.wordpress.com/2017/10/01/overachieving-september-arctic-ice/.

36. John Holdren, "Climate Wars," CBC, 2009, http://tomnelson.blogspot.com/2009/02/complete-barking-madness-from-john.html.

37. Interview of John Holdren, Morano and Curran, *Climate Hustle*, 2016.

38. "Arctic Warming Reverse! New Study Finds Winter Arctic Sea Ice 'to Increase Towards 2020,'" June 23, 2017, No Tricks Zone, http://notrickszone.com/2017/06/23/arctic-warming-reverse-new-study-says-winter-arctic-sea-ice-to-increase-towards-2020/.

39. Interview of Tony Heller, Morano and Curran, *Climate Hustle*.

40. "Greenland's Climate Becoming Milder," *Brisbane Courier-Mail*, May 6, 1940, https://stevengoddard.wordpress.com/2013/07/06/1940-shock-news-greenland-glaciers-melting-at-catastrophic-rates-2/.

41. "Scientific American 1904: Alpine and Greenland Glaciers Retreating 'Very Considerably,'" Real Climate Science, https://realclimatescience.com/2016/08/scientific-american-1904-alpine-and-greenland-glaciers-retreating-very-considerably/.

42. Interview of Don Easterbrook, Morano and Curran, *Climate Hustle*.

43. B. M. Vinther, et al., "Extending Greenland Temperature Records into the Late Eighteenth Century," *Journal of Geophysical Research* vol. 111, D11105, June 6, 2006, https://crudata.uea.ac.uk/cru/data/greenland/vintheretal2006.pdf.

44. "Latest Scientific Studies Refute Fears of Greenland Melt," Senate Environment and Public Works Committee, July 30, 2007, https://www.epw.senate.gov/public/index.cfm/press-releases-all?ID=175b568a-802a-23ad-4c69-9bdd978fb3cd.

45. "Greenland Warming of 1920–1930 and 1995–2005," *Geophysical Research Letters* 33, L11707, June 13, 2006, http://meteo.lcd.lu/globalwarming/Chylek/greenland_warming.html.

46. "Study finds ice isn't being lost from Greenland's interior," UIC Today, May 4, 2016, http://today.uic.edu/study-finds-ice-isnt-being-lost-from-greenlands-interior.

47. Patrick J. Michaels and Paul C. "Chip" Knappengerger, "Taming the Greenland Melting Global Warming Hype," Cato, June 10, 2016, https://www.cato.org/blog/taming-greenland-melting-global-warming-hype.

48. "Latest Scientific Studies Refute Fears of Greenland Melt," Senate Environment and Public Works Committee, July 30, 2007, https://www.epw.senate.gov/public/index.cfm/press-releases-all?ID=175b568a-802a-23ad-4c69-9bdd978fb3cd.

49. Leshan, "Sea Level Rise."

50. Interview of Michael Oppenheimer, Morano and Curran, *Climate Hustle*.

51. Interview of Bob Carter, Morano and Curran, *Climate Hustle*.

52. John Marchese, "Science: Al Gore Is a Greenhouse Gasbag," *Philadelphia*, February 2, 2007, http://www.phillymag.com/articles/science-al-gore-is-a-greenhouse-gasbag/?all=1.

53. Roy W. Spencer, "Hillary Clinton Boards the Climate Crisis Train to Nowhere," *Forbes*, October 25, 2016, https://www.forbes.com/sites/realspin/2016/10/25/hillary-clinton-boards-the-climate-crisis-train-to-nowhere/#12d38548371e.

54. "New Paper Finds Global Sea Level Rise Has Decelerated 31% since 2002 along with the 'Pause' of Global Warming," The Hockey Schtick, March 23, 2014, http://hockeyschtick.blogspot.com/2014/03/new-paper-finds-global-sea-level-rise.html.

55. Ibid.

56. Interview of Don Easterbrook.

57. Interview of Patrick Moore, Morano and Curran, *Climate Hustle*.

58. Interview of Nils Axel Morner, *Climate Hustle*.

59. Kate Sheppard, "Former Top NASA Scientist Predicts Catastrophic Rise in Sea Levels," HuffPost, July 21, 2015, https://www.huffingtonpost.com/entry/james-hansen-sea-level-rise_us_55aecb02e4b0a9b94852e7f5.

60. Ryan Maue, @RyanMaue, Twitter, July 23, 2015, https://twitter.com/RyanMaue/status/624412278520217600.

61. Michael E Mann, Facebook, July 23, 2015, https://www.facebook.com/MichaelMannScientist/posts/932041946852008?comment_id=932375396818663&reply_comment_id=932380823484787&total_comments=4&comment_tracking=%7B"tn"%3A"R9"%7D; Marc Morano, "NASA's James Hansen Gets Dissed by Global Warming Establishment! Warmists Say Sea Level Rise Study Based on 'Flimsy Evidence' & 'Rife with Speculation,'" Climate Depot, July 24, 2015, http://www.climatedepot.com/2015/07/24/nasas-ex-con-james-hansen-gets-dissed-by-global-warming-establishment-warmists-say-sea-level-rise-study-based-on-flimsy-evidence-rife-with-speculation/.

62. "Understanding Sea Level Observations," NASA, 2017, https://sealevel.nasa.gov/understanding-sea-level/observations/sea-level.

63. Klaus-Eckart Puls, "Global Sea Level Rise from Tide Gauges Is Half of That Claimed from Satellites. Which Is Right?" The Hockey Schtick, July 14, 2014, http://hockeyschtick.blogspot.com/2014/07/global-sea-level-rise-from-tide-gauges.html.

64. Michael Oppenheimer, "Years of Living Dangerously," *True Colors*, 2017, http://yearsoflivingdangerously.com/story/christie-and-the-storm/; "Years of Living Dangerously, Season 1 Episode 5: True Colors," January 20, 2017, https://www.youtube.com/watch?v=pd2f1rq1I_E.

65. Interview of Michael Oppenheimer.

66. Bjørn Lomborg, "About Those Non-Disappearing Pacific Islands," *Wall Street Journal*, October 13, 2016, https://www.wsj.com/articles/about-those-non-disappearing-pacific-islands-1476400840?mod=djemMER.

67. Nils-Axel Mörner, "Sea Level Manipulation," *International Journal of Engineering Science Invention*, August 2017, http://www.ijesi.org/papers/Vol(6)8/Version-1/G0608014851.pdf.

68. Puls, "Global Sea Level Rise."

69. Marchese, "Science: Al Gore."

70. Marc Morano, "Sea Level Expert Rips Study Claiming Fastest Rise in 2800 Years: Study 'Full of Very Bad Violations of Observational Facts,'" Climate Depot, February 23, 2016, http://www.climatedepot.com/2016/02/23/sea-level-expert-rips-study-claiming-fastest-rise-in-2800-years-study-full-of-very-bad-violations-of-observational-facts/.

71. Marc Morano, "Special Report: More than 1000 International Scientists Dissent over Man-Made Global Warming Claims—Challenge UN IPCC &

Gore," Climate Depot, December 8, 2010, http://www.climatedepot.com/2010/12/08/special-report-more-than-1000-international-scientists-dissent-over-manmade-global-warming-claims-challenge-un-ipcc-gore-2/.

72. Morano, "Sea Level Expert Rips Study."

73. Philippa McDonald, "Pacific islands growing, not sinking," ABC News (Australia), June 2, 2010, http://www.abc.net.au/news/2010-06-03/pacific-islands-growing-not-sinking/851738.

74. Interview of Bob Carter.

75. Susan Crockford, "Polar Bears Have Not Been Harmed by Sea Ice Declines in Summer—the Evidence," Polar Bear Science, August 18, 2013, https://polarbearscience.com/2013/08/18/polar-bears-have-not-been-harmed-by-sea-ice-declines-in-summer-the-evidence/.

76. "Impact: New York Times Features EPW Polar Bear Report," U.S. Senate Environment and Public Works Committee, February 1, 2008, https://www.epw.senate.gov/public/index.cfm/press-releases-all?ID=D6C356AF-802A-23AD-407D-FF2680A65F1B.

77. Valerie Richardson, "Polar Bear Zoologist Blasts Obama's Climate Alarmism: 'Sensationalized Nonsense,'" Washington Times, January 9, 2017, http://www.washingtontimes.com/news/2017/jan/9/polar-bear-conservation-group-blasts-obama-admin/.

78. Susan Crockford, "Ten Dire Polar Bear Predictions That Have Failed As Global Population Hits 22–31k," Polar Bear Science, February 25, 2016, https://polarbearscience.com/2016/02/25/ten-dire-polar-bear-predictions-that-have-failed-as-global-population-hits-20-31k/.

79. Interview of Judith Curry, Morano and Curran, Climate Hustle.

80. Interview of Don Easterbrook.

81. Susan Crockford, "Survey Results: Svalbard Polar Bear Numbers Increased 42% over Last 11 Years," Polar Bear Science, December 23, 2015, https://polarbearscience.com/2015/12/23/survey-results-svalbard-polar-bear-numbers-increased-42-over-last-11-years/.

82. "U.S. Senate Report Debunks Polar Bear Extinction Fears," U.S. Senate Environment and Public Works Committee, January 30, 2008, https://www.epw.senate.gov/public/index.cfm/press-releases-all?ID=CB2FAA9C-802A-23AD-4BCC-29BB94CEB993.

83. Jonathan Amos, "Ancient Polar Bear Jawbone Found," BBC News, December 10, 2007, http://news.bbc.co.uk/2/hi/science/nature/7132220.stm.

84. "U.S. Senate Report Debunks Polar Bear Extinction Fears."

85. Ibid.

86. Ibid.

87. Christopher Booker, "Polar Bear Expert Barred by Global Warmists," Telegraph, June 27, 2009, http://www.telegraph.co.uk/comment/columnists/christopherbooker/5664069/Polar-bear-expert-barred-by-global-warmists.html.

88. "Polar Bears Not at Risk: Nunavut," CBC News, May 28, 2010, http://www.cbc.ca/news/canada/north/polar-bears-not-at-risk-nunavut-1.914601.

89. Kim Murphy, "Canada's Inuit Roar in Protest over Move to Protect Polar Bears," Los Angeles Times, July 08, 2012, http://articles.latimes.com/2012/jul/08/nation/la-na-polar-bears-20120708.

90. "Churchill Guide Chronicles Adventures with Polar Bears in New Book," CBC News, October 21, 2015, http://www.cbc.ca/news/canada/manitoba/churchill-guide-chronicles-adventures-with-polar-bears-in-new-book-1.3281636.

91. "U.S. Senate Report Debunks Polar Bear Extinction Fears," U.S. Senate Environment and Public Works Committee, January 30, 2008, https://www.epw.senate.gov/public/index.cfm/press-releases-all?ID=CB2FAA9C-802A-23AD-4BCC-29BB94CEB993.

92. Ibid.

93. Susan Crockford, "Death of a Climate Icon," GWPF TV (Global Warming Policy Foundation), August 31, 2017, https://www.thegwpf.com/death-of-a-climate-icon/.

Chapter 6:
The Hottest Temperatures in a Thousand Years! Michael Mann's Hockey Schtick

1. Matt Ridley, "The Case against the Hockey Stick," *Prospect*, March 10, 2010, https://www.prospectmagazine.co.uk/magazine/the-case-against-the-hockey-stick.

2. Figure 7.1c from the 2990 IPCC First Assessment Report (FAR): IPCC Report 1990, p. 202, http://wwwFigure.ipcc.ch/ipccreports/far/wg_I/ipcc_far_wg_I_chapter_07.pdf.

3. James E. Hansen, Andrew A. Lacis, David H. Rind, and Gary L. Russel, "Climate Sensitivity to Increasing Greenhouse Gases," Research Gate, 1984, chapter two, https://www.researchgate.net/publication/237429936_Climate_Sensitivity_to_Increasing_Greenhouse_Gases. And https://web.archive.org/web/20070404001809/http://www.epa.gov/climatechange/effects/downloads/Challenge_chapter2.pdf.

4. Tony Heller, "Hansen Confirmed the MWP in 1981," Real Climate Science, January 14, 2016, https://realclimatescience.com/2016/01/hansen-confirmed-the-mwp-in-1981/.

5. Natalie Angier, "In Unexpected Places, Clues to Ancient and Future Climate; Warming? Tree Rings Say Not Yet," *New York Times*, December 1, 1992, http://www.nytimes.com/1992/12/01/news/unexpected-places-clues-ancient-future-climate-warming-tree-rings-say-not-yet.html.

6. Tom Bethell, "The False Alert of Global Warming," *American Spectator*, July 19, 2013, https://spectator.org/55208_false-alert-global-warming/.

7. William M. Connolley, "We have to get rid of the Medieval Warm Period?" Stoat, October 17, 2012, http://scienceblogs.com/stoat/2012/10/17/1943/.

8. Tom Nelson, "Warmist Overpeck Writes to Jones/Trenberth/Mann/Solomon/Santer about the 'Get Rid of the Warm Medieval Period' Email; He's Worried that Deming May Be 'Taking the Quote Out of Context,'" December 22, 2011, http://tomnelson.blogspot.com/2011/12/warmist-overpeck-writes-to.html.

9. Karl Blankenship, "98 Was Warmest Year in Hot Century," *Bay Journal*, January 1, 1999, http://www.bayjournal.com/article/98_was_warmest_year_in_hot_century._.

10. "Project Overview," Center for the Study of Carbon Dioxide and Global Change, 2017, http://co2science.org/data/mwp/description.php.

11. "SPPI Monthly CO_2 Report: May," Science and Public Policy Institute, June 8, 2009, http://scienceandpublicpolicy.org/science-papers/monthly-report/sppi-monthly-co2-report-may.

12. "20th Century Climate Not So Hot," Harvard-Smithsonian Center for Astrophysics, March 31, 2003, https://www.cfa.harvard.edu/news/archive/pr0310.html.

13. "Mid- to Late Holocene Environmental and Climatic Changes in New Caledonia, Southwest Tropical Pacific, Inferred from the Littoral Plain Gouaro-Déva," *Quaternary Research* 76: 229–42 (September 2011), http://www.sciencedirect.com/science/article/pii/S0033589411000627.

14. "Little Ice Age Cold Interval in West Antarctica: Evidence from Borehole Temperature at the West Antarctic Ice Sheet (WAIS)," *Geophysical Research Letters* 39 (2012), http://hockeyschtick.blogspot.com/2012/05/new-paper-confirms-little-ice-age-.

15. "More Evidence the Medieval Warm Period was Global," Elsevier, March 22, 2012, https://wattsupwiththat.com/2012/03/22/more-evidence-the-medieval-warm-period-was-global/.

16. Interview of Robert Giegengack, Marc Morano and Mick Curran, *Climate Hustle*, 2016, www.ClimateHustle.com.

17. Interview of Ross Mckitrick, Morano and Curran, *Climate Hustle*.

18. John R. Christy, "Examining the Process concerning Climate Change Assessments," House Science, Space and Technology Committee, March 31, 2011, http://science.house.gov/sites/republicans.science.house.gov/files/documents/hearings/ChristyJR_written_110331_all.pdf.

19. Interview of Ross Mckitrick, Morano and Curran, *Climate Hustle*.

20. "Gerald North Caught Fibbing about Hockey," Real Science, October 18, 2010, https://stevengoddard.wordpress.com/2010/10/18/gerald-north-caught-fibbing-about-hockey/.

21. P. Gosselin, "German Professor Slams Global Warming Science—Calls Mann's Hockey Stick 'A Very Very Nasty Fabrication,'" No Tricks Zone, November 13, 2011, http://notrickszone.com/2011/11/13/german-professor-slams-global-warming-science-calls-manns-hockey-stick-a-very-very-nasty-fabrication/#sthash.ntj2MKmm.dpbs.

22. Kenneth Richard, "'Hide the Decline' Unveiled: 50 Non-Hockey Stick Graphs Quash Modern 'Global' Warming Claims," No Tricks Zone, September 27, 2016, http://notrickszone.com/2016/09/27/hide-the-decline-unveiled-50-non-hockey-stick-graphs-quash-modern-global-warming-claims/#sthash.Ess7KkSO.dpbs.

23. Andrew Orlowski, "The Scandal We See and the Scandal We Don't," *Register*, November 30, 2009, http://www.theregister.co.uk/Print/2009/11/30/crugate_analysis/.

24. Interview of Ross Mckitrick, Morano and Curran, *Climate Hustle*.

25. "Guest Post by Bo Christiansen: On Temperature Reconstructions, Past Climate Variability, Hockey Sticks and Hockey Teams," Die Klimazwiebel, May 30, 2010, http://klimazwiebel.blogspot.com/2010/05/guest-post-by-bo-christiansen-on.html.

26. Matt Ridley, "The Case against the Hockey Stick," *Prospect*, March 10, 2010, https://www.prospectmagazine.co.uk/magazine/the-case-against-the-hockey-stick.

27. "There He Goes Again: Mann Claims His Hockey Stick Was Affirmed by the NAS," Hockey Schtick, March 29, 2011, http://hockeyschtick.blogspot.com/2011/03/there-he-goes-again-mann-claims-his.html.

28. Louise Gray, "'Hockey Stick' Graph Was Exaggerated," *Telegraph*, April 14, 2010, http://www.telegraph.co.uk/news/earth/environment/climatechange/7589897/Hockey-stick-graph-was-exaggerated.html.

29. "Hide the Decline," Just the Facts, 2017, http://www.justfacts.com/globalwarming.hidethedecline.asp.

30. David Rose, "Special Investigation: Climate Change Emails Row deepens As Russians Admit They DID Come from Their Siberian Server," *Daily Mail*, December 13, 2009, http://www.dailymail.co.uk/news/article-1235395/SPECIAL-INVESTIGATION-Climate-change-emails-row-deepens—Russians-admit-DID-send-them.html.

31. Tom Nelson "ClimateGate FOIA Grepper!—1999 Briffa Email," Tom Nelson (blog), December 1, 2011, http://tomnelson.blogspot.com/2011/12/climategate-foia-grepper-1999-briffa.html.

32. Tom Nelson, "2004 Email: Phil Jones on Why He Thought the Last 20 Years Was Warmer than the Medieval Warm Period: 'This Is All Gut Feeling, No Science'"; Warmist Tom Wigley Also Calls the Hockey Stick 'a Very Sloppy Piece of Work,' Tom Nelson (blog), December 26, 2011, http://tomnelson.blogspot.com/2011/12/2004-email-phil-jones-on-why-he-thought.html.

33. Ronald Bailey, "Climate Scientist Michael Mann's Defamation Lawsuit against Critics Can Proceed, Rules Court," Hit and Run, December 22, 2016, http://reason.com/blog/2016/12/22/climate-scientist-michael-manns-defamati.

34. "Variability and Extremes of Northern Scandinavian Summer Temperatures over the Past Two Millennia," *Global and Planetary Change*, January 24, 2012, http://us4.campaign-archive1.com/?u=c920274f2a364603849bbb505&id=6946bd70cb&e=f4e33fdd1e.

35. "Paper finds Medieval Warm Period in Arctic was much warmer than the present," The Hockey Schtick, August 8, 2012, http://hockeyschtick.blogspot.com/2012/08/paper-finds-medieval-warm-period-in.html.

36. 36 "New paper finds Medieval Warming Period was ~1°C warmer than current temperatures," The Hockey Schtick, July 16, 2012, http://hockeyschtick.blogspot.com/2012/07/new-paper-finds-medieval-warming-period.html.

37. "Climate in northern Europe reconstructed for the past 2,000 years," Phys.org, July 9, 2012, https://phys.org/news/2012-07-climate-northern-europe-reconstructed-years.html.

38. "Tree-Rings Prove Climate Was WARMER in Roman and Medieval Times Than It Is Now—and World Has Been Cooling for 2,000 Years," *Daily Mail*, July 11, 2012, http://www.dailymail.co.uk/sciencetech/article-2171973/Tree-ring-study-proves-climate-WARMER-Roman-Medieval-times-modern-industrial-age.html.

39. Andrew Revkin, "New Study Finds 10,000 Years of Deep Pacific Ocean Cooling Preceded Fast Recent Warming," Oct 31, 2013, https://www.youtube.com/watch?v=ta-0mo7UjUE.

40. Anthony Watts, "New Paper Shows Medieval Warm Period Was Global in Scope," Watts Up with That, October 31, 2013, https://wattsupwiththat.com/2013/10/31/new-paper-shows-medieval-warm-period-was-global-in-scope/.

Chapter 7:
A Long Cool Pause

1. Doyle Rice, "The Last Time the Earth Was This Warm Was 125,000 Years Ago," *USA Today*, January 18, 2017, https://www.usatoday.com/story/weather/2017/01/18/hottest-year-on-record/96713338/.

2. Al Gore, "2016 Was the Hottest Year on Record: Confirmed by @NASA & @NOAA," January 18, 2017, https://twitter.com/algore/status/821745205057638402.

3. Justin Gillis, "Earth Sets a Temperature Record for the Third Straight Year," *New York Times*, January 18, 2017, https://www.nytimes.com/2017/01/18/science/earth-highest-temperature-record.html.

4. Robert Tracinski, "Why NYT Hid the Numbers for the 'Hottest Year on Record,'" The Federalist, January, 2017, http://thefederalist.com/2017/01/18/nyt-hid-numbers-hottest-year-record/.

5. The 0.01 Celsius increase was reported by the British Met Office. According to the U.S. National Oceanic and Atmospheric Association, it was 0.04, still well within the margin of error. James Varney, "Scientists Criticize 'Hottest Year on Record' Claim as Hype," Real Clear Investigations, January 29, 2017, http://www.realclearinvestigations.com/articles/2017/01/29/scientists_criticize_hottest_year_on_record_claim_as_hype.html.

6. David Whitehouse, "2016 Not Statistically Warmer than 1998, Satellites Data Shows," Global Warming Policy Foundation, January 5, 2017, http://us4.campaign-archive2.com/?u=c920274f2a364603849bbb505&id=dcd5a228f8&e=f4e33fdd1e.

7. Varney, "Scientists Criticize."

8. Ibid.

9. "Despite Subtle Differences, Global Temperature Records in Close Agreement," NASA, January 13, 2011, https://www.giss.nasa.gov/research/news/20110113/.

10. Jason Samenow, "Scientists React to Warmest Year: 2014 Underscores 'Undeniable Fact' of Human-Caused Climate Change," *Washington*

Post, January 16, 2015, https://www. washingtonpost.com/news/capital-weather-gang/wp/2015/01/16/scientists-react-to-warmest-year-2014-underscores-undeniable-fact-of-human-caused-climate-change/?utm_ term=.31de5c3509f4.

11. Marc Morano, "Prominent Scientists Declare Climate Claims Ahead of UN Summit 'Irrational'—'Based On Nonsense'—'Leading Us Down a False Path,'" Climate Depot, November 19, 2015, http://www.climatedepot. com/2015/11/19/scientists-declare-un-climate-summit-goals-irrational-based-on-nonsense-leading-us-down-a-false-path/.

12. James Varney, "Scientists Criticize."

13. Marc Morano, "MIT Climate Scientist on 'Hottest Year': 'The Hysteria over This Issue Is Truly Bizarre'—Warns of Return 'Back to the Dark Ages,'" Climate Depot, January 18, 2017, http://www.climatedepot.com/2017/01/18/mit-climate-scientist-on-hottest-year-the-hysteria-over-this-issue-is-truly-bizarre-warns-of-return-back-to-the-dark-ages/.

14. "Selling climate policy on 'hottest year ever' hard bc 2016 also had v low disasters & record high crop productivity. It doesn't scare people." Roger Pielke Jr., Twitter, January 18, 2017, https:// twitter.com/RogerPielkeJr/ status/821746921769467904.

15. Tony Heller, "NOAA September Temperature Fraud," Real Climate Science, November 17, 2016, https://realclimatescience.com/2016/11/ noaa-september-temperature-fraud/.

16. Paul Homewood, "BBC Ignore the Satellite Record," January 18, 2017, https:// notalotofpeopleknowthat.wordpress. com/2017/01/18/bbc-ignore-the-satellite-record/.

17. "A Report Issued by the U.S. Space Agency NASA," *Canberra Times*, April 1, 1990, http:// www.climatedepot.com/2014/10/20/ flashabck-1990-nasa-report-satellite-analysis-of-upper-atmosphere-is-more-accurate-should-be-adopted-as-the-standard-way-to-monitor-temp-change/; Marc Morano, "Flashback 1990: AP Calls Satellite Temperatures 'More Accurate'—but in 2016: AP Ignores Satellite's Showing 18 Year Plus 'Standstill' in Global Temps," Climate Depot, January 20, 2016, http://www. climatedepot.com/2016/01/20/flashback-1990-ap-calls-satellite-temperatures-more-accurate-but-in-2016-ap-ignores-satellites-showing-18-year-plus-standstill-in-global-temps/.

18. Roger Pielke Sr., "Load of Bollocks: 2016 Allegedly 'Hottest Year' by Unmeasureable 1/100 of a Degree—While Satellites Show 'Pause' Continues," Climate Depot, January 18, 2017, http://www.climatedepot.com/2017/01/18/load-of-bollocks-2016-allegedly-hottest-year-by-immeasurable-1100-of-a-degree-while-satellites-show-pause-continues/.

19. Lubos Motl, "GISS: 1998–2016 Comparison Suggests a Trend of 2 °C per Century," The Reference Frame, January 18, 2017, https://motls. blogspot.com/2017/01/giss-1998-2016-comparison-suggests.html.

20. Marc Morano, "Adjust the Satellite 'Pause' Away! RSS Temperature Data Revised with 'Improved Adjustments'—Have Now 'Found' the Missing Warming," March 2, 2016, Climate Depot, http:// www.climatedepot.com/2016/03/02/adjust-the-satellite-pause-away-rss-temperature-data-revised-with-improved-adjustments-have-now-found-the-missing-warming/.

21. Paul C. "Chip" Knappenberger, "Comment on New Satellite-Observed Temperature Dataset," Cato at Liberty, March 7, 2016, https://www.cato. org/blog/comment-new-satellite-observed-temperature-dataset.

22. Christopher Monckton, "How They Airbrushed Out the Inconvenient Pause," Watts Up with That, July 7, 2017, https://wattsupwiththat. com/2017/07/07/how-they-airbrushed-out-the-inconvenient-pause/.

23. Interview of Patrick Michaels, Marc Morano and Mick Curran, *Climate Hustle*, 2016, www. ClimateHustle.com,

24. Interview of Anthony Watts, Morano and Curran, *Climate Hustle*.

25. James Hansen, Reto Ruedy, Jay Glascoe and Makiko Sato, "Whither U.S. Climate?" NASA Science Briefs, August 1999, https://www.giss. nasa.gov/research/briefs/hansen_07/.

26. Ibid.

27. Philip Shabecoff, "Global Warmth in '88 Is Found to Set a Record," *New York Times*, February 4, 1989, http://www.nytimes. com/1989/02/04/us/global-warmth-in-88-is-found-to-set-a-record.html.

28. Tony Heller, "NOAA/NASA Dramatically Altered US Temperatures After the Year 2000," Real Science, June 23, 2014, https:// stevengoddard.wordpress.com/2014/06/23/ noaanasa-dramatically-altered-us-temperatures-after-the-year-2000/.

29. Tony Heller, "NOAA/NASA Dramatically Altered US Temperatures After the Year 2000," Real Science, June 23, 2014, https:// stevengoddard.wordpress.com/2014/06/23/ noaanasa-dramatically-altered-us-temperatures-after-the-year-2000/.

30. Heller, "NOAA/NASA."

31. C. Essex, R. McKitrick, and B. Andresen, "Researchers Question Validity of a 'Global Temperature,'" Science Daily, March 18, 2007, https://www.sciencedaily.com/ releases/2007/03/070315101129.htm., http:// www.uoguelph.ca/~rmckitri/research/ globaltemp/GlobTemp.JNET.pdf.

32. Interview of Lamar Smith, Morano and Curran, *Climate Hustle*.

33. Gayathri Vaidyanathan, "NOAA Declines to Provide GOP with Emails Related to 'Pause' Study," Climatewire, October 28, 2015, http:// junkscience.com/2015/10/wow-noaa-refuses-to-provide-congress-with-e-mails-on-study-claiming-climate-pause-never-happened/.

34. David Whitehouse, "2016 Global Temperature: The Pause Never Went Away," Global Warming Policy Foundation, January 19, 2017, https:// www.thegwpf.com/2016-global-temperature-the-pause-never-went-away/.

35. "*Science* Publishes New NOAA Analysis: Data Show No Recent Slowdown in Global Warming," National Oceanographic and Atmospheric Administration, June 4, 2015, http://www. noaanews.noaa.gov/stories2015/noaa-analysis-journal-science-no-slowdown-in-global-warming-in-recent-years.html.

36. "'Pause-Buster?' Scientists Challenge New Study Attempting to Erase the 'Pause': Warmists Rewrite Temperature History to Eliminate the 'Pause,'" Climate Depot, June 4, 2015, http:// www.climatedepot.com/2015/06/04/scientists-challenge-new-study-attempting-to-erase-the-pause-warmists-rewrite-temperature-history-to-eliminate-the-pause/.

37. "It's Official—There Are Now 66 Excuses for Temp 'Pause'—Updated List of 66 Excuses for the 18–26 Year 'Pause' in Global Warming," Climate Depot, November 20, 2014, http://www. climatedepot.com/2014/11/20/its-official-there-are-now-66-excuses-for-temp-pause-updated-list-of-66-excuses-for-the-18-26-year-pause-in-global-warming/.

38. Interview of Roy Spencer, Morano and Curran, *Climate Hustle*.

39. Patrick J. Michaels, "The Current Wisdom: The Lack of Recent Warming and the State of Peer Review," Cato, August 3, 2011, https://www.cato. org/publications/commentary/current-wisdom-lack-recent-warming-state-peer-review.

40. Justin Gillis, "In the Ocean, Clues to Change," *New York Times*, August 11, 2014, https://www. nytimes.com/2014/08/12/science/in-the-ocean-clues-to-change.html?mcubz=0.

41. Lubos Motl, "Did Chinese Coal Cause the Cooling since 1998?" The Reference Frame, July 06, 2011, https://motls.blogspot.com/2011/07/ did-chinese-coal-cause-cooling-since.html.

42. "Volcanic Aerosols, Not Pollutants, Tamped Down Recent Earth Warming, Says CU Study," University of Colorado, Boulder, March 1, 2013, http://www.colorado.edu/today/2013/03/01/volcanic-aerosols-not-pollutants-tamped-down-recent-earth-warming-says-cu-study.

43. Victoria Woolaston, "Government Measures 'May Have Slowed Down Global Warming': Energy Minister Claims Policies Are Playing a Role in Curbing Rising Temperatures," *Daily Mail*, October 27, 2014, http://www.dailymail.co.uk/sciencetech/article-2809995/Government-measures-slowed-global-warming-Energy-minister-claims-policies-playing-role-curbing-rising-temperature.html.

44. Paul Homewood, "Baroness Verma In Wonderland," Not A Lot of People Know That, October 27, 2014, https://notalotofpeopleknowthat.wordpress.com/2014/10/27/baroness-verma-in-wonderland/.

45. Willie Soon, Istvan Marko, et al., "To Put America First Is to Put Our Planet's Climate First," Breitbart News, June 16, 2017, http://www.breitbart.com/big-government/2017/06/16/america-first-climate/.

46. Interview of Judith Curry, Morano and Curran, *Climate Hustle*.

47. Interview of Robert Giegengack, Morano and Curran, *Climate Hustle*.

48. Clip of Al Gore, Morano and Curran, *Climate Hustle*.

49. Interview of Patrick Moore, Morano and Curran, *Climate Hustle*.

50. Speech by Ivar Giaever, Morano and Curran, *Climate Hustle*.

51. Interview of Patrick Michaels, Morano and Curran, *Climate Hustle*.

52. Patrick J. Michaels, "When Will Climate Scientists Say They Were Wrong?" Cato, May 29, 2015, https://www.cato.org/publications/commentary/when-will-climate-scientists-say-they-were-wrong?utm_medium=twitter&utm_source=twitterfeed.

53. Lennart Bengtsson, quoted in Peter Ferrara, "As the Economy Recesses, Obama's Global Warming Delusions Are Truly Cruel," Forbes, February 22, 2013, https://www.forbes.com/sites/peterferrara/2013/02/22/as-the-economy-recesses-obamas-global-warming-delusions-are-truly-cruel/#53f78db3a779.

54. Interview of Robert M. "Bob" Carter, Morano and Curran, *Climate Hustle*.

55. John R. Christy, House Energy and Power Subcommittee. September 20, 2012, http://energycommerce.house.gov/sites/republicans.energycommerce.house.gov/files/Hearings/EP/20120920/HHRG-112-IF03-WState-ChristyJ-20120920.pdf.

56. "Climate Change Indicators: High and Low Temperatures," EPA, August 2016, https://www.epa.gov/climate-indicators/climate-change-indicators-high-and-low-temperatures.

57. Ibid.

Chapter 8:
Models Do Not Equal Evidence

1. Interview of Patrick J. Michaels, Marc Morano and Mick Curran, *Climate Hustle*, 2016, www.ClimateHustle.com.

2. Michael E. Mann, "Climate Catastrophe Is a Choice," *Foreign Affairs*, April 21, 2017, https://www.foreignaffairs.com/articles/2017-04-21/climate-catastrophe-choice.

3. Willie Soon, István Markó, et al., "To Put America First Is to Put Our Planet's Climate First," Breitbart News, June 16, 2017, http://www.breitbart.com/big-government/2017/06/16/america-first-climate/.

4. "Public Release: Antarctic Temperatures Disagree with Climate Model Predictions," Ohio State University, February 15, 2007, https://www.eurekalert.org/pub_releases/2007-02/osu-atd021207.php.

5. Roger Pielke, "Another Publication of An Unverifiable Multi-Decadal Climate Prediction: 'Cold Spells In a Warming World,'" April 27, 2011, https://pielkeclimatesci.wordpress.

com/2011/04/27/another-publication-of-an-unverifiable-multi-decadal-climate-prediction-cold-spells-in-a-warming-world/; Evan Kodra, Karsten Steinhaeuser, et al., "Persisting Cold Extremes under21st-Century Warming Scenarios," *Geophysical Research Letters* 38 (2011), http://onlinelibrary.wiley.com/doi/10.1029/2011GL047103/pdf.

6. Rocky Barker, "Study Finds Idaho Trout Face Climate Trouble," *Idaho Statesman*, August 16, 2011, http://www.heliogenic.net/2011/08/16/climate-model-output-is-now-data/.

7. Roger Pielke, "Another Publication of an Unverifiable Multi-Decadal Climate Prediction: 'Cold Spells in A Warming World,'" April 27, 2011, https://pielkeclimatesci.wordpress.com/2011/04/27/another-publication-of-an-unverifiable-multi-decadal-climate-prediction-cold-spells-in-a-warming-world/.

8. Interview of Robert M. "Bob" Carter, Morano and Curran, *Climate Hustle*.

9. Sue Dremann, "U.S. Energy Secretary Chu Looks to Innovation to Solve Warming Crisis," Palo Alto Online, June 28, 2009, https://www.paloaltoonline.com/news/2009/06/28/us-energy-secretary-chu-looks-to-innovation-to-solve-warming-crisis.

10. Peter Griffiths, "Arctic to Be Ice-Free in Summer in 20 Years: Scientist," Reuters, October 15, 2009, https://www.reuters.com/article/us-climate-britain-arctic-science/arctic-to-be-ice-free-in-summer-in-20-years-scientist-idUSTRE59E18W20091015.

11. Gary Strand, "The Secret of the Rahmstorf 'Non-Linear Trend Line,'" Climate Audit, July 5, 2009, https://climateaudit.org/2009/07/04/march-2106/#comment-187196.

12. Anthony Watts, "Quote of the Week #13," Watts Up with That, July 6, 2009, https://wattsupwiththat.com/2009/07/06/quote-of-the-week-13/.

13. Kevin Trenberth, "Predictions of Cimate," Climate Feedback, June 4, 2007, http://blogs.nature.com/climatefeedback/2007/06/predictions_of_climate.html.

14. "The Physics That We Know: A Conversation with Gavin Schmidt: Introduction by Russell Weinberger," Edge, July 2009, https://www.edge.org/conversation/the-physics-that-we-know.

15. "World Climate Predictors Right Only Half the Time," The New Zealand Climate Science Coalition, June 7, 2007, http://nzclimatescience.net/index.php?option=com_content&task=view&id=23&Itemid=32.

16. "IPCC Report Slammed as 'Dangerous Nonsense,'" New Zealand Climate Science Coalition, April 10, 2007, http://www.scoop.co.nz/stories/SC0704/S00023.htm.

17. Marc Morano, "UN IPCC Scientist: 'No Convincing Scientific Arguments to Support Claim That Increases in Greenhouse Gases Are Harmful to the Climate,'" Climate Depot, May 5, 2009, http://www.climatedepot.com/2009/05/05/un-ipcc-scientist-no-convincing-scientific-arguments-to-support-claim-that-increases-in-greenhouse-gases-are-harmful-to-the-climate/.

18. "Climatologist Slams RealClimate.org for 'Erroneously Communicating the Reality of How the Climate System Is Actually Behaving,'" Climate Depot, June 30, 2009, http://www.climatedepot.com/2009/06/30/climatologist-slams-realclimateorg-for-erroneously-communicating-the-reality-of-the-how-climate-system-is-actually-behaving/.

19. "U.S. Senate Report: Over 400 Prominent Scientists Disputed Man-Made Global Warming Claims in 2007: Scientists Debunk 'Consensus,'" U.S. Senate Environment and Public Works Committee Minority Staff Report, December 20, 2007, https://www.epw.senate.gov/public/_cache/files/b/b/bba2ebce-6d03-48e4-b83c-44fe321a34fa/01AFD79733D77F24A71FEF9DAFCCB056.consensusbusterscompletedocument.pdf.

20. Marc Morano, "U.S. Government Scientist's Shock Admission: 'Climate Model Software Doesn't Meet the Best Standards Available,'" Climate Depot, July 6, 2009, http://www.climatedepot.com/2009/07/06/us-government-scientists-shock-admission-climate-model-software-doesnt-meet-the-best-standards-available/.

21. Marc Morano, "Gore Losing: No Cause for Alarm at 5-Year Mid-point of Armstrong-Gore Climate 'Bet'—'Gore Should Be Pleased to Find concerns about a 'Tipping Point' Have Turned Out to Be Unfounded,'" January 18, 2013, http://www.climatedepot.com/2013/01/18/gore-l; Anthony Watts, "Forecasting Guru Announces: 'No Scientific Basis for Forecasting Climate,'" What's Up With That?, January 28, 2009, https://wattsupwiththat.com/2009/01/28/forecasting-guru-announces-no-scientific-basis-for-forecasting-climate/.

22. Andrew Orlowski, "Dyson: Climate Models Are Rubbish," *Register*, August 14, 2007, http://www.theregister.co.uk/2007/08/14/freeman_dyson_climate_heresies/.

23. Interview of Robert Giegengack, Morano and Curran, *Climate Hustle*.

24. Cornelia Dean, "The Problems in Modeling Nature with Its Unruly Natural Tendencies," *New York Times*, February 20, 2007, http://www.nytimes.com/2007/02/20/science/20book.html.

25. Interview of Michael Oppenheimer, Morano and Curran, *Climate Hustle*.

Chapter 9:
The Eroding "Consensus"

1. Interview of Roger Pielke Sr., Morano and Curran, *Climate Hustle*.

2. Interview of William "Matt" Briggs, Marc Morano and Mick Curran, *Climate Hustle*, 2016, www.ClimateHustle.com.

3. Interview of Christopher Monckton, Morano and Curran, *Climate Hustle*.

4. Marc Morano, "Special Report: More than 1000 International Scientists Dissent Over Man-Made Global Warming Claims—Challenge UN IPCC & Gore," Climate Depot, December 8, 2010, http://www.climatedepot.com/2010/12/08/special-report-more-than-1000-international-scientists-dissent-over-manmade-global-warming-claims-challenge-un-ipcc-gore-2/.

5. "Renowned Scientist Defects from Belief in Global Warming—Caps Year of Vindication for Skeptics," Senate Environment and Public Works Committee, October 17, 2006, https://www.epw.senate.gov/public/index.cfm/2006/10/post-e58dff04-5a65-42a4-9f82-87381de894cd.

6. Marc Morano, "Et Tu, Francois? Skeptical Scientist Who Mocked Gore's Nobel Prize as 'Political Gimmick,' May Be Appointed to French Super-Ministry Post," Climate Depot, May 27, 2009, http://www.climatedepot.com/2009/05/27/et-tu-francois-skeptical-scientist-who-mocked-gores-nobel-prize-as-political-gimmick-may-be-appointed-to-french-superministry-post/.

7. Ibid.

8. "Claude Allègre, Geochemist and Former French Minister," *The Interview*, France 24, May 18, 2011, http://www.france24.com/en/20110517-2011-05-16-1815-wb-en-interview.

9. "James Lovelock: The Earth Is about to Catch a Morbid Fever That May Last as Long as 100,000 Years," *Independent*, January 16, 2006, http://www.independent.co.uk/voices/commentators/james-lovelock-the-earth-is-about-to-catch-a-morbid-fever-that-may-last-as-long-as-100000-years-5336856.html.

10. "Naked Science" videoclip featured in Morano and Curran, *Climate Hustle*.

11. Andrew Orlowski, "Gaia Scientist Lovelock: 'I Was Wrong and Alarmist on Climate,'" *Register*, April 24, 2012 http://www.theregister.co.uk/2012/04/24/lovelock_clangers/.

12. Marc Morano, "Green Guru James Lovelock on Climate Change: 'I Don't Think Anybody Really Knows What's Happening. They Just Guess'— Lovelock Reverses Himself on Global Warming," Climate Depot, April 3, 2014, http://www. climatedepot.com/2014/04/03/green-guru-james-lovelock-on-climate-change-i-dont-think-anybody-really-knows-whats-happening-they-just-guess-lovelock-reverses-himself-on-global--warming/.

13. Leo Hickman, "James Lovelock on the Value of Sceptics and Why Copenhagen Was Doomed," Guardian, March 29, 2010, https://www. theguardian.com/environment/blog/2010/mar/29/james-lovelock.

14. Morano and Curran, Climate Hustle.

15. Decca Aitkenhead, "James Lovelock: 'Before the End of This Century, Robots Will Have Taken Over,'" Guardian, September 30, 2016, https:// www.theguardian.com/environment/2016/sep/30/james-lovelock-interview-by-end-of-century-robots-will-have-taken-over.

16. Adam Vaughan, "James Lovelock: Environmentalism Has Become a Religion," Guardian, March 30, 2014, https://www. theguardian.com/environment/2014/mar/30/james-lovelock-environmentalism-religion.

17. Hickman, "James Lovelock."

18. Vaughan, "James Lovelock."

19. Marc Morano, "Scientist Dr. Daniel Botkin Tells Congress Why He Reversed His Belief in Global Warming to Become a Skeptic: 'There Are Several Lines of Evidence Suggesting That It (AGW) Is a Weaker Case Today, Not a Stronger Case'—Rips Obama Climate Report As 'Filled with Misstatements Contradicted by Well-Established and Well-Known Scientific Papers'" Climate Depot, May 29, 2014, http://www. climatedepot.com/2014/05/29/scientist-dr-daniel-botkin-tells-congress-why-he-reversed-his-belief-in-global-warming-to-become-a-

skeptic-there-are-several-lines-of-evidence-suggesting-that-it-agw-is-a-weaker-case-today-not/.

20. Interview of Daniel Botkin, Morano and Curran, Climate Hustle.

21. Morano, "Special Report: More than 1,000 International Scientists."

22. Botkin, "Full Committee Hearing."

23. Richard Tol, "Examining the UN Intergovernmental Panel on Climate Change Process for the Fifth Assessment Report Committee on Science, Space and Technology, U.S. House of Representatives, May 29, 2014, https://nofrakkingconsensus.files.wordpress. com/2014/05/tol_written_testimony_may2014. pdf.

24. Interview of Richard Tol, Morano and Curran, Climate Hustle.

25. Ben Spencer, "UK Professor Refuses to Put His Name to 'Apocalyptic' UN Climate Change Survey That He Claims Is Exaggerating the Effects," Daily Mail, March 26, 2014, http://www. dailymail.co.uk/news/article-2589424/UK-professor-refuses-apocalyptic-UN-climate-change-survey.html#ixzz2x3avUrY3.

26. Interview of Richard Tol, Morano and Curran, Climate Hustle.

27. Marc Morano, "Former UN IPCC Lead Author Dr. Richard Tol: 'One of the Startling Facts about Climate Change Is That There Are Very Few Facts about Climate Change. Climate Change Is Mainly Something of the Future So We Are Really Talking about Model Projections,'" Climate Depot, May 29, 2014, http://www. climatedepot.com/2014/05/29/former-un-ipcc-lead-author-dr-richard-tol-one-of-the-startling-facts-about-climate-change-is-that-there-are-very-few-facts-about-climate-change-climate-change-is-mainly-something-of-the-future-s/.

28. "An Open Letter to the American People," 2008, https://globalhighered.files.wordpress. com/2008/11/nobel_letter_v6.pdf.

29. Marc Morano, "Nobel Prize–Winning Scientist Who Endorsed Obama Now Says Prez. is 'Ridiculous' & 'Dead Wrong' on 'Global Warming,'" Climate Depot, July 6, 2015, http://www.climatedepot.com/2015/07/06/nobel-prize-winning-scientist-who-endorsed-obama-now-says-prez-is-ridiculous-dead-wrong-on-global-warming/.

30. Interview of Robert Giegengack, Marc Morano and Mick Curran, Climate Hustle, 2016, www.ClimateHustle.com.

31. Paul Mulshine, "Climatologists Are No Einsteins, Says His Successor," Star Ledger, April 3, 2013, http://blog.nj.com/njv_paul_mulshine/2013/04/climatologists_are_no_einstein.html.

32. Andrew Orlowski, "Top Boffin Freeman Dyson on Climate Change, Interstellar Travel, Fusion, and More," Register, October 11, 2015, http://www.theregister.co.uk/2015/10/11/freeman_dyson_interview/.

33. "Video: Earth Is Actually Growing Greener," Vancouver Sun, April 20, 2015, http://www.vancouversun.com/technology/Conversations+that+matter+Earth+actually+growing+greener/10944052/story.html.

34. Mulshine, "Climatologists Are No Einsteins."

35. Thomas Lin, "At 90, Freeman Dyson Ponders His Next Challenge," Wired, March 31, 2014, https://www.wired.com/2014/03/quanta-freeman-dyson-qa/.

36. "Video: Earth Is Actually Growing Greener."

37. Marc Morano, "Renowned Princeton Physicist Freeman Dyson: 'I Like Carbon Dioxide'—Warns 'Climate Experts Have Set Themselves Up as Being the Guardians of the Truth…& That Is a Dangerous Situation,'" June 30, 2014, https://www.climatedepot.com/2015/02/19/renowned-princeton-physicist-freeman-dyson-i-like-carbon-dioxide-warns-climate-excerpts-have-set-themselves-up-as-being-the-guardians-of-the-truth-that-is-a-dangerous-situation/.

38. Interview of Judith Curry, Morano and Curran, Climate Hustle.

39. Michael D. Lemonick, "Climate Heretic: Judith Curry Turns on Her Colleagues,' Scientific American, November 2010, https://www.scientificamerican.com/article/climate-heretic/.

40. David Rose, "'I Was Tossed Out of the Tribe': Climate Scientist Judith Curry Interviewed," Spectator, November 25, 2015, http://new.spectator.co.uk/2015/11/i-was-tossed-out-of-the-tribe-climate-scientist-judith-curry-interviewed/.

41. Interview of Judith Curry, Morano and Curran, Climate Hustle.

42. Lemonick, "Climate Heretic."

43. Interview of Judith Curry, Morano and Curran, Climate Hustle.

44. Judith Curry, "IPCC Diagnosis—Permanent Paradigm Paralysis," Climate Etc., September 28, 2013, https://judithcurry.com/2013/09/28/ipcc-diagnosis-permanent-paradigm-paralysis/.

45. Marc Morano, "Climatologist Dr. Judith Curry's at Senate Hearing: 'Attempts to Modify the Climate through Reducing CO_2 Emissions May Turn Out to be Futile'—UN IPCC Now Making 'a Weaker Case for Anthropogenic Global Warming,'" Climate Depot, January 16, 2014 http://www.climatedepot.com/2014/01/16/climatologist-dr-judith-currys-testimony-at-senate-climate-hearing-attempts-to-modify-the-climate-through-reducing-co2-emissions-may-turn-out-to-be-futile-un-ipcc-now-making-a-weaker-case/.

46. Interview of Patrick Moore, Morano and Curran, Climate Hustle.

47. "Exclusive: Former Greenpeace Founder's 'Reality Check' for Liberals," Hannity, Fox News, February 27, 2014, http://www.foxnews.com/on-air/hannity/2014/02/28/exclusive-former-greenpeace-founders-reality-check-liberals.

48. EPW (blog), "Scientist Who Was Fmr. Greenpeace Member Says 'No Proof' CO_2 Is Driving Global Temps!" Canada Free Press,

October 15, 2008, http://canadafreepress.com/article/scientist-who-was-fmr-greenpeace-member-says-no-proof-co2-is-driving-global.

49. Marc Morano, "Left-Wing Env. Scientist Bails Out of Global Warming Movement: Declares It a 'Corrupt Social Phenomenon...Strictly an Imaginary Problem of the 1st World Middleclass,'" Climate Depot, July 26, 2010, http://www.climatedepot.com/2010/07/26/leftwing-env-scientist-bails-out-of-global-warming-movement-declares-it-a-corrupt-social-phenomenonstrictly-an-imaginary-problem-of-the-1st-world-middleclass/.

50. Interview of Denis Rancourt, Morano and Curran, *Climate Hustle*.

51. Morano, "Left-Wing Env. Scientist."

52. Denis Rancourt, "On the Gargantuan Lie of Climate Change Science," Activist Teacher, March 21, 2011, http://activistteacher.blogspot.com/2011/03/on-gargantuan-lie-of-climate-change.html.

53. Morano, "Left-Wing Env. Scientist."

54. Interview of Caleb Rossiter, Morano and Curran, *Climate Hustle*.

55. Marc Morano, "Politically Left Scientist Dissents—Calls President Obama 'Delusional' on Global Warming," Climate Depot, September 23, 2014, http://www.climatedepot.com/2014/09/23/politically-left-scientist-dissents-calls-president-obama-delusional-on-global-warming/.

56. Interview of Caleb Rossiter, Morano and Curran, *Climate Hustle*.

57. Morano, "Politically Left Scientist Dissents."

58. Interview of Caleb Rossiter, Morano and Curran, *Climate Hustle*.

59. Caleb Rossiter, "Sacrificing Africa for Climate Change," *Wall Street Journal*, May 4, 2014, https://www.wsj.com/articles/caleb-rossiter-sacrificing-africa-for-climate-change-1399243547.

60. Philip Stott, "Major Reductions in Carbon Emissions Are Not Worth the Money,"

Intelligence Squared Debate, January 13, 2009, https://www.intelligencesquaredus.org/debates/major-reductions-carbon-emissions-are-not-worth-money.

61. Philip Stott, "It's All Unraveling," Global Warming Politics, May 2, 2008, https://www.epw.senate.gov/public/index.cfm/press-releases-all?ID=A17DEFA8-802A-23AD-4912-8AB7138A7C3F.

62. Stott, "Major Reductions."

63. Philip Stott, "Global Warming Is Not a Crisis," ABC News, March 9, 2007, http://abcnews.go.com/International/story?id=2938762&page=1.

64. Morano, "Politically Left Scientist Dissents."

65. Interview of Leighton Steward, Morano and Curran, *Climate Hustle*.

66. Denis G. Rancourt, "Dr. Jean-Louis Pinault Explains His Idea about Global Climate Forcing and His Experience with Climate Politics," Denis Rancourt on Climate (blog), May 19, 2014, http://climateguy.blogspot.ca/2014/05/dr-jean-louis-pinault-explains-his-idea.html.

67. P. Gosselin, "The Belief That CO_2 Can Regulate Climate Is 'Sheer Absurdity' Says Prominent German Meteorologist," May 9, 2012, http://notrickszone.com/2012/05/09/the-belief-that-co2-can-regulate-climate-is-sheer-absurdity-says-prominent-german-meteorologist/#sthash.ULhVwrl8.dpbs.

68. David Kear, "Global Warming Alias Climate Change—the Non-Existent, Incredibly Expensive Threat to Us All, including to Our Grandchildren," New Zealand Center for Political Research, April 6, 2014, http://www.nzcpr.com/global-warming-alias-climate-change/.

69. Ibid.

70. Marc Morano, "Another Prominent Scientist Dissents! Fmr. NASA Scientist Dr. Les Woodcock 'Laughs' at Global Warming—'Global Warming Is Nonsense' Top Prof. Declares," Climate Depot, April 3, 2014, http://www.climatedepot.

com/2014/04/03/another-prominent-scientist-dissents-fmr-nasa-scientist-dr-les-woodcock-laughs-at-global-warming-top-prof-declares-global-warming-is-nonsense/.

71. James Varney, "Scientists Criticize 'Hottest Year on Record' Claim as Hype," Real Clear Investigations, January 29, 2017, http://www.realclearinvestigations.com/articles/2017/01/29/scientists_criticize_hottest_year_on_record_claim_as_hype.html.

72. Chris White, "Former Obama Official: Bureaucrats Manipulate Climate Stats to Influence Policy," The Daily Caller, April 24, 2017, http://dailycaller.com/2017/04/24/former-obama-official-says-climate-data-was-often-misleading-and-wrong/.

73. Steven Koonin, "A 'Red Team' Exercise Would Strengthen Climate Science," *Wall Street Journal*, April 21, 2017, https://www.wsj.com/articles/a-red-team-exercise-would-strengthen-climate-science-1492728579.

74. Marc Morano, "More than 700 International Scientists Dissent Over Man-Made Global Warming Claims," U.S. Senate Environment and Public Works Committee, March 16, 2009, https://www.epw.senate.gov/public/index.cfm/press-releases-all?ID=10fe77b0-802a-23ad-4df1-fc38ed4f85e3.

75. Morano, "Special Report: More than 1,000 International Scientists."

Chapter 10:
Climategate: The UN IPCC Exposed

1. Declan McCullagh, "Physicists Stick to Warming Claim Post-ClimateGate," CBS News, December 8, 2009, https://www.cbsnews.com/news/physicists-stick-to-warming-claim-post-climategate/.

2. Ibid.

3. Brent Baker, "ABC and NBC Acknowledge 'Climategate' but Remain Undeterred: 'Science Is Solid,'" Newsbusters, December 7, 2009, https://www.newsbusters.org/blogs/nb/

brent-baker/2009/12/07/abc-and-nbc-acknowledge-climategate-remain-undeterred-science-solid.

4. James Taylor, "Climategate 2.0: New E-Mails Rock the Global Warming Debate," Forbes, November 23, 2011, https://www.forbes.com/sites/jamestaylor/2011/11/23/climategate-2-0-new-e-mails-rock-the-global-warming-debate/#69075ca27ba6.

5. Christopher Horner, "Penn State Whitewashed ClimateGate," The Daily Caller, March 8, 3022, http://dailycaller.com/2011/03/08/penn-state-whitewashed-climategate/.

6. Christopher Horner, "A Reply to Michael Mann and Eugene Wahl," The Daily Caller, March 9, 2011, http://dailycaller.com/2011/03/09/a-reply-to-michael-mann-and-eugene-wahl/.

7. Chris Irvine, "Climategate: Phil Jones Accused of Making Error of Judgment by Colleague," *The Telegraph*, December 3, 2009, http://www.telegraph.co.uk/news/earth/copenhagen-climate-change-confe/6718183/Climategate-Phil-Jones-accused-of-making-error-of-judgment-by-colleague.html.

8. "Climate of Denial" (editorial), *Washington Post*, November 25, 2009, http://www.washingtonpost.com/wp-dyn/content/article/2009/11/24/AR2009112403549.html.

9. John R. Lott, "Why You Should Be Hot and Bothered about 'Climate-Gate,'" Fox News, November 24, 2009, http://www.foxnews.com/opinion/2009/11/24/john-lott-climate-change-emails-copenhagen.html.

10. Tom Nelson, "ClimateGate—See 'Outed' Emails Here," Tom Nelson (blog), 2012, http://tomnelson.blogspot.com/p/climategate_05.html.

11. Marc Morano, "UN Scientists Turn on Each Other: UN Scientist Declares Climategate colleagues Mann, Jones and Rahmstorf 'Should Be Barred from the IPCC Process'—They Are 'Not Credible Any More,'" Climate Depot, November 27, 2009, http://www.climatedepot.com/2009/11/27/

un-scientists-turn-on-each-other-un-scientist-
declares-climategate-colleagues-mann-jones-and-
rahmstorf-should-be-barred-from-the-ipcc-process-
they-are-not-credible-any-more/.

12. Ibid.

13. "Richard Tol: IPCC Was 'Captured,'" The Global
Warming Policy Forum, April 22, 2010, https://
www.thegwpf.com/richard-tol-ipcc-was-
captured/.

14. Richard Tol, "IPCC: This Time Will be Different
(Not), A Guest Post by Richard Tol," Roger Pielke
Jr."s Blog, June 11, 2010, http://rogerpielkejr.
blogspot.com/2010/06/ipcc-time-will-be-
different-not-guest.html.

15. Andrew Revkin, "A Climate Scientist Who
Engages Skeptics," New York Times, November
27, 2009, https://dotearth.blogs.nytimes.
com/2009/11/27/a-climate-scientist-on-climate-
skeptics/?mtrref=www.climatedepot.com&gwh=
AC78DCF78BF9FEB8D07494E6A139A115&gwt=
pay&assetType=opinion#more-11377.

16. Keith Johnson, "Climate Emails Stoke Debate,"
Wall Street Journal, November 23, 2009, https://
www.wsj.com/articles/SB12588340529485921.5.

17. Vincent Gray, "Vincent Gray on Climategate:
'There Was Proof of Fraud All Along' (PJM
Exclusive)" PJ Media, November 27, 2009,
https://pjmedia.com/blog/vincent-gray-on-
climategate-there-was-proof-of-fraud-all-along-
pjm-exclusive/.

18. Marc Morano, "Another F-Bomb: S. African UN
Scientist Declares UN IPCC a 'Worthless Carcass'
As Pachauri Is in 'Disgrace' and 'Fraudulent
Science Continues to Be Exposed,'" Climate
Depot, February 26, 2010, http://www.
climatedepot.com/2010/02/26/another-fbomb-s-
african-un-scientist-declares-un-ipcc-a-
worthless-carcass-as-pachauri-is-in-disgrace-
and-fraudulent-science-continues-to-be-expos-
ed/.

19. Marc Morano, "Special Report: More than 1000
International Scientists Dissent Over Man-Made
Global Warming Claims—Challenge UN IPCC &

Gore," Climate Depot, December 8, 2010, http://
www.climatedepot.com/2010/12/08/special-
report-more-than-1000-international-scientists-
dissent-over-manmade-global-warming-claims-
challenge-un-ipcc-gore-2/.

20. Ibid.

21. Richard Muller, "Climategate 'Hide the Decline'
Explained by Berkeley Professor Richard A.
Muller," YouTube, February 27, 2011, https://
www.youtube.com/watch?feature=player_
embedded&v=8BQpciw8suk.

22. Charles Cooper, "Tempers Flare in Climate
Change Flap," CBS News, December 5, 2009,
https://www.cbsnews.com/news/tempers-flare-
in-climate-change-flap/.

23. John Lott, "As Climategate Becomes Pressgate,
Questions for the Media," Breitbart, January 8,
2010, http://www.breitbart.com/big-
journalism/2010/01/08/as-climategate-becomes-
pressgate-questions-for-the-media/.

24. Marc Morano, "Newsweek Profile: Climate
Depot 'Is Quickly Becoming King of the
Skeptics!'—'Most Popular Denial Site'—Morano
Thanks 'Mainstream Media for Ignoring
Climategate,'" Climate Depot, December 15, 2009,
http://www.climatedepot.com/2009/12/15/
newsweek-profile-climate-depot-is-quickly-
becoming-king-of-the-skeptics-most-popular-
denial-site-morano-thanks-mainstream-media-
for-ignoring-climategate/.

25. "Michael Mann Threatens Legal Action over
Steyn Comment," Australian Climate Madness,
July 25, 2012, https://australianclimatemadness.
com/2012/07/25/michael-mann-threatens-legal-
action-over-steyn-comment/.

26. Clive Crook, "Climategate and the Big Green
Lie," Atlantic, July 14, 2010, https://www.
theatlantic.com/politics/archive/2010/07/
climategate-and-the-big-green-lie/59709/.

27. "Michael Mann's Climate Stimulus," Wall Street
Journal, January 20, 2010, https://www.wsj.com/
articles/SB10001424052748704541004575010931344
004278.

28. Willis Eschenbach, "Exonerated? Not.," Watts Up with That, October 19, 2010, https://wattsupwiththat.com/2010/10/19/exonerated-not/.

29. Steve McIntyre, "Another Tainted Inquiry," Climate Audit, March 23, 2010, https://climateaudit.org/2010/03/23/another-tainted-inquiry/.

30. Andrew Orlowski, "Climategate Report: 'Campaign to Win Hearts and Minds' Needed," *Register*, July 7, 2010, http://www.theregister.co.uk/2010/07/07/muir_russell_climategate_report/.

31. Steve McIntyre, "Lord Russell of Holyrood," Climate Audit, July 7, 2010, https://climateaudit.org/2010/07/07/lord-russell-of-holyrood/.

32. Andrew Orlowski, "Oops: Chief Climategate Investigator Failed to Declare Eco Directorship," *Register*, March 24, 2010, http://www.theregister.co.uk/2010/03/24/climategate_oxburgh_globe/.

33. Judith Curry, "The Legacy of Climategate: 5 Years Later," Climate Etc., December 1, 2014, https://judithcurry.com/2014/12/01/the-legacy-of-climategate-5-years-later/.

34. Marc Morano, "CBC's Rex Murphy Unloads about ClimateGate: It 'Pulls Back the Curtain on Pettiness, Turf Protection, Manipulation, Defiance of FOIA, Loss [sic] or Destroyed Data and Attempts to Blacklist,'" Climate Depot, December 3, 2009, http://www.climatedepot.com/2009/12/04/cbcs-rex-murphy-unloads-about-climategate-it-pulls-back-the-curtain-on-pettiness-turf-protection-manipulation-defiance-of-foia-loss-or-destroyed-data-and-attempts-to-blacklist/.

35. Roger Pielke Jr., "Your Politics Are Showing," Roger Pielke Jr.'s Blog, December 4, 2009, http://rogerpielkejr.blogspot.com/2009/12/your-politics-are-showing.html.

36. Catherine Brahic, "Major Climate Change Report Looks Set to Alarm," *New Scientist*, January 29, 2007, https://www.newscientist.com/article/dn11049-major-climate-change-report-looks-set-to-alarm/.

37. John Brignell, "On Refereeing," Number Watch, July 23, 2008, http://www.numberwatch.co.uk/2008%20July.htm#refereeing.

38. John McLean, "IPCC Peer Review Process an Illusion, Finds SPPI Analysis," Science and Public Policy Institute, September 10, 2007, http://scienceandpublicpolicy.org/press-releases/press/ipccprocessillusion.

39. John McLean, "Prejudiced Authors, Prejudiced Findings: Did the UN Bias Its Attribution of 'Global Warming' to Humankind?" Science and Public Policy Institute, July 2008, http://scienceandpublicpolicy.org/images/stories/papers/originals/McLean_IPCC_bias.pdf.

40. McLean, "IPCC Peer Review Process an Illusion."

41. Marc Morano, "UN Scientists Who Have Turned on the UN IPCC & Man-Made Climate Fears—A Climate Depot Flashback Report," Climate Depot, March 30, 2014, http://www.climatedepot.com/2013/08/21/un-scientists-who-have-turned-on-unipcc-man-made-climate-fears-a-climate-depot-flashback-report/.

42. Ibid.

43. Ibid.

44. Ibid.

45. Marc Morano, "Inhofe Debunks So-Called 'Consensus' On Global Warming," U.S. Senate Committee on Environment and Public Works, October 26, 2007, https://www.epw.senate.gov/public/index.cfm/press-releases-all?ID=595F6F41-802A-23AD-4BC4-B364B623ADA3; "IPCC Factsheet: How Does the IPCC Approve Reports?," Intergovernmental Panel on Climate Change, https://www.ipcc.ch/news_and_events/docs/factsheets/FS_ipcc_approve.pdf.

46. Morano, "Inhofe Debunks So-Called 'Consensus.'"

47. Suzanne Goldenberg, "IPCC Report: Climate Change Felt 'on All Continents and across the

Oceans,'" *Guardian*, March 28, 2014, https://www.theguardian.com/environment/2014/mar/28/ipcc-report-climate-change-report-human-natural-systems?CMP=twt_gu.

48. Steve McIntyre, "IPCC Schedule: WG1 Report Available Only to Insiders Until May 2007," Climate Audit, January 24, 2007, https://climateaudit.org/2007/01/24/ipcc-4ar/.

49. Morano, "Inhofe Debunks So-Called 'Consensus.'"

50. Morano, "UN Scientists Who Have Turned."

51. Suzanne Goldenberg, "IPCC Chairman Dismisses Climate Report Spoiler Campaign," *Guardian*, September 19, 2013, https://www.theguardian.com/environment/2013/sep/19/ipcc-chairman-climate-report?CMP=twt_fd.

52. Marc Morano, "Outgoing UN IPCC Chief Reveals Global Warming 'Is My Religion and my Dharma,'" Climate Depot, February 24, 2015, http://www.climatedepot.com/2015/02/24/un-ipcc-chief-admits-global-warming-is-religious-issue-it-is-my-religion-and-my-dharma/.

53. Donna Laframboise, "Rajendra Pachauri's Resignation Letter," February 24, 2015, Big Pic News, https://nofrakkingconsensus.com/2015/02/24/rajendra-pachauris-resignation-letter/.

54. Donna Laframboise, "US Scientific Integrity Rules Repudiate the UN Climate Process," January 29, 2017, https://nofrakkingconsensus.com/2017/01/29/us-scientific-integrity-rules-repudiate-the-un-climate-process/.

55. Marc Morano, "UN IPCC 'Altered' Climate Reports Violate U.S. Science Policy Guidelines," Climate Depot, January 31, 2017, http://www.climatedepot.com/2017/01/31/u-s-scientific-integrity-rules-repudiate-the-un-climate-process/.

56. Morano, "UN Scientists Who Have Turned on the UN IPCC."

57. Ibid.

58. Ibid.

59. Morano, "Inhofe Debunks So-Called 'Consensus.'"

60. "IPCC Conclusions Were Reached Three Years before the 'Research' Was Done," Real Science, November 22, 2010, https://stevengoddard.wordpress.com/2013/10/30/ipcc-conclusion-were-reached-three-years-before-the-research-was-done/.

61. Tony Heller, "IPCC Gets Caught—Digs Their Hole Deeper," Real Climate Science, September 17, 2017, https://realclimatescience.com/2017/09/ipcc-gets-caught-digs-their-hole-deeper/.

62. "Ethics Are 'Missing Dimension' in Climate Debate, says IPCC head," Bahá'í World News Service, September 23, 2009, http://news.bahai.org/story/729/.

63. Roger Harrabin, "Human Role in Warming More Certain'—UN Climate Chief," BBC, October 17, 2013, http://www.bbc.com/news/science-environment-24204323.

64. Peter Hannam, "Former UN Official Says Climate Report Will Shock Nations into Action," *Age*, November 7, 2012, http://www.theage.com.au/environment/climate-change/former-un-official-says-climate-report-will-shock-nations-into-action-20121106-28w5c.html.

65. Marc Morano, "Climate Depot's Morano Statement on New UN IPCC Report: 'This New IPCC Report Represents the Culmination of Years of Pre-Determined Science,'" Climate Depot, March 31, 2014, http://www.climatedepot.com/2014/03/31/climate-depots-morano-statement-on-new-un-ipcc-report-this-new-ipcc-report-represents-the-culmination-of-years-of-pre-determined-science/.

66. Morano, "UN Scientists Who Have Turned on the UN IPCC."

Chapter 11:
What's in a Name? How "Global Warming" Became "Climate Change"

1. Eric Scheiner, "Obama's Science Adviser: Don't Call it 'Global Warming,'" CNSNews.com,

September 14, 2010, https://www.cnsnews.com/news/article/obamas-science-adviser-dont-call-it-global-warming.

2. Conservation International, "Lost There, Felt Here: A Message from Tom Friedman—Conservation International (CI)," YouTube, December 8, 2008, https://www.youtube.com/watch?v=W1Li3O81uDs.

3. Ibid.

4. Marc Morano, "New Report: 'Extreme Weather Report 2012': Latest Peer-Reviewed Studies, Data & Analyses Undermine Claims That Current Weather Is 'Unprecedented' or a 'New Normal,'" Climate Depot, December 6, 2012, http://www.climatedepot.com/2012/12/06/new-report-extreme-weather-report-2012-latest-peerreviewed-studies-data-analyses-undermine-claims-that-current-weather-is-unprecedented-or-a-new-normal/.

5. Ibid.

6. "President Obama on Energy Policy," CSPAN, May 9, 2014, https://www.c-span.org/video/?319284-1/president-obama-energy-policy.

7. Kerry Jackson, "Global Warming Alarmist Reveals the Anti-Science Con," Investor's Business Daily, July 6, 2016, http://www.investors.com/politics/commentary/global-warming-alarmist-reveals-the-anti-science-con/.

8. Charles Onians, "Snowfalls Are Now Just a Thing of the Past," Independent, March 20, 2000, https://wattsupwiththat.files.wordpress.com/2015/11/snowfalls-are-now-just-a-thing-of-the-past-the-independent.pdf.

9. Marc Morano, "Meteorologist: 2010s Officially the Snowiest Decade in the East Coast in the NOAA Record—Surpassing the 1960s," Climate Depot, January 26, 2015, http://www.climatedepot.com/2015/01/26/meteorologist-2010s-officially-the-snowiest-decade-in-the-east-coast-in-the-noaa-record-surpassing-the-1960s/.

10. nextdrink97, "Barbara Boxer & Antonio Villaraigosa—Both Sounding Like IDIOTS!!!!" YouTube, May 22, 2013, https://www.youtube.com/watch?v=clORMerYQN0.

11. George Monbiot, "That Snow Outside Is What Global Warming Looks Like," Guardian, December 20, 2010, https://www.theguardian.com/commentisfree/2010/dec/20/uk-snow-global-warming.

12. Richard Alleyne, "Snow Is Consistent with Global Warming, Say Scientists," February 3, 2009, Telegraph, http://www.telegraph.co.uk/news/weather/4436934/Snow-is-consistent-with-global-warming-say-scientists.html.

13. "Why Global Warming Means…More Snow," Financial Times, January 20, 2012, http://www.aer.com/news-events/in-the-news/why-global-warming-means-more-snow.

14. Judah Cohen, "Bundle Up, It's Global Warming," New York Times, December 25, 2010, http://www.nytimes.com/2010/12/26/opinion/26cohen.html?_r=2&ref=opinion&mtrref=www.climatedepot.com&gwh=1F2AA2BBBE20B17C5E4E811F5FA50F2D&gwt=pay&assetType=opinion.

15. Michael Batasch, "An Inconvenient Review: After 10 Years Al Gore's Film Is Still Alarmingly Inaccurate," The Daily Caller, May 3, 2016, http://dailycaller.com/2016/05/03/an-inconvenient-review-after-10-years-al-gores-film-is-still-alarmingly-inaccurate/.

16. "Gore Reports Snow and Ice Across the World Vanishing Quickly," Environmental News Service, December 14, 2009, http://www.ens-newswire.com/ens/dec2009/2009-12-14-02.html.

17. Marc Morano, "Internet 'Creator' Gore's Website Crashed by Drudge—Gore Now Claims 'Increased Heavy Snowfalls Are Completely Consistent with Man-Made Global Warming,'" Climate Depot, February 1, 2011, http://www.climatedepot.com/2011/02/01/internet-creator-gores-website-crashed-by-drudge-gore-now-claims-increased-heavy-snowfalls-are-completely-consistent-with-manmade-global-warming/.

18. Barbara Stewart, "Winter in New York: Something's Missing; Absence of Snow Upsets Rhythms of Urban Life and Natural World," New

York Times, January 15, 2000, http://www.nytimes.com/2000/01/15/nyregion/winter-new-york-something-s-missing-absence-snow-upsets-rhythms-urban-life.html?scp=23&sq=global+warming+snow&st=nyt.

19. "Al Roker: Extreme Snow Storms Are Due To Climate Change/Al Roker/Larry King Now on OratV," YouTube, February 24, 2015, https://www.youtube.com/watch?v=-ZZVP2iwN0Y.

20. Interview of Michael Oppenheimer, Marc Morano and Mick Curran, *Climate Hustle*, 2016, www.ClimateHustle.com.

21. "The Ed Show" clip in Morano and Curran, *Climate Hustle*.

22. Clip of CNN's *American Morning* in Morano and Curran, *Climate Hustle*.

23. Paul Eccleston, "Penguins Now Threatened by Global Warming," *Telegraph*, December 11, 2007, http://www.telegraph.co.uk/news/earth/earthnews/3318079/Penguins-now-threatened-by-global-warming.html.

24. "Antarctica is Cold? Yeah, We Knew That," Real Climate, February 21, 2008, http://www.realclimate.org/index.php/archives/2008/02/antarctica-is-cold/.

25. AAP, "Antarctic Sea Ice Is Expanding: Study," *Herald Sun*, March 31, 2013, http://www.heraldsun.com.au/news/breaking-news/antarctic-sea-ice-is-expanding-study/news-story/dce551b4dc9c1cc19d5e4ae1fa2386f6.

26. Andrew Bolt, "Less Ice, More Ice, Whatever. It's Global Warming," *Herald Sun*, March 31, 2013, http://www.heraldsun.com.au/blogs/andrew-bolt/less-ice-more-ice-whatever-its-global-warming/news-story/16609bc3423e9220f387bf31449f42dd.

27. Marc Morano, "AP's Seth Borenstein at It Again! Claims 'Global Warming Means More Antarctic Ice'—Meet the New Consensus, the Opposite of the Old Consensus," Climate Depot, October 10, 2012, http://www.climatedepot.com/2012/10/10/aps-seth-borenstein-at-it-again-claims-global-warming-means-more-antarctic-ice-meet-the-

new-consensus-the-opposite-of-the-old-consensus/.

28. Interview of Patrick Michaels, Morano and Curran, *Climate Hustle*.

29. John Roach, "Hurricanes Have Doubled Due to Global Warming, Study Says," National Geographic News, July 30, 2007, http://news.nationalgeographic.com/news/2007/07/070730-hurricane-warming.html.

30. "Robust Science! More than 30 Contradictory Pairs of Peer-Reviewed Papers," No Tricks Zone, March 30, 2011, http://notrickszone.com/2011/03/30/robust-science-more-than-30-contradictory-pairs-of-peer-reviewed-papers/#sthash.DLCY61Nu.dpbs.

31. "Amazon Rainforests Green-Up with Sunlight in Dry Season," *Geophysical Research Letters*, March 22, 2006, http://onlinelibrary.wiley.com/doi/10.1029/2005GL025583/abstract.

32. "Amazon Forests Did Not Green-up during the 2005 Drought," *Geophysical Research Letters*, March 5, 2010, http://onlinelibrary.wiley.com/doi/10.1029/2009GL042154/abstract.

33. "Climate Change and Geomorphological Hazards in the Eastern European Alps," The Royal Society, April 19, 2010, http://rsta.royalsocietypublishing.org/content/368/1919/2461.

34. "Impact of a Climate Change on Avalanche Hazard," *Annals of Glaciology*, September 14, 2017, https://www.cambridge.org/core/journals/annals-of-glaciology/article/impact-of-a-climate-change-on-avalanche-hazard/929FAF1908AF3358884A24D502A29D2D.

35. "Potential Impacts of Climatic Change on the Breeding and Non-Breeding Ranges and Migration Distance of European Sylvia Warblers," *Journal of Biogeography*, March 19, 2009, http://onlinelibrary.wiley.com/doi/10.1111/j.1365-2699.2009.02086.x/abstract.

36. "Climate Change Leads to Decreasing Bird Migration Distances," *Global Change Biology*, January 15, 2009, http://onlinelibrary.wiley.com/doi/10.1111/j.1365-2486.2009.01865.x/abstract.

37. Eric S. Kasischke, et al., "Fire, Global Warming, and the Carbon Balance of Boreal Forests," *Ecological Applications* 5:2 (May 1995), http://www.jstor.org/stable/1942034.

38. "Future Wildfire in Circumboreal Forests in Relation to Global Warming," *Journal of Vegetation Science*, August 1998, http://onlinelibrary.wiley.com/doi/10.2307/3237261/abstract.

39. "Impacts of Climate Change on Historical Locust Outbreaks in China," *Journal of Geophysical Research*, September 17, 2009, http://onlinelibrary.wiley.com/doi/10.1029/2009JD011833/abstract.

40. "Thousand-Year-Long Chinese Time Series Reveals Climatic Forcing of Decadal Locust Dynamics" *Proceedings of the National Academy of Sciences*, March 16, 2007, http://www.pnas.org/content/104/41/16188.

41. "Climatic Change and Wetland Desiccation Cause Amphibian Decline in Yellowstone National Park," *Proceedings of the National Academy of Sciences*, January 15, 2008, http://www.pnas.org/content/105/44/16988.full.

42. "Decreased Winter Severity Increases Viability of a Montane Frog Population," *Proceedings of the National Academy of Sciences*, April 6, 2010, http://www.pnas.org/content/107/19/8644.

43. "Sea-Level Rise at Tropical Pacific and Indian Ocean Islands," *Global and Planetary Change*, 155–68, September 2006, http://www.sciencedirect.com/science/article/pii/S0921818106000877?via%3Dihub.

44. "The Dynamic Response of Reef Islands to Sea-Level Rise: Evidence from Multi-Decadal Analysis of Island Change in the Central Pacific," *Global and Planetary Change*, June 3, 2010, http://www.sciencedirect.com/science/article/pii/S0921818110001013?via%3Dihub.

45. "Effect of Global Warming on the Length-of-Day," *Geophysical Research Letters*, April 12, 2002, http://onlinelibrary.wiley.com/doi/10.1029/2001GL013672/abstract.

46. "Ocean Bottom Pressure Changes Lead to a Decreasing Length-of-Day in a Warming Climate," *Geophysical Research Letters*, March 28, 2007, http://onlinelibrary.wiley.com/doi/10.1029/2006GL029106/abstract.

47. "A Westward Extension of the Warm Pool Leads to a Westward Extension of the Walker Circulation, Drying Eastern Africa," *Climate Dynamics*, December 2011, https://link.springer.com/article/10.1007%2Fs00382-010-0984-y

48. "African Climate Change: 1900–2100," *Inter-Research Climate Research*, August 15, 2001, http://www.int-res.com/abstracts/cr/v17/n2/p145-168/.

49. "Climate Change Scenarios for Great Lakes Basin Ecosystem Studies," *Limnology and Oceanography*, July 1996 http://www.jstor.org/stable/2838781?seq=1#page_scan_tab_contents.

50. "Increasing Great Lake–Effect Snowfall during the Twentieth Century: A Regional Response to Global Warming?" American Meteorological Society, November 2003, http://journals.ametsoc.org/doi/abs/10.1175/1520-0442(2003)016<3535:IGLSDT>2.0.CO;2.

51. "Slowing of the Atlantic Meridional Overturning Circulation at 25°N," *Nature*, December 1, 2005, https://www.nature.com/nature/journal/v438/n7068/full/nature04385.html.

52. Josh K. Willis, "Can in Situ Floats and Satellite Altimeters Detect Long-Term Changes in Atlantic Ocean Overturning?" *Geophysical Research Letters*, March 25, 2010, http://onlinelibrary.wiley.com/doi/10.1029/2010GL042372/abstract.

53. "Weakening of North Indian SST Gradients and the Monsoon Rainfall in India and the Sahel," American Meteorological Society, May 15, 2006, http://journals.ametsoc.org/doi/full/10.1175/JCLI3820.1.

54. *Climate Change 2007*: *Working Group I: The Physical Science Basis* (Fourth Assessment Report), Intergovernmental Panel on Climate Change, http://www.ipcc.ch/publications_and_data/ar4/wg1/en/ch10s10-3-5-2.html.

55. "Rice Yields in Tropical/Subtropical Asia Exhibit Large but Opposing Sensitivities to Minimum and Maximum Temperatures," *Proceedings of the National Academy of Sciences*, July 6, 2010, http://www.pnas.org/content/107/33/14562.

56. "Climate Change and Rice Yields in Diverse Agro-Environments of India. II. Effect of Uncertainties in Scenarios and Crop Models on Impact Assessment," *Climatic Change*, February 2002, https://link.springer.com/article/10.1023%2FA%3A1013714506779.

57. "Climate Change Consequences on the Biome Distribution in Tropical South America," *Geophysical Research Letters*, May 12, 2007, http://onlinelibrary.wiley.com/doi/10.1029/2007GL029695/abstract.

58. "Effects of Rapid Global Warming at the Paleocene-Eocene Boundary on Neotropical Vegetation," *Science*, November 12, 2010, http://science.sciencemag.org/content/330/6006/957.

59. "Global Synthesis of Leaf Area Index Observations: Implications for Ecological and Remote Sensing Studies," *Global Ecology and Biogeography*, April 11, 2003, http://onlinelibrary.wiley.com/doi/10.1046/j.1466-822X.2003.00026.x/abstract.

60. "Spatial and Temporal Variation of Global LAI during 1981–2006," *Journal of Geographical Sciences*, 2010, http://cat.inist.fr/?aModele=afficheN&cpsidt=22575167.

61. "Climate Change and Future Populations at Risk of Malaria," *Global Environmental Change*, October 1999, http://www.sciencedirect.com/science/article/pii/S0959378099000205?via%3Dihub.

62. "Climate Change and the Global Malaria Recession," *Nature*, April 16, 2010, https://www.nature.com/nature/journal/v465/n7296/full/nature09098.html.

63. "North Sea Cod and Climate Change—Modelling the Effects of Temperature on Population Dynamics," *Global Change Biology*, October 17, 2003, http://onlinelibrary.wiley.com/doi/10.1046/j.1365-2486.2003.00685.x/abstract.

64. "The Response of Atlantic Cod (Gadus Morhua) to Future Climate Change," *ICES Journal of Marine Science*, October 1, 2005, https://academic.oup.com/icesjms/article-lookup/doi/10.1016/j.icesjms.2005.05.015.

65. "Tropical Cyclones (Hurricanes)," *Climate Change 2007*.

66. "Simulated Reduction in Atlantic Hurricane Frequency under Twenty-First-Century Warming Conditions," *Nature Geoscience*, May 18, 2008, https://www.nature.com/ngeo/journal/v1/n6/full/ngeo202.html.

67. "Dilution of the Northern North Atlantic Ocean in Recent Decades," *Science*, June 17, 2005, http://science.sciencemag.org/content/308/5729/1772.

68. "Changes in freshwater content in the North Atlantic Ocean 1955–2006," Geophysical Research Letters, August 21, 2007, http://onlinelibrary.wiley.com/doi/10.1029/2007GL030126/abstract.

69. "Mountains and Sub-Arctic Regions," *Climate Change 2007*.

70. "Will Greenhouse Warming Lead to Northern Hemisphere Ice-Sheet Growth?" *Nature*, January 16, 1992, https://www.nature.com/nature/journal/v355/n6357/abs/355244a0.html.

71. "Impacts of Climate Change on the Tree Line," *Annals of Botany* 90:4 (October 1, 2002), https://academic.oup.com/aob/article-lookup/doi/10.1093/aob/mcf222.

72. "Changes in Climatic Water Balance Drive Downhill Shifts in Plant Species' Optimum

Elevations," *Science*, January 21, 2011, http://science.sciencemag.org/content/331/6015/324.

73. "Simulation of Sahel Drought in the 20th and 21st Centuries," *Proceedings of the National Academy of Sciences*, October 17, 2005, http://www.pnas.org/content/102/50/17891.

74. "Sahel Rainfall Variability and Response to Greenhouse Warming," *Geophysical Research Letters*, September 10, 2005, http://onlinelibrary.wiley.com/doi/10.1029/2005GL023232/abstract.

75. "Climatic Context and Ecological Implications of Summer Fog Decline in the Coast Redwood Region," *Proceedings of the National Academy of Sciences*, August 10, 2009, http://www.pnas.org/content/107/10/4533.

76. Janny Hu, "Get Ready for Even Foggier Summers," *SF Gate*, July 6, 2009, http://www.sfgate.com/science/article/Get-ready-for-even-foggier-summers-3226235.php.

77. "A 20th Century Acceleration in Global Sea-Level Rise," *Geophysical Research Letters*, January 6, 2006, http://onlinelibrary.wiley.com/doi/10.1029/2005GL024826/abstract.

78. "Sea-Level Acceleration Based on U.S. Tide Gauges and Extensions of Previous Global-Gauge Analyses," *Journal of Coastal Research*, February 23, 2011, http://www.bioone.org/doi/abs/10.2112/JCOASTRES-D-10-00157.1.

79. "Predominant Role of Water in Regulating Soil and Microbial Respiration and Their Responses to Climate Change in a Semiarid Grassland," *Global Change Biology*, October 15, 2008, http://onlinelibrary.wiley.com/doi/10.1111/j.1365-2486.2008.01728.x/abstract.

80. "Plants Reverse Warming Effect on Ecosystem Water Balance," *Proceedings of the National Academy of Sciences*, June 16, 2003, http://www.pnas.org/content/100/17/9892.

81. "The Potential Impacts of Climate Change on Inshore Squid: Biology, Ecology and Fisheries," *Reviews in Fish Biology and Fisheries*, November 2008, https://link.springer.com/article/10.1007%2Fs11160-007-9077-3.

82. "Spatial and Temporal Variation in Growth Rates and Maturity in the Indo-Pacific Squid Sepioteuthis Lessoniana (Cephalopoda: Loliginidae)," *Marine Biology*, April 2002, https://link.springer.com/article/10.1007%2Fs00227-001-0746-9.

83. "Climatic Change and Debris Flows in High Mountain Regions: The Case Study of the Ritigraben Torrent (Swiss Alps)," *Climatic Change*, July 1997, https://link.springer.com/article/10.1023%2FA%3A1005356130392.

84. "On the Incidence of Debris Flows from the Early Little Ice Age to a Future Greenhouse Climate: A Case Study from the Swiss Alps," *Geophysical Research Letters*, August 24, 2006, http://onlinelibrary.wiley.com/doi/10.1029/2006GL026805/abstract.

85. "An Extreme Value Analysis of UK Drought and Projections of Change in the Future," *Journal of Hydrology*, June 25, 2010, http://www.sciencedirect.com/science/article/pii/S0022169410002349?via%3Dihub.

86. "Multi-Model Ensemble Estimates of Climate Change Impacts on UK Seasonal Precipitation Extremes," *International Journal of Climatology*, January 6, 2009, http://onlinelibrary.wiley.com/doi/10.1002/joc.1827/abstract.

87. "Changing Cyclones and Surface Wind Speeds over the North Atlantic and Europe in a Transient GHG Experiment," *Inter-Research Climate Research*, July 20, 2000, http://www.int-res.com/abstracts/cr/v15/n2/p109-122/.

88. "Northern Hemisphere Atmospheric Stilling Partly Attributed to an Increase in Surface Roughness," *Nature Geoscience*, October 17, 2010, https://www.nature.com/ngeo/journal/v3/n11/full/ngeo979.html.

89. "Simulation of Recent Northern Winter Climate Trends by Greenhouse-Gas Forcing," *Nature*, June 3, 1999, https://www.nature.com/nature/journal/v399/n6735/full/399452a0.html.

90. "A Link between Reduced Barents-Kara Sea Ice and Cold Winter Extremes over Northern

Continents," *Journal of Geophysical Research*, November 5, 2010, http://onlinelibrary.wiley. com/doi/10.1029/2009JD013568/abstract.

91. Associated Press, "Study: Climate Change Not behind Hurricanes," NBC News, May 18, 2008, http://www.nbcnews.com/id/24694854/ns/us_ news-environment/t/study-climate-change-not-behind-hurricanes/#.WaRuINWPLrc.

92. Eric Berger, "Hurricane Expert Reconsiders Global Warming's Impact," *Houston Chronicle*, April 11, 2008, http://www.chron.com/disp/ story.mpl/tech/news/5693436.html.

93. Shalini Saxena, "Global Warming Means Fewer—but More Powerful—Hurricanes," ARS Technica, May 21, 2015, http://arstechnica.com/ science/2015/05/global-warming-means-fewer-but-more-powerful-hurricanes/.

94. Ethan Weston and Caitlin Balwin, "Climate Change Might Make Intense Hurricanes like Harvey More Common," ABC News 2, August 27, 2017.

95. Alister Doyle, "Extreme Weather Calls for Action, U.N. Climate Chief Says," Reuters, November 28, 2012, http://uk.reuters.com/ article/us-climate-talks-2012/extreme-weather-calls-for-action-u-n-climate-chief-says-idUKBRE8AR0DY20121128.

96. "Climate Change Is Real, Experts Say," News 24 (Capetown), February 2, 2013, http://www. news24.com/Green/News/Climate-change-is-real-experts-say-20130213.

97. Philip Stott, "Global Warming Is Not a Crisis," ABC News, March 9, 2007, http://abcnews. go.com/International/ story?id=2938762&page=1#.UB_XbNn5k74.

98. Bill McGuire, *Waking the Giant*: *How a Changing Climate Triggers Earthquakes, Tsunamis, and Volcanoes*, (Oxford University Press, 2013).

99. Alex Renton, "More Fatal Earthquakes to Come, Geologists Warn," *Newsweek*, April 28, 2015, http://www.newsweek.com/nepal-earthquake-could-have-been-manmade-disaster-climate-change-brings-326017.

100. Helen Lawson, "Earthquakes 'Contribute to Global Warming by Releasing Greenhouse Gas from the Ocean Floor,'" *Daily Mail*, July 28, 2013, http://www.dailymail.co.uk/news/ article-2380434/Earthquakes-contribute-global-warming-releasing-greenhouse-gas-ocean-floor. html.

101. "Which Is It? Time Mag. Goes Both Ways on Volcanoes! TIME Magazine: 'Climate Change Leads to Volcanoes'—OR Flashback Time Magazine 2014: 'Volcanoes May Be Slowing Down Climate Change,'" Climate Depot, January 31, 2015, http://www.climatedepot. com/2015/01/31/which-is-it-time-mag-goes-both-ways-on-volcanoes-time-mag-climate-change-leads-to-volcanoes-or-flashback-time-mag-2014-volcanoes-may-be-slowing-down-climate-change/; Jeffrey Kluger, "How Climate Change Leads to Volcanoes (Really)," *Time*, January 29, 2015, http://time.com/3687893/volcanoes-climate-change/; Francesca Trianni, "Volcanoes May Be Slowing Down Climate Change," *Time*, February 25, 2014, http://time.com/9717/ volcanoes-may-be-slowing-down-climate-change/.

102. Marc Morano, "Dem Sen. Boxer Blames Tornadoes on Global Warming—Plugs Her Carbon Tax Bill to Fix Bad Weather: 'This Is Climate Change. We Were Warned about Extreme Weather…We Need to Protect Our People'—'Carbon Could Cost Us the Planet,'" Climate Depot, May 21, 2013, http://www. climatedepot.com/2013/05/21/dem-sen-boxer-blames-tornadoes-on-global-warming-plugs-her-carbon-tax-bill-to-fix-bad-weather-this-is-climate-change-we-were-warned-about-extreme--weather-we-need-to-protect-our/.

103. Jeff Poor, "Democratic Senator Uses Okla. Tornado for Anti-GOP Rant over Global Warming," Daily Caller, May 20, 2013, http:// dailycaller.com/2013/05/20/democratic-senator-goes-on-anti-gop-rant-over-climate-change-as-tornadoes-hit-oklahoma/#ixzz2TsyeamLC.

104. Interview of Roger Pielke Sr., Morano and Curran, *Climate Hustle.*

105. Roger Pielke Jr., "It Has Been Foretold," Roger Pielke Jr.'s Blog, August 12, 2010, http://rogerpielkejr.blogspot.com/2010/08/it-has-been-foretold.html.

106. Ben Geman, "EPA, Dems Cheer Climate Ruling," *Hill*, June 26, 2012, http://thehill.com/policy/energy-environment/234801-epa-dems-cheer-climate-ruling.

107. "NBC Nightly News: NBC: January 16, 2015 6:30PM–7:01PM EST," NBC, January 16, 2015, https://archive.org/details/WCAU_20150116_233000_NBC_Nightly_News.

108. Bruce Leshan, "Sea Level Rise Could Lap at DC Memorials," WUSA 9, May 14, 2014, http://www.wusa9.com/news/local/dc/sea-level-rise-could-lap-at-dc-memorials/285179702.

109. Marc Morano, "2013 Conflict in Syria Blamed on Drought Caused by Global Warming—Flashback 1933: 'YO-YO BANNED IN SYRIA—Blamed for Drought by Moslems,'" Climate Depot, September 6, 2013, http://www.climatedepot.com/2013/09/06/2013-conflict-in-syria-blamed-on-drought-caused-by-global-warming-flashback-1933-yo-yo-banned-in-syria-blamed-for-drought-by-moslems/.

110. Ibid.

111. Randy Olson, "A Professional Foe of Climate Campaigners Gets His Big-Screen Moment," *New York Times*, April 17, 2016, https://dotearth.blogs.nytimes.com/2016/04/18/a-professional-foe-of-climate-campaigners-gets-his-big-screen-moment/.

112. Marc Morano, "Watch Now: Morano v. Bill Nye Climate Debate on Fox with Stossel—Nye Cites 'Hockey Stick' as Proof—Says Politicians Can Fix Potholes & the Climate—Morano Denounces as 'Medieval Witchcraft'—Full Transcript," Climate Depot, January 28, 2014, http://www.climatedepot.com/2014/01/28/watch-now-morano-v-bill-nye-climate-debate-on-fox-with-stossel-nye-cites-hockey-stick-as-proof-says-politicians-can-fix-potholes-the-climate-morano/.

113. Marc Morano, "SPECIAL REPORT: More than 1000 International Scientists Dissent Over Man-Made Global Warming Claims—Challenge UN IPCC & Gore," Climate Depot, December 8, 2010, http://www.climatedepot.com/2010/12/08/special-report-more-than-1000-international-scientists-dissent-over-manmade-global-warming-claims-challenge-un-ipcc-gore-2/.

114. Barack Obama, "Obama Speech Oceans Receding, Planet Healing," YouTube, June 8, 2008, https://www.youtube.com/watch?v=u2pZSvq9bto.

115. Marc Morano, "Promise Kept—Planet Healer Obama Calls It: In 2008, He Declared His Presidency Would Result in 'the Rise of the Oceans Beginning to Slow'—and by 2011, Sea Level Drops!" Climate Depot, September 21, 2011, http://www.climatedepot.com/2011/09/21/promise-kept-planet-healer-obama-calls-it-in-2008-he-declared-his-presidency-would-result-in-the-rise-of-the-oceans-beginning-to-slow-and-by-2011-sea-level-drops/.

116. League of Conservation Voters, "President Obama: Climate Change Is Not a Hoax," YouTube, September 6, 2012, https://www.youtube.com/watch?v=21qRfPLK-84.

117. Mike Lillis, "Pelosi: Obama Can't Control the Weather," *Hill*, September 5, 2012, http://thehill.com/conventions-2012/dem-convention-charlotte/247641-pelosi-obamas-good-but-cant-control-the-weather.

118. Tony Heller, "Yes We Can't," Real Science, September 7, 2012, https://stevengoddard.wordpress.com/2012/09/07/yes-we-cant/.

119. P. Gosselin, "University College London, Professor of Climatology Claims Mankind Now Able to Control Climate!" No Tricks Zone, January 26, 2015, http://notrickszone.com/2015/01/26/university-college-london-professor-of-climatology-mark-maslin-claims-

mankind-now-able-to-control-climate/#sthash. HoQsNhZN.nI2z1oNS.dpbs.

120. Interview of Sheldon Whitehouse, Morano and Curran, *Climate Hustle.*

121. Richard S. Dunham, "Obama Defends Climate Bill's Cap-and-Trade Plan," SF Gate, June 25, 2009, http://www.sfgate.com/green/article/ Obama-defends-climate-bill-s-cap-and-trade-plan-3294664.php.

122. "EPA's Jackson and Energy Sec. Chu on the Senate Hot Seat," Watts Up with That?, July 7, 2009, https://wattsupwiththat.com/2009/07/07/ epas-jackson-and-energy-sec-chu-on-the-senate-hot-seat/.

123. Michael Moore, @MMFlint, Twitter, March 28, 2017, https://twitter.com/mmflint/status/846797 843193106432?lang=en.

124. Scott Whitlock, "Fear-Mongering ABC Hypes Latest Global Warming Disaster: The End of Coffee!," MRC Newsbusters, November 13, 2012, http://www.newsbusters.org/blogs/nb/scott-whitlock/2012/11/13/fear-mongering-abc-hypes-latest-global-warming-disaster-end.

125. Michael Oppenheimer, NBC Nightly News, March 31, 2014, http://www.nbcnews.com/ video/nightly-news-netcast/54832241.

126. Mark Joseph Ubalde, "'Climate Change Pushes Poor Women to Prostitution, Dangerous Work,'" GMA News TV, November 19, 2009, http://www. gmanetwork.com/news/news/nation/177346/ climate-change-pushes-poor-women-to-prostitution-dangerous-work/story/.

127. "House Dem Warns Climate Change Will Force Millions of Poor Women to Engage in 'Transactional Sex,'" CBS DC, March 26, 2015, http://washington.cbslocal.com/2015/03/26/ house-dem-warns-climate-change-will-force-millions-of-poor-women-to-engage-in-transactional-sex/.

128. Interview of Judith Curry, Morano and Curran, *Climate Hustle.*

129. Anthony Watts, "Global warming creates volatility. I feel it when I'm flying," Watts Up

with That?, August 11, 2009, https:// wattsupwiththat.com/2009/08/11/global-warming-creates-volatility-i-feel-it-when-i'm-flying/.

130. Bjørn Lomborg, "The Climate-Change Distraction," *Wall Street Journal*, September 7, 2017, https://www.wsj.com/articles/the-climate-change-distraction-1504802476.

131. "Fasten Seat Belts: Climate Change Could Mean More Turbulence," NBC News, February 19, 2014, https://www.nbcnews.com/science/science-news/fasten-seat-belts-climate-change-could-mean-more-turbulence-n33956.

132. Interview of Judith Curry, Morano and Curran, *Climate Hustle*, 2016.

133. James West, "One Reason It May Be Harder to Find Flight 370: We Messed Up the Currents," *Mother Jones*, March 21, 2014, http://www. motherjones.com/environment/2014/03/climate-change-malaysia-airlines-370-search/.

134. "Did Global Warming Help Bring Down Air France Flight 447?" Russia Today, June 4, 2009, https://www.rt.com/news/did-global-warming-help-bring-down-air-france-flight-447/.

135. Jeremy Schulman, "Study: Global Warming Will Cause 180,000 More Rapes by 2099," *Mother Jones*, February 27, 2014, http://www. motherjones.com/environment/2014/02/climate-change-murder-rape/.

136. Justin Doom, "Global Warming Sparks Fistfights and War, Researchers Say," Bloomberg News, August 1, 2013, https://www.bloomberg.com/ news/articles/2013-08-01/global-warming-sparks-fistfights-and-war-researchers-say.

137. Louis Sahagun, "Climate Change Brings More Crime," *Los Angeles Times*, February 19, 2014, http://www.latimes.com/science/sciencenow/ la-sci-sn-climate-change-crime-20140219-story. html.

138. Lindsay Abrams, "Increased Murder and War Linked to Climate Change," *Salon*, August 1, 2013, https://www.salon.com/2013/08/01/

increased_murder_and_war_linked_to_climate_change/.

139. Marc Morano, "Flashback 1941: Scientist Claimed Global Warming Caused Hitler—Warmer Temps 'May Produce a Trend toward Dictatorial Govts'—'People Are More Docile & Easily Led in Warm Weather,'" Climate Depot, November 22, 2015, http://www.climatedepot.com/2015/11/22/flashback-1941-scientist-claims-global-warming-caused-hitler-warmer-temps-may-produce-a-trend-toward-dictatorial-govts-people-are-are-more-docile-easily-led-in-warm-weather/.

140. Peter Hoffmann, *The History of the German Resistance 1933–1945* (McGill-Queens University Press, 1996), 655.

141. "War Blamed for Weather Extremes" *Barrier Miner*, August 19, 1941, http://www.climatedepot.com/2012/03/28/flashback-1941-war-caused-weather-extremes-and-global-weirding-the-ancient-belief-that-years-of-war-tend-to-bring-abnormal-vagaries-in-the-weather/.

142. Bruno Tertrais, "The Climate Wars Myth," June 21, 2011, Journal The Washington Quarterly, http://www.tandfonline.com/doi/abs/10.1080/0163660X.2011.587951?src=recsys&journalCode=rwaq20.

143. Tatiana Schlossberg, "How Lowering Crime Could Contribute to Global Warming," *New York Times*, August 3, 2016, https://www.nytimes.com/2016/08/04/science/climate-change-rebound-effect.html?_r=0&mtrref=www.climatedepot.com&gwh=DA65DE6F63B4ADD9CE26D5CA1143DC9F&gwt=pay.

144. "Climate Change Threatens Norway's Moose," *Aftenposten*, May 15, 2008, https://article.wn.com/view/2008/05/15/Climate_change_threatens_Norways_moose/; http://www.climatedepot.com/2016/05/11/flashback-say-what-moose-threatened-by-climate-change-and-causing-it-note-sounds-like-the-problem-will-resolve-itself/.

145. "Norway's Moose Population in Trouble for Belching," *Spiegel*, August 21, 2007, http://www.spiegel.de/international/zeitgeist/global-warming-fears-norway-s-moose-population-in-trouble-for-belching-a-501145.html.

146. Clip from Miles O'Brien, "Gore, UN Panel Lauded For Global Warming Awareness," *Newsroom*, CNN, www.ClimateHustle.com.

147. Ted Turner; Global Warming Will Create Cannibals," Charlie Rose Show, PBS, April 1, 2008, YouTube, https://www.youtube.com/watch?v=DSlB1nW4S54; Brent Baker, "Turner: Global Warming Will Cause Mass Cannibalism, Insurgents Are Patriots," MRC News Busters, April 2, 2008, https://www.newsbusters.org/blogs/nb/brent-baker/2008/04/02/turner-global-warming-will-cause-mass-cannibalism-insurgents-are.

148. Roger Pielke Jr., "It Has Been Foretold," Roger Pielke Jr.'s Blog, August 12, 2010, http://rogerpielkejr.blogspot.com/2010/08/it-has-been-foretold.html.

149. Marc Morano, "MIT Climate Scientist: 'Ordinary People See through Man-Made Climate Fears—but Educated People Are Very Vulnerable,'" Climate Depot, http://www.climatedepot.com/2009/07/06/mit-climate-scientist-ordinary-people-see-through-manmade-climate-fears-but-educated-people-are-very-vulnerable/.

150. "What Climate Change Looks Like: A Race to Save Frozen Mummies," *New York Times*, December 9, 2015, https://www.nytimes.com/interactive/projects/cp/climate/2015-paris-climate-talks/what-climate-change-looks-likea-race-to-save-frozen-mummies.

151. Lizzie Wade, "Climate Change Turns Mummies into Black Ooze," *Science*, March 9, 2015, http://www.sciencemag.org/news/2015/03/climate-change-turns-mummies-black-ooze.

152. "The Change of Climate," *Maitland Mercury & Hunter River General Advertiser*, March 11, 1846, https://stevengoddard.wordpress.com/2013/05/21/was-captain-cook-the-first-

climate-climate-wrecking-white-man/ and
http://trove.nla.gov.au/newspaper/article/67978
7?searchTerm=climate%20
change&searchLimits=.

153. Bill McKibben, "Immigration Reform—for the
Climate," *Los Angeles Times*, March 14, 2013,
http://articles.latimes.com/2013/mar/14/
opinion/la-oe-mckibben-immigration-
environment-20130314.

154. John Brignell, "A Complete List of Things Caused
by Global Warming," Number Watch, 2015,
numberwatch.co.uk/warmlist.htm.

155. "A (Not Quite Complete) List of Things
Supposedly Caused by Global Warming," What
Really Happened, http://whatreallyhappened.
com/WRHARTICLES/globalwarming2.html.

156. Marc Morano, "'The Dead Rising from the
Grave!' Global Warming Claims Imitate Scene
from 1984 Comedy 'Ghostbusters'—'A Disaster of
Biblical Proportions...Real Wrath of God Type
Stuff,'" Climate Depot, March 10, 2014, http://
www.climatedepot.com/2014/03/10/the-dead-
rising-from-the-grave-global-warming-claims-
imitate-scene-from-1984-comedy-ghostbusters-a-
disaster-of-biblical-proportions-real-wrath-of-
-god-type-stuff/.

157. Rebecca Savransky, "Gore: 'TV News Is like a
Nature Hike through the Book of Revelations,"
The Hill, August 6, 2017, http://thehill.com/
policy/energy-environment/345515-gore-tv-
news-is-like-a-nature-hike-through-the-book-of-
revelation.

158. Rose Marie Berger, "God's Wrath Caused East
Coast Earthquake," HuffPost, October 24, 2011,
http://www.huffingtonpost.com/rose-marie-
berger/keystone-pipeline-protest-
earthquake_b_934776.html.

159. Anthony Watts, "Quote of the Week: Dr. James
Hansen of NASA GISS, Unhinged," Watts Up
with That, January 12, 2012, https://
wattsupwiththat.com/2012/01/12/quote-of-the-
week-dr-james-hansen-of-nasa-giss-unhinged/.

160. Geoffrey Lean, "Hay Festival 2012: Government
Adviser Bill McGuire Says Global Warming Is
Causing Earthquakes and Landslides,"
Telegraph, June 5, 2012, http://www.telegraph.
co.uk/culture/hay-festival/9312347/Hay-Festival-
2012-Government-adviser-Bill-McGuire-says-
global-warming-is-causing-earthquakes-and-
landslides.html.

161. Sarah Griffiths, "Will Climate Change Bring
Back SMALLPOX? Siberian Corpses Could Ooze
Contagious Virus If Graveyards Thaw Out, Claim
Scientists," *Daily Mail*, March 10, 2014, http://
www.dailymail.co.uk/sciencetech/
article-2551664/Will-climate-change-bring-
SMALLPOX-Siberian-corpses-ooze-contagious-
virus-graveyards-thaw-claim-scientists.html.

162. Morano, "'The Dead Rising from the Grave!'"

Chapter 12:
Not So Extreme

1. Michael E. Mann, "It's a Fact: Climate Change
Made Hurricane Harvey More Deadly," *Guardian*, August 28, 2017, https://www.theguardian.
com/commentisfree/2017/aug/28/climate-
change-hurricane-harvey-more-deadly; Marc
Morano, "What Hurricane 'Consensus'?! 'Global
Warming' Causes MORE Hurricanes—Except
When It Causes FEWER Hurricanes," Climate
Depot, August 28, 2017, http://www.climatede-
pot.com/2017/08/28/what-hurricane-consensus-
global-warming-causes-more-hurricanes-except-
when-it-causes-less-hurricanes/.

2. Roger Pielke Jr., "Guest Commentary: Climate
Spin Is Rampant," *Denver Post*, October 11, 2012,
http://www.denverpost.com/2012/10/11/guest-
commentary-climate-spin-is-rampant/.

3. Brian K. Sullivan, "There Was Nothing Normal
About America's Freakish Winter Weather,"
Bloomberg, April 10, 2017, https://www.
bloomberg.com/news/features/2017-04-10/there-
was-nothing-normal-about-america-s-freakish-
winter-weather?bcomANews=true.

4. Marc Morano, "Scientist to Congress: 'No Evidence' That Hurricanes, Floods, Droughts, Tornadoes Are Increasing," Climate Depot, March 29, 2017, http://www.climatedepot.com/2017/03/29/scientist-to-congress-no-evidence-that-hurricanes-floods-droughts-tornadoes-are-increasing/.

5. Susan Brooks Thistlethwaite, "'Super' Typhoon Haiyan: Suffering and the Sin of Climate Change Denial," *Washington Post*, November 12, 2013, http://www.washingtonpost.com/blogs/on-faith/wp/2013/11/12/super-typhoon-haiyan-suffering-and-the-sin-of-climate-change-denial/ and https://www.onfaith.co/onfaith/2013/11/12/super-typhoon-haiyan-suffering-and-the-sin-of-climate-change-denial.

6. Roger Pielke Jr., "Weather-Related Natural Disasters: Should We Be Concerned about a Reversion to the Mean?" Risk Frontiers, July 31, 2017, https://riskfrontiers.com/weather-related-natural-disasters-should-we-be-concerned-about-a-reversion-to-the-mean/.

7. Interview of Tony Heller, Marc Morano and Curran, *Climate Hustle*, 2016, www.ClimateHustle.com.

8. Eric Holthaus, "Harvey Is What Climate Change Looks Like," *Politico*, August 28, 2017, http://www.politico.com/magazine/story/2017/08/28/climate-change-hurricane-harvey-215547.

9. John D. Sutter, "Hurricane Matthew Looks a Lot like the Future of Climate Change," CNN, October 7, 2016, http://www.cnn.com/2016/10/07/opinions/sutter-hurricane-matthew-climate-change/index.html.

10. "Superstorm Sandy Is 'What Global Warming Looks Like,'" Environmental News Service, October 30, 2012, http://ens-newswire.com/2012/10/30/superstorm-sandy-is-what-global-warming-looks-like/.

11. Ross Gelbspan, "Katrina's Real Name: 'Its Real Name Is Global Warming," *Boston Globe*, August 30, 2005, http://archive.boston.com/news/weather/articles/2005/08/30/katrinas_real_name/.

12. John R. Christy, Testimony before the U.S. House of Representatives Subcommittee on Energy and Power, September 20, 2012, https://archives-energycommerce.house.gov/sites/republicans.energycommerce.house.gov/files/Hearings/EP/20120920/HHRG-112-IF03-WState-ChristyJ-20120920.pdf.

13. "Al Gore 2017: Was That Science or Gratuitous Random Weather-Porn to Fuel Superstitious Belief," Jo Nova, July 14th, 2017, http://joannenova.com.au/2017/07/al-gore-2017-was-that-science-or-gratuitous-random-weather-porn-to-fuel-superstitious-belief/.

14. Roy W. Spencer, "Hillary Clinton Boards the Climate Crisis Train to Nowhere," *Forbes*, October 25, 2016, https://www.forbes.com/sites/realspin/2016/10/25/hillary-clinton-boards-the-climate-crisis-train-to-nowhere/#46b71bbc371e.

15. Interview of Tom Wysmuller, Marc Morano and Mick Curran, *Climate Hustle*, 2016, www.ClimateHustle.com.

16. Tony Heller, "1000 Days Without Hurricane Landfall?" Real Climate Science, October 24, 2010, https://stevengoddard.wordpress.com/2010/10/24/1000-days-without-hurricane-landfall/.

17. Marc Morano, "Flashback 1981: Climatologists Blame Recurring Droughts & Floods on a Global Cooling Trend That Could Trigger Massive Tragedies for Mankind," Climate Depot, September 9, 2016, http://www.climatedepot.com/2016/09/09/flashback-1981-climatologists-blame-recurring-droughts-floods-on-a-global-cooling-trend-that-could-trigger-massive-tragedies-for-mankind/.

18. Patrick Hughes, "Climate: A Key to the World's Food Supply," National Oceanographic and Atmospheric Administration, October 1974, https://docs.lib.noaa.gov/rescue/journals/noaa/QC851U461974oct.pdf#page=5.

19. "The Cooling World," *Newsweek*, April 28, 1975, http://www.denisdutton.com/cooling_world.htm.

20. "No Major U.S. Hurricane Landfalls in Nine Years: Luck?" National Aeronautics and Space Administration, May 13, 2015, https://www.giss.nasa.gov/research/news/20150513/.

21. National Hurricane Center, "U.S. Hurricane Strikes by Decade," National Oceanographic and Atmospheric Administration, http://www.nhc.noaa.gov/pastdec.shtml.

22. "UN IPCC AR5 Working Group I, 2013, chapter 2, http://www.ipcc.ch/pdf/assessment-report/ar5/wg1/WG1AR5_Chapter02_FINAL.pdf.

23. P. Gosselin, "US Atmospheric Scientist Sees No Link Between Accumulated Cyclone Energy and Global Warming over Past 30 Years," No Tricks Zone, September 20, 2016, http://notrickszone.com/2016/09/20/us-atmospheric-scientist-sees-no-link-between-accumulated-cyclone-energy-and-global-warming-over-past-30-years/#sthash.DzmD9uYt.D7i7ycuc.dpbs.

24. Michael Chenoweth, "A New Compilation of North Atlantic Tropical Cyclones, 1851–1898," *Journal of Climate*, December 5, 2014, http://journals.ametsoc.org/doi/abs/10.1175/JCLI-D-13-00771.1?af=R; http://hockeyschtick.blogspot.com/2014/09/new-paper-finds-strong-hurricanes-were.html.

25. NOAA website, "What Is the Difference between a Hurricane, a Cyclone, and a Typhoon?," https://oceanservice.noaa.gov/facts/cyclone.html, accessed December 20, 2017.

26. Kenneth Richard, "Scientific 'Consensus' Says 'Global Warming' Leads to Less Intense, Less Frequent Hurricanes," No Tricks Zone, October 10, 2016, http://notrickszone.com/2016/10/10/scientific-consensus-30-papers-global-warming-leads-to-less-intense-less-frequent-hurricanes/#sthash.KYYZYa6A.dpbs.

27. Huang XiaoyanHe Li and Zhao Huasheng Huang Ying, "Characteristics of Tropical Cyclones Generated in South China Sea and Their Landfalls over China and Vietnam," published in *Natural Hazards* 88:2 (September 2017, 1043–1057, https://link.springer.com/article/10.1007/s11069-017-2905-4.

28. Kim Stephens, "Cyclone Blanche Is Latest to Cross Land in Second Consecutive Quiet Season in Australian History," news.com.au, March 10, 2017, http://www.news.com.au/technology/environment/climate-change/cyclone-blanche-is-latest-to-cross-land-in-second-consecutive-quiet-season-in-australian-history/news-story/220bd07cbd24d1db32cfd2175d3ec2ac.

29. Paul Homewood, "Pacific Ocean Seeing One of the Quietest Typhoon Seasons on Record," Not A Lot of People Know That, October 11, 2017, https://notalotofpeopleknowthat.wordpress.com/2017/10/11/pacific-ocean-seeing-one-of-the-quietest-typhoon-seasons-on-record/.

30. Darren Goode, "Senate Dems Bash Climate Deniers," *Politico*, November 30, 2012, http://www.politico.com/story/2012/11/senate-panel-democrats-bash-climate-change-deniers-084412.

31. Andrew Freedman, "NWS Confirms Sandy Was Not a Hurricane at Landfall," Climate Central, February 12, 2013, http://www.climatecentral.org/news/nws-confirms-sandy-was-not-a-hurricane-at-landfall-15589.

32. Kerry Jackson, "Warming Alarmists Redefine What a Hurricane Is so We'll Have More of Them," Investor's Business Daily, October 17, 2016, http://www.investors.com/politics/commentary/warming-alarmists-redefine-what-a-hurricane-is-so-well-have-more-of-them/.

33. Interview of Patrick J. Michaels, Marc Morano and Mick Curran, *Climate Hustle*, 2016, www.ClimateHustle.com

34. Joe Bastardi, "Hermine a Poor Example to Push Man-Made Global Warming," The Patriot Post, September. 8, 2016, https://patriotpost.us/opinion/44728.

35. "Historic Hurricane Harvey's Recap," Weather Channel, September 2, 2017, https://weather.com/storms/hurricane/news/tropical-storm-harvey-forecast-texas-louisiana-arkansas.

36. Judith Curry, "Hurricane Harvey: Long-Range Forecasts," Climate Etc., August 27, 2017, https://judithcurry.com/2017/08/27/hurricane-harvey-long-range-forecasts/.

37. Valerie Richardson, "Not Unprecedented: Hurricanes Irma and Harvey Rank Seventh, 18th after Making Landfall," *Washington Times*, September 11, 2017, http://www.washingtontimes.com/news/2017/sep/11/hurricanes-irma-and-harvey-rank-seventh-18th-after/.

38. Jonathan Erdman, "Hurricane Maria Was One of the 10 Most Intense Atlantic Basin Hurricanes on Record," September 20, 2017, https://weather.com/storms/hurricane/news/hurricane-maria-one-of-most-intense-atlantic-hurricanes.

39. Richardson, "Not Unprecedented."

40. Jess Shankleman and Stefan Nicola, "Hurricane Irma Made Worse by Climate Change, Scientists Say," Bloomberg News, September 6, 2017, https://www.bloomberg.com/news/articles/2017-09-06/hurricane-irma-was-made-worse-by-climate-change-scientists-say.

41. Marc Morano, "Prof. Michael Mann Warns Hurricanes May 'Literally Force Us to Relocate the Major Coastal Cities of the World'—Recipe for Climate 'Catastrophe,'" Climate Depot, September 7, 2017, http://www.climatedepot.com/2017/09/07/prof-michael-mann-warns-hurricanes-may-literally-force-us-to-relocate-the-major-coastal-cities-of-the-world/.

42. Seth Borenstein, "Winds, Fire, Floods and Quakes: A Nutty Run of Nature," Associated Press, September 08, 2017, https://apnews.com/b0860812593744c588dd9db9f6a8b599/Winds,-fire,-floods-and-quakes:-A-nutty-run-of-nature.

43. Ryan Maue, @RyanMaue, Twitter, September 13, 2017, https://twitter.com/RyanMaue/status/908030951975989261.

44. Andrew Griffin, "Hurricane Irma Likely to be Followed by More Extreme Weather Events so We Should Prepare for Horror of Global Warming Now, Say Experts," *Independent*, September 7, 2017, http://www.independent.co.uk/environment/irma-hurricane-extreme-weather-events-climate-change-global-warming-experts-a7934211.html.

45. Marc Morano, "Climate Depot's Point-by-Point Rebuttal to Warmist Claims on Extreme Weather Events," Climate Depot, September 7, 2017, http://www.climatedepot.com/2017/09/07/climate-depots-point-by-point-rebuttal-to-warmist-claims-on-extreme-weather-events/.

46. Ryan Maue, "Climate Change Hype Doesn't Help," *Wall Street Journal*, September 17, 2017, https://www.wsj.com/articles/climate-change-hype-doesnt-help-1505672774.

47. Ryan Maue, @RyanMaue, Twitter, September 14, 2017, https://twitter.com/RyanMaue/status/908367291133435904.

48. Roger A. Pielke Sr ., @RogerAPielkeSr, Twitter, September 12, 2017, https://twitter.com/RogerAPielkeSr/status/907774541622951938.

49. Steve Goddard, @SteveSGoddard, Twitter, September 12, 2017, https://twitter.com/SteveSGoddard/status/907804829602291716.

50. Marc Morano, "Weather Channel Hypes Report Claiming 'Climate Change Will Mean More Stress, Anxiety, PTSD in the Future'—Growing 'Substance Abuse' due to AGW—Expect 'Broad Psychological Impacts,'" Climate Depot, June 10, 2014, http://www.climatedepot.com/2014/06/10/weather-channel-hypes-report-claiming-climate-change-will-mean-more-stress-anxiety-ptsd-in-the-future-growing-substance-abuse-due-to-agw-expect-broad-psychological-impacts/.

51. Bjorn Lomborg, @Bjorn Lomborg, Twitter, September 7, 2017, https://twitter.com/BjornLomborg/status/905820404765179904.

52. Marc Morano, "Chart: As CO_2 Has risen, Major Landfalling US Hurricanes Declining over Past 140 Years—'Maybe We Need MORE CO_2,'" Climate Depot, September 7, 2017, http://www.climatedepot.com/2017/09/07/chart-as-co2-has-risen-major-landfalling-us-hurricanes-declining-over-past-140-years-maybe-we-need-more-co2/.

53. "Alex Epstein praises oil & gas industry: 'Thousands upon thousands of lives saved in Texas thanks to fossil fuels and the development they make possible,'" August 30, 2017, http://

www.climatedepot.com/2017/08/30/alex-epstein-praises-oil-gas-industry-thousands-upon-thousands-of-lives-saved-in-texas-thanks-to-fossil-fuels-and-the-development-they-make-possible/.

54. Valerie Richardson, "Climate Scientist Rebuts Hollywood Hurricane Hype: 'This Is What Weather Looks Like,'" *Washington Times*, September 21, 2017, http://www.washingtontimes.com/news/2017/sep/21/climate-scientist-rebuts-hollywood-hurricane-hype/.

55. Ibid.

56. Roy W. Spencer, "Texas Major Hurricane Intensity Not Related to Gulf Water Temperatures," Roy Spencer, Ph.D. (blog), August 29th, 2017, http://www.drroyspencer.com/2017/08/texas-major-hurricane-intensity-not-related-to-gulf-water-temperatures/.

57. Richardson, "Climate Scientist Rebuts Hollywood."

58. Roger Pielke Jr. "Pielke on Climate #5," The Climate Fix, September 18, 2017, https://theclimatefix.wordpress.com/2017/09/18/pielke-on-climate-5/.

59. Clip from Dan Rather, *CBS Evening News*, February 19, 2001, in Morano and Curran, *ClimateHustle*.

60. Michael Mogil, "On Misrepresenting Hurricane Statistics (H. Michael Mogil, CCM, CBM, NWA-DS*)" H. Michael Mogil's Lifelong Learning Blog, September 19, 2017, http://www.weatherworks.com/lifelong-learning-blog/?p=1411.

61. James Rosen, "NOAA Scientist Rejects Global Warming Link to Tornadoes," Fox News, April 28, 2011, http://www.foxnews.com/politics/2011/04/28/noaa-scientist-rejects-global-warming-link-tornadoes.html.

62. Interview of Caleb Rossiter, Morano and Curran, *Climate Hustle*.

63. Roy W. Spencer, "Today's Tornado Outlook: High Risk of Global Warming Hype," Roy Spencer, Ph.D. (blog), May 24th, 2011, http://www.drroyspencer.com/2011/05/todays-tornado-outlook-high-risk-of-global-warming-hype/.

64. Rosen, "NOAA Scientist Rejects Global Warming Link."

65. Marc Morano, "Prof Roger Pielke Jr: Testimony on the Current State of Weather Extremes: "It Is Misleading, and Just Plain Incorrect, to Claim That Disasters Associated with Hurricanes, Tornadoes, Floods or Droughts Have Increased on Climate Timescales Either in the United States or Globally," Climate Depot, July 2013, http://www.climatedepot.com/2013/07/18/prof-roger-pielke-jr-testimony-on-the-current-state-of-weather-extremes-it-is-misleading-and-just-plain-incorrect-to-claim-that-disasters-associa-ted-with-hurricanes-tornadoes-floods-or-d/.

66. Roger Pielke Jr., "Guest Commentary: Climate Spin Is Rampant," *Denver Post*, October 11, 2012, http://www.denverpost.com/2012/10/11/guest-commentary-climate-spin-is-rampant/.

67. Marc Morano, "'Floods Are Not Increasing': Dr. Roger Pielke Jr. Slams 'Global Warming' Link to Floods & Extreme Weather—How Does Media 'Get Away with This?" Climate Depot, August 23, 2016, http://www.climatedepot.com/2016/08/23/floods-are-not-increasing-dr-roger-pielke-jr-slams-global-warming-link-to-floods-extreme-weather-how-does-media-get-away-with-this/.

68. Tony Heller, "Tornado Outbreak of April 3–4, 1974," Real Climate Science, April 3, 2017, https://realclimatescience.com/2017/04/tornado-outbreak-of-april-3-4-1974/.

69. "Review of the Draft Climate Science Special Report," National Academy of Sciences, 2017, https://www.nap.edu/catalog/24712/review-of-the-draft-climate-science-special-report; Roger Pielke Jr., "Pielke on Climate #3," The Climate Fix, June 15, 2017, https://theclimatefix.wordpress.com/2017/06/15/pielke-on-climate-3/.

70. Joe Bastardi, @BigJoeBastardi, Twitter, November 8, 2016, https://twitter.com/BigJoeBastardi/status/796031582654107649.

71. Marc Morano, "NOAA Tornado Data: 2016 'One of the Quietest Years Since Records Began in 1954'—Below Average for 5th Year in a Row," Climate Depot, November 12, 2016, http://www.climatedepot.com/2016/11/12/noaa-tornado-data-2016-one-of-the-quietest-years-since-records-began-in-1954-below-average-for-5th-year-in-a-row/.

72. Paul Homewood, "NOAA's Tornado Fraud," Not A Lot of People Know That, January 15, 2017, https://notalotofpeopleknowthat.wordpress.com/2017/01/15/noaas-tornado-fraud/.

73. Nicholas Kristof, "As Donald Trump Denies Climate Change, These Kids Die of It," *New York Times*, January 6, 2017, https://www.nytimes.com/2017/01/06/opinion/sunday/as-donald-trump-denies-climate-change-these-kids-die-of-it.html?emc=eta1&mtrref=www.climatedepot.com&gwh=57E1FA982DAAB0B86A770E7B95663143&gwt=pay&assetType=opinion.

74. Justin Sheffield, et al., "Little Change in Global Drought over the Past 60 Years," *Nature*, November 15, 2012, http://www.nature.com/nature/journal/v491/n7424/abs/nature11575.html.

75. Morano, "Prof Roger Pielke Jr: Testimony."

76. Doyle Rice, "U.S. Drought Reaches Record Low as Rain Reigns," *USA Today*, April 27, 2017, https://www.usatoday.com/story/weather/2017/04/27/us-drought-record-low/100971018/.

77. Nick Stockton, "Thanks El Niño, but California's Drought Is Probably Forever," Wired, May 13, 2016, https://www.wired.com/2016/05/thanks-el-nino-californias-drought-probably-forever/.

78. Paul Rogers, "California Storms Add 350 Billion Gallons to Parched Reservoirs," *Mercury News*, January 9, 2017, http://www.mercurynews.com/2017/01/09/california-storms-fill-drought-parched-reservoirs/; Mark Gomez, "California Storms: Wettest Water Year, So Far, in 122 Years of Records," *Mercury News*, March 8, 2017, http://www.mercurynews.com/2017/03/08/california-storms-wettest-water-year-so-far-in-122-years-of-records/.

79. "Review of the Draft Climate Science Special Report."

80. The Earth Institute at Columbia University, "New Drought Atlas Maps 2,000 Years of Climate in Europe: Completes the First Big-Picture View across Northern Hemisphere," Science Daily, November 6, 2015. <www.sciencedaily.com/releases/2015/11/151106144515.htm.

81. Jo Nova, "Megadroughts in Past 2000 Years Worse, Longer, than Current Droughts," November 6, 2015, http://joannenova.com.au/2015/11/megadroughts-in-past-2000-years-worse-longer-than-current-droughts/.

82. Gregory J. McCabe, et al., "Variability of Runoff-Based Drought Conditions in the Conterminous United States," *International Journal of Climatology*, May 6, 2016, http://onlinelibrary.wiley.com/doi/10.1002/joc.4756/full.

83. Paul Rogers, "California Drought: Past Dry Periods Have Lasted More than 200 Years, Scientists Say," *Mercury News*, January 25, 2014, http://www.mercurynews.com/2014/01/25/california-drought-past-dry-periods-have-lasted-more-than-200-years-scientists-say/.

84. Ibid.

85. Brantley Hargrove, "Chatting with a NOAA Meteorologist about This Drought: What It Is and What It Ain't," *Dallas Observer*, September 16, 2011, http://www.dallasobserver.com/news/chatting-with-a-noaa-meteorologist-about-this-drought-what-it-is-and-what-it-aint-7133049.

86. Marc Morano, "Flashback: Experts Blamed 'Global Cooling' for fhe Widespread Droughts of the 1970s," Climate Depot, September 9, 2016, http://www.climatedepot.com/2016/09/09/flashback-experts-blamed-global-cooling-for-the-widespread-droughts-of-the-1970s/.

87. James P. Sterba, "Climatologists Forecast Stormy Economic Future," *New York Times*, July 12, 1976, http://www.nytimes.com/1976/07/12/archives/climatologists-forecast-stormy-economic-future-climatologists.html.

88. Linyin Cheng, et al., "How Has Human-Induced Climate Change Affected California Drought Risk?" *Journal of Climate*, January 2016, http://journals.ametsoc.org/doi/10.1175/JCLI-D-15-0260.1.

89. Marc Morano, "Climatologist Dr. David Legates Tells U.S. Senate: 'Droughts in the U.S. Are More Frequent and More Intense during Colder Periods,'" Climate Depot, David Legates, "Statement to the Environment and Public Works Committee of the United States Senate," June 3, 2014, http://www.climatedepot.com/2014/06/03/climatologist-dr-david-legates-droughts-in-the-united-states-are-more-frequent-and-more-intense-during-colder-periods/.

90. Roger Pielke Jr., "Are US Floods Increasing? The Answer Is Still No," Roger Pielke Jr.'s Blog, October 24, 2011, http://rogerpielkejr.blogspot.com/2011/10/are-us-floods-increasing-answer-is.html.

91. Marc Morano, "Prof. Roger Pielke Jr. Rips Flooding Claims Using UN IPCC Quotes: 'No Gauge-Based Evidence Has Been Found for a Climate-Driven, Globally Widespread Change in the Magnitude and Frequency of Floods,'" Climate Depot, August 15, 2014, http://www.climatedepot.com/2014/08/15/prof-roger-pielke-jr-rips-flooding-claims-using-un-ipcc-quotes-no-gauge-based-evidence-has-been-found-for-a-climate-driven-globally-widespread-change-in-the-magnitude-and-frequency-of-floods/.

92. Roger Pielke Jr., @RogerPielkeJr, Twitter, August 22, 2016, https://twitter.com/RogerPielkeJr/status/767712040840409088.

93. "Review of the Draft Climate Science Special Report."

94. Marc Morano, "All of the World's Deadliest Floods Occurred with CO_2 Well Below 350 PPM," Climate Depot, December 5, 2011 http://www. climatedepot.com/2011/12/05/all-of-the-worlds-deadliest-floods-occurred-with-co2-well-below-350-ppm/.

95. R. M. Hirsch and K. R. Ryberg, "Has the Magnitude of Floods across the USA Changed with Global CO_2 levels?" *Hydrological Sciences Journal*, October 24, 2011, http://www.tandfonline.com/doi/abs/10.1080/02626667.2011.621895.

96. Andrew Restuccia, "Study Finds No Evidence that Climate Change Caused More Severe Flooding," *Hill*, October 24, 2011, http://thehill.com/policy/energy-environment/189349-study-finds-no-evidence-that-climate-change-has-caused-more-severe-floods.

97. Robert Holmes, "Flood Frequency Q&A," U.S. Geological Survey, 2015, https://water.usgs.gov/floods/events/2015/Joaquin/HolmesQA.html.

98. "Climate-Driven Variability in the Occurrence of Major Floods across North America and Europe," *Journal of Hydrology* 552 (September 2017), 704–17, http://www.sciencedirect.com/science/article/pii/S002216941730478X#%21

99. Holmes, "Flood Frequency Q&A."

100. Richardson, "Climate Scientist Rebuts Hollywood."

101. "Another Ice Age?" *Time*, June 24, 1974, https://stevengoddard.wordpress.com/2014/05/01/rain-used-to-be-caused-by-global-cooling-but-now-caused-by-global-warming/.

102. Roy W. Spencer, "South Carolina Flooding is NOT a 1 in 1,000 Year Event," Roy Spencer, Ph.D. (blog), October 6, 2015, http://www.drroyspencer.com/2015/10/south-carolina-flooding-is-not-a-1-in-1000-year-event/.

103. Topper Shutt, Tenacious Topper Shutt, Facebook, August 28, 2017, https://www.facebook.com/permalink.php?story_fbid=1783008711728131&id=161366410559044.

104. Sean Long, "Networks Blame Wildfires, Droughts on Climate Change, Despite Fact They've Declined," Newsbusters, June 5, 2014, https://www.newsbusters.org/blogs/nb/

sean-long/2014/06/05/networks-blame-wildfires-droughts-climate-change-despite-fact-theyve.

105. Yvonne Gonzalez, "In Las Vegas, Gore Says Global Warming Worsening Hurricanes, Drought," *Las Vegas Sun*, October 13, 2017, https://lasvegassun.com/news/2017/oct/13/in-las-vegas-gore-says-global-warming-worsening-hu/.

106. Stefan H. Doerr and Cristina Santín, "Global Trends in Wildfire and Its Impacts: Perceptions versus Realities in a Changing World," *Philosophical Transactions of the Royal Society B*, May 23, 2016, http://rstb.royalsocietypublishing.org/content/371/1696/20150345.

107. Juli G. Pausas amd Santiago Fernández Muñoz, "Fire Regime Changes in the Western Mediterranean Basin: From Fuel-Limited to Drought-Driven Fire Regime," *Climatic Change*, March 21, 2011, http://portal.uc3m.es/portal/page/portal/actualidad_cientifica/noticias/forest_fires.

108. Michael Bastasch, "Report: Global Warming Not Causing More Wildfires," The Daily Caller, June 3, 2014, http://dailycaller.com/2014/06/03/report-global-warming-not-causing-more-wildfires/.

109. Paige St. John, "Gov. Brown's Link between Climate Change and Wildfires Is Unsupported, Fire Experts Say," *Los Angeles Times*, October 18, 2015, http://www.latimes.com/local/politics/la-me-pol-ca-brown-wildfires-20151019-story.html.

110. Tom Banse, "Is Wildfire Severity Really Getting Worse?" NW News Network, August 8, 2014, http://nwnewsnetwork.org/post/wildfire-severity-really-getting-worse.

111. "Colorado's Front Range Fire Severity Today Not Much Different than in Past, Says CU-Boulder Study," University of Colorado Boulder, September 24, 2014, https://www.colorado.edu/today/2014/09/24/colorados-front-range-fire-severity-today-not-much-different-past-says-cu-boulder-study.

112. NonGovernmental International Panel on Climate Change, "Why Scientists Disagree About Global Warming," Heartland Institute, http://climatechangereconsidered.org.

113. Interview of Christopher Monckton, Morano and Curran, *Climate Hustle*.

114. Interview of Robert Giegengack, Morano and Curran, *Climate Hustle*.

115. Osborne, K., Dolman, et al, "Disturbance and the Dynamics of Coral Cover on the Great Barrier Reef (1995–2009)," *PLoS ONE* 6: 10.1371/journal.pone.0017516, http://www.co2science.org/articles/V14/N26/B1.php.

116. "Great Barrier Reef Scare: Exaggerated Threats Says Head of GBR Authority," Jo Nova, June 6, 2016, http://joannenova.com.au/2016/06/great-barrier-reef-scare-exaggerated-threats-says-head-of-gbr-authority/.

117. Patrick Moore, "Ocean 'Acidification' Alarmism in Perspective," Frontier Centre for Public Policy, November 2015, https://fcpp.org/sites/default/files/documents/Moore%20-%20Ocean%20Acidification%20Alarmism.pdf.

Chapter 13:
The Ever-Receding Tipping Point

1. Interview of Christiana Figurines, Marc Morano and Mick Curran, *Climate Hustle*, 2016, www.ClimateHustle.com.

2. Al Gore, Copenhagen Climate Summit, 2009, www.ClimateHustle.com.

3. Al Gore, *Charlie Rose Show*, PBS, 2006, www.ClimateHustle.com.

4. Laurie David, *Today Show*, NBC, February 3, 2007, www.ClimateHustle.com

5. "Planet Earth Is a Sick Patient Due to Climate Change, Says Prince Charles," *Guardian*, February 26, 2015, https://www.theguardian.com/environment/2015/feb/26/planet-earth-is-a-sick-patient-due-to-climate-change-says-prince-charles.

6. Prince Charles, 2014 video, www.ClimateHustle.com.

7. *Nightline*, ABC News, 2007, www.ClimateHustle. com.

8. Marc Morano, "Earth 'Serially Doomed': UN Issues New 15 Year Climate Tipping Point—but UN Issued Tipping Points in 1982 & Another 10-Year Tipping Point in 1989!" Climate Depot, April 16, 2014, http://www.climatedepot. com/2014/04/16/earth-serially-doomed-un-issues-new-15-year-climate-tipping-point-but-un-first-issued-10-year-tipping-point-in-1989/.

9. Marc Morano, "Alert: NASA's James Hansen Declared Obama Has One Week Left to Save The Planet!—'On Jan. 17, 2009 Hansen Declared Obama Only 'Has Four Years to Save Earth'— Only 7 Days Left!" Climate Depot, January 10, 2013, http://www.climatedepot.com/2013/01/10/ alert-nasas-james-hansen-declared-obama-has-one-week-left-to-save-the-planet-on-jan-17-2009-hansen-declared-obama-only-has-four-years-to-save-earth-only-7-days-left/.

10. Robert Verkaik, "Just 96 Months to Save World, Says Prince Charles," *Independent*, July 8, 2009, http://www.independent.co.uk/environment/ green-living/just-96-months-to-save-world-says-prince-charles-1738049.html.

11. "PM Warns of Climate 'Catastrophe,'" BBC News, October 19, 2009, http://news.bbc.co.uk/2/hi/ uk_news/8313672.stm.

12. Elizabeth May, "'We Have Hours' to Prevent Climate Disaster," *Star*, March 24, 2009, https:// www.thestar.com/news/gta/2009/03/24/we_ have_hours_to_prevent_climate_disaster.html.

13. Scot Lehigh, "Slogging Forward on Climate Change," *Boston Globe*, April 16, 2014, http:// www.bostonglobe.com/opinion/2014/04/15/ climate-change-skeptics-don-deserve-veto/ NsUEz1Epj9SjQchg2GIqoN/story.html.

14. "Ecological Disaster Feared," *Vancouver Sun*, May 11, 1982, https://stevengoddard.wordpress. com/2011/01/19/un-1982-world-to-end-before-the-year-2000/.

15. Leo Marx, "American Institutions and Ecological Ideals: Scientific and Literary Views of Our Expansionary Life-Style Are Converging," in *Readings in Environmental Impact*, Peter E. Black and Lee P. Herrington, eds. (SUNY at Syracuse College of Environmental Science & Forestry, 1974), 111; Steven Goddard, "1864 Expert Warning: Man's Climate Excess Rapidly Destroying the Human Race," Real Climate, April 25, 1864, https://stevengoddard.wordpress. com/2013/04/25/1864-expert-warning-mans-climate-excess-rapidly-destroying-the-human-race/.

16. "Grim Forecast," *San Jose Mercury News*, June 30, 1989, http://nl.newsbank.com/nl-search/we/ Archives?p_product=SJ&s_ site=mercurynews&p_multi=SJ&p_ theme=realcities&p_action=search&p_ maxdocs=200&p_topdoc=1&p_text_direct-0=0EB7304FF9A84273&p_field_direct-0=document_id&p_perpage=10&p_sort=YMD_ date:D&s_trackval=GooglePM.

17. Morano, "Earth 'Serially Doomed.'"

18. Ibid.

19. Ibid.

20. Andrew Alderson, "Global Warming Has Reached a 'Defining Moment,' Prince Charles Warns," *Telegraph*, March 12, 2009, http://www. telegraph.co.uk/news/earth/environment/ climatechange/4980347/Global-warming-has-reached-a-defining-moment-Prince-Charles-warns.html.

21. Valerie Richardson, "Prince Charles Extends Climate Doomsday Deadline by 33 Years," *Washington Times*, July 28, 2015, http://www. washingtontimes.com/news/2015/jul/28/prince-charles-extends-climate-doomsday-deadline/.

22. Robert Verkaik, "Just 96 Months to Save world, Says Prince Charles," *Independent*, July 8, 2009, http://www.independent.co.uk/environment/ green-living/just-96-months-to-save-world-says-prince-charles-1738049.html.

23. "Prince Charles Copenhagen Speech: 'The Eyes of the World Are upon You,'" *Guardian*, December 15, 2009, https://www.theguardian. com/environment/2009/dec/15/prince-charles-speech-copenhagen-climate.

24. Kevin Ball, video of Prince Charles, 2010, www. ClimatHustle.com.

25. Prince Charles, "Prince Charles, the British Monarchy," www.ClimatHustle.com.

26. Michael Bastasch, "We Just Passed Prince Charles's 96-Month Deadline to Save the World from 'Ecosystem Collapse,'" The Daily Caller, July 9, 2017, http://dailycaller.com/2017/07/09/we-just-passed-prince-charless-96-month-deadline-to-save-the-world-from-ecosystem-collapse/?utm_source=site-share.

27. Marc Morano, "Prince Charles Gives World Reprieve: Extends '100-Month' Climate 'Tipping Point' to 35 More Years," Climate Depot, July 28, 2015, http://www.climatedepot.com/2015/07/28/prince-charles-gives-world-reprieve-on-global-warming-extends-100-month-tipping-point-to-35-more-years/.

28. Harry McGee and Suzanne Lynch, "Robinson Calls for Climate Agreement by 2015," *Irish Times*, April 14, 2014, https://www.irishtimes.com/news/environment/robinson-calls-for-climate-agreement-by-2015-1.1761501.

29. "Gore Delivers Remarks on Energy and the Climate," *Washington Post*, July 17, 2008, http://www.washingtonpost.com/wp-dyn/content/article/2008/07/17/AR2008071701896.html.

30. Roy Spencer, "Al Gore's 10-Year Warning—Only 2 Years Left, Still No Warming," January 10, 2014, http://www.drroyspencer.com/2014/01/al-gores-10-year-warning-only-2-years-left-still-no-warming/.

31. Michael Mann, "Climate Danger Threshold Approaching," RT TV, 2014, www.ClimateHustle.com.

32. "Mann's Latest Propaganda: 'Global Warming Will Cross a Dangerous Threshold in 2036,'" The Hockey Schtick, March 18, 2014, http://hockeyschtick.blogspot.com/2014/03/manns-latest-propaganda-global-warming.html;. Michael E. Mann, "Earth Will Cross the Climate Danger Threshold by 2036," *Scientific American*, April 1, 2014, https://www.scientificamerican. com/article/earth-will-cross-the-climate-danger-threshold-by-2036/.

33. Marc Morano, "Global Warming Activist Scientists May Not Be the First to Proclaim a Doomsday Year of 2047 as the End of Time!—2047 Is the New 2012—but Global Warming Activists Were Beaten to Armageddon!" Climate Depot, October 10, 2013, http://www.climatedepot.com/2013/10/10/global-warming-activist-scientists-may-not-be-the-first-to-proclaim-a-doomsday-year-of-2047-as-the-end-of-time-2047-is-the-new-2012-but-global-warming-activists-were-beaten-to-armageddon/.

34. Nina Chestney, "100 Mln Will Die by 2030 If World Fails to Act on Climate—Report," Reuters, September 25, 2012, http://www.reuters.com/article/climate-inaction/100-mln-will-die-by-2030-if-world-fails-to-act-on-climate-report-idINDEE88O0HH20120925.

35. Geoffrey Lean, "Why Antarctica Will Soon Be the Only Place to Live—Literally," Independent, May 2, 2004, http://web.archive.org/web/20100817023019/http://www.independent.co.uk/environment/why-antarctica-will-soon-be-the-ionlyi-place-to-live—literally-561947.html.

36. "The Cooling World," *Newsweek*, April 28, 1975, http://www.denisdutton.com/cooling_world.htm.

37. Ron Bailey, "Tipping Points in Environmentalist Rhetoric: An Unscientific Survey of Nexis," Hit and Run (blog), June 8, 2012, http://reason.com/blog/2012/06/08/tipping-points-in-environmentalist-rheto.

38. "Paris Is Probably the Last Chance" Klimaretter. Info, November 6, 2015, https://translate.google.com/translate?hl=en&sl=de&tl=en&u=http%3A%2F%2Fwww.klimaretter.info%2Fforschung%2Fhintergrund%2F19986-qparis-ist-die-wahrscheinlich-letzte-chanceq&sandbox=1.

39. "Every UN Climate Summit Hailed as 'Last Chance' To Stop 'Global Warming' Before It's Too Late," Climate Depot, November 6, 2015, http://www.climatedepot.com/2015/11/06/

every-un-climate-summit-hailed-as-last-chance-to-stop-global-warming-before-its-too-late/.

40. Marc Morano, "Every UN Climate Summit Hailed as 'Last Chance' to Stop 'Global Warming' Before It's Too Late," Climate Depot, November 6, 2015, http://www.climatedepot.com/2015/11/06/every-un-climate-summit-hailed-as-last-chance-to-stop-global-warming-before-its-too-late/.

41. Tom Nelson, "Hunting the Climate Change Snark," Tom Nelson (blog), November 18, 2007, http://tomnelson.blogspot.com/2007/11/hunting-climate-change-snark.html.

42. Donna Laframboise, "50 to 1 Project—a Donna Laframboise Interview," YouTube, September 1, 2013, https://www.youtube.com/watch?v=U5weFQYBL5w.

43. Interview of Patrick Moore, Morano and Curran, *Climate Hustle*.

44. Interview of Don Easterbrook, Morano and Curran, *Climate Hustle*.

45. "Environmental Hysteria," *Penn & Teller Bullshit*," April 18, 2003, http://www.imdb.com/title/tt0672528/.

46. Eduardo Porter, "Earth Isn't Doomed Yet. The Climate Could Survive Trump," November 29, 2016, https://www.newsbusters.org/blogs/nb/scott-whitlock/2016/11/30/nyt-trump-hasnt-doomed-earth-yet-climate-might-survive-four-years.

47. Noel Sheppard, "New Zealand Weatherman on Global Warming: 'It's All Going to be a Joke in 5 Years,'" Newsbusters, https://www.newsbusters.org/blogs/nb/noel-sheppard/2007/05/18/new-zealand-weatherman-global-warming-its-all-going-be-joke-5.

Chapter 14:
Controlling Climate…or You?

1. "Inhofe Slams New Cap-and-Trade Bill as All 'Economic Pain for No Climate Gain,'" U.S. Senate Committee on Environment and Public Works, October 18, 2007, https://www.epw.

senate.gov/public/index.cfm/press-releases-all?ID=b4f81115-802a-23ad-4e54-f0137d7a406f&Issue_id=.

2. Bruno Waterfield, "EU Policy on Climate Change Is Right Even If Science Was Wrong, Says Commissioner," *Telegraph*, September 16, 2013, http://www.telegraph.co.uk/news/earth/environment/climatechange/10313261/EU-policy-on-climate-change-is-right-even-if-science-was-wrong-says-commissioner.html.

3. John P. Holdren, "Too Much Energy, Too Soon, a Hazard," *Windsor Star*, August 11, 1975, https://hauntingthelibrary.wordpress.com/2011/01/31/flashback-1975-holdren-says-real-threat-to-usa-is-cheap-energy/.

4. Anthony Watts, "Quote of the Week—Still 'Wirthless' After All These Years Edition," Watts Up with That, June 26, 2011, https://wattsupwiththat.com/2011/06/26/quote-of-the-week-still-wirthless-after-all-these-years-edition/.

5. Noel Sheppard, "UN IPCC Official Admits 'We Redistribute World's Wealth by Climate Policy,'" Newsbusters, November 18, 2010, https://www.newsbusters.org/blogs/nb/noel-sheppard/2010/11/18/un-ipcc-official-admits-we-redistribute-worlds-wealth-climate.

6. Roger Pielke Jr., "Ehrlich and Holdren Senate Testimony," Roger Pielke Jr.'s Blog, September 3, 2013, http://rogerpielkejr.blogspot.de/2013/09/1974-ehrlich-and-holdren-senate.html.

7. Marc Morano, "Global Carbon Tax Urged at UN Climate Conference," U.S. Senate Committee on Environment and Public Works, December 13, 2007, https://www.epw.senate.gov/public/index.cfm/press-releases-all?ID=D5C3C93F-802A-23AD-4F29-FE59494B48A6.

8. "MIT Professor Urging Climate Change Activists to 'Slow Down,'" CBS Boston, January 14, 2014, http://boston.cbslocal.com/2014/01/14/

mit-professor-urging-climate-change-activists-to-slow-down/.

9. Marc Morano, "Flashback: Gore: U.S. Climate Bill Will Help Bring About 'Global Governance,'" Climate Depot, July 10, 2009, http://www.climatedepot.com/2009/07/10/flashback-gore-us-climate-bill-will-help-bring-about-global-governance-2/.

10. Ban Ki-Moon, "We Can Do It," *New York Times*, October 25, 2009, http://www.nytimes.com/2009/10/26/opinion/26iht-edban.html.

11. Adam Brickley, "Global 'Ecological Board of Directors' Envisioned by State Department's Climate Czar," CNSNews, September 4, 2009, https://www.cnsnews.com/news/article/global-ecological-board-directors-envisioned-state-department-s-climate-czar.

12. Christopher C. Horner, "Chirac: Kyoto 'First Step Toward Global Governance,'" Competitive Enterprise Institute, November 19, 2000, https://cei.org/content/chirac-kyoto-first-step-toward-global-governance.

13. Leo Marx, "American Institutions and Ecological Ideals," Science vol. 170 (November 27, 1970), 945–52, http://nomadism.org/pdf/ecology.pdf.

14. Morano, "Global Carbon Tax Urged."

15. Marc Morano, "Flashback: ALERT: German Climate Advisor 'Proposes Creation of a CO_2 Budget for Every Person on Planet!,'" Climate Depot, September 6, 2009, http://www.climatedepot.com/2009/09/06/flashback-alert-german-climate-advisor-proposes-creation-of-a-co2-budget-for-every-person-on-planet/.

16. David Derbyshire, "Every Adult in Britain Should Be Forced to Carry 'Carbon Ration Cards,' Say MPs," *Daily Mail*, May 26, 2008, http://www.mailonsunday.co.uk/news/article-1021983/Every-adult-Britain-forced-carry-carbon-ration-cards-say-MPs.html.

17. Ben Webster, "Staff in Carbon Footprint Trial Face £100 Fines for High Emissions," *Times*, September 14, 2009, https://www.thetimes.co.uk/article/staff-in-carbon-footprint-trial-face-pound100-fines-for-high-emissions-7h87tdmfbqv.

18. Ibid.

19. Felicity Barringer, "California Seeks Thermostat Control," *New York Times*, January 11, 2008, http://www.nytimes.com/2008/01/11/us/11control.html?_r=1.

20. Marc Morano, "Special Report: 'Unholy Alliance'—Exposing the Radicals Advising Pope Francis on Climate," Climate Depot, September 24, 2015, http://www.climatedepot.com/2015/09/24/special-report-unholy-alliance-exposing-the-radicals-advising-pope-francis-on-climate/.

21. Thomas L. Friedman, "Our One-Party Democracy," *New York Times*, September 8, 2009, http://www.nytimes.com/2009/09/09/opinion/09friedman.htm

22. Sangwon Yoon, "Biggest Emitter China Best on Climate, Figueres Says," Bloomberg, January 14, 2014, https://www.bloomberg.com/news/articles/2014-01-13/top-global-emitter-china-best-on-climate-change-figueres-says.

23. Elizabeth Kolbert, "Global Warming Talks Progress Is 'Slow but Steady'—UN Climate Chief," *Guardian*, November 21, 2012, https://www.theguardian.com/environment/2012/nov/21/global-warming-talks-progress-un-climate-chief.

24. Ed Morrissey, "Video: Inventorying Pelosi," Hot Air, May 29, 2009, https://hotair.com/archives/2009/05/29/video-inventorying-pelosi/.

25. S. Matthew Liao, "Human Engineering and Climate Change," S. Matthew Liao (blog), February 9, 2012, http://www.smatthewliao.com/2012/02/09/human-engineering-and-climate-change/; Marc Morano, "NYU Prof. S. Matthew Liao of Center for Bioethics Promotes 'Solution of Human Engineering. It Involves the Biomedical Modification of Humans to Make Them Better at Mitigating Climate Change,'"

Climate Depot, http://www.climatedepot.com/2012/03/12/nyu-prof-s-matthew-liao-of-center-for-bioethics-promotes-solution-of-human-engineering-it-involves-the-biomedical-modification-of-humans-to-make-them-better-at-mitigating-climate-change/.

26. Marc Morano, "Climate Depot Round Up: Flashback 2012: Meet Man Who Wants to Engineer a Master Climate Race?! NYU Prof. Matthew Liao: Humans Genetically Engineered to Combat Global Warming—'Pharmacological Enhancement,'" Climate Depot, February 28, 2014, http://www.climatedepot.com/2014/02/28/climate-depot-round-up-flashback-2012-meet-man-who-wants-to-engineer-a-master-climate-race-nyu-prof-matthew-liao-humans-genetically-engineered-to-combat-global-warming-pharmac/; Sydney Opera House Talks & Ideas, "Matthew Liao—Engineer Humans to Stop Climate Change (Ideas at the House)," YouTube, June 25, 2013, https://www.youtube.com/watch?v=KzBVtmyN6_Y.

27. Marc Morano, "NYU Bioethicist Prof. Liao on Eating Meat: Seeks to 'Make Ourselves Allergic to Those Proteins…Unpleasant Reaction…the Way We Can Do That Is to Create Some Sort of Meat Patch,'" Climate Depot, October 20, 2012, http://www.climatedepot.com/2012/10/20/nyu-bioethicist-prof-liao-on-eating-meat-seeks-to-make-ourselves-allergic-to-those-proteinsunpleasant-reactionthe-way-we-can-do-that-is-to-create-some-sort-of-meat-patch/.

28. Bill McKibben, "Worst Climate Change Solutions of All Time," Atlantic, http://www.theatlantic.com/technology/archive/2012/03/how-engineering-the-human-body-could-combat-climate-change/253981/; Bill McKibben, @billmckibben, Twitter, March 12, 2012, https://twitter.com/billmckibben/status/179338270180786176.

29. Marc Morano, "'A Planned Economic Recession': Global Warming Prof. Kevin Anderson—Who Has 'Cut Back on Showering' to Save Planet—Asserts Economic 'De-Growth' Is Needed to Fight Climate Change," Climate Depot, November 25, 2013, http://www.climatedepot.com/2013/11/25/a-planned-economic-recession-global-warming-prof-kevin-anderson-who-has-cut-back-on-showering-to-save-planet-asserts-economic-de-growth-is-needed-continuing-with-ec/.

30. Marc Morano, "Global Warming Professor Kevin Anderson 'Cuts Back on Washing and Showering' to Fight Climate Change—Admits at UN Climate Summit: 'That Is Why I Smell'—Defends His Call for 'a Planned Economic Recession,'" Climate Depot, November 19, 2013, http://www.climatedepot.com/2013/11/19/global-warming-professor-kevin-anderson-cuts-back-on-washing-and-showering-to-fight-climate-change-admits-at-un-climate-summit-that-is-why-i-smell/.

31. Marc Morano, "Sierra Club Touts Economic 'De-Growth': 'We Have to De-Grow Our Economy' to 'Temper Climate Disruption, and Foster a Stable, Equitable World Economy'—'WORK LESS TO LIVE MORE,'" February 21, 2014, http://www.climatedepot.com/2014/02/21/sierra-club-touts-study-we-have-to-de-grow-our-economy-to-temper-climate-disruption-and-foster-a-stable-equitable-world-economy-work-less-to-live-more/.

32. Marc Morano, "Warmist Prof. Alice Bows-Larkin Calls for 'Planned Recessions' to Fight 'Global Warming': 'Economic Growth Needs to Be Exchanged' for 'Planned Austerity'—'Whole System Change,'" Climate Depot, October 19, 2015, http://www.climatedepot.com/2015/10/19/warmist-prof-alice-bows-larkin-calls-for-planned-recessions-to-fight-global-warming-economic-growth-needs-to-be-exchanged-for-planned-austerity-whole-system-change/.

33. Pielke, "Ehrlich and Holdren Senate Testimony."

34. Emily Gosden, "Prince Charles: Reform Capitalism to Save the Planet," *Telegraph*, May 27, 2014, http://www.telegraph.co.uk/news/uknews/prince-charles/10859230/Prince-Charles-reform-capitalism-to-save-the-planet.html.

35. Joseph Hall, "Facing Climate Change Head-On Means Changing Capitalism: Naomi Klein," *Star*, September 13, 2014, https://www.thestar.com/news/insight/2014/09/13/facing_climate_change_headon_means_changing_capitalism_naomi_klein.html.

36. Grist Staff, "Naomi Klein Says Climate Activists Need to Get Comfortable Attacking Capitalism," Grist, October 9, 2014, http://grist.org/climate-energy/naomi-klein-says-climate-activists-need-to-get-comfortable-attacking-capitalism/.

37. Marc Morano, "Capitalism in Crosshairs as Socialism Promoted at Opening Event of People's Climate March," Climate Depot, September 21, 2014, http://www.climatedepot.com/2014/09/21/capitalism-in-crosshairs-as-socialism-promoted-at-opening-event-of-people-climate-march/.

38. Marc Morano, "U. of Arizona Professor: Climate Change Now 'Irreversible'—Humans Must 'Terminate Industrial Civilization,'" Climate Depot, October 21, 2014, http://www.climatedepot.com/2014/10/21/u-of-arizona-professor-climate-change-now-irreversible-humans-must-terminate-industrial-civilization/.

39. Keith Farnish, *Time's Up! An Uncivilized Solution to a Global Crisis* (UIT Cambridge Ltd., 2009).

40. "Top NASA Scientist Arrested (Again) in White House Protest," Fox News, February 13, 2013, http://www.foxnews.com/science/2013/02/13/top-nasa-climate-scientist-arrested-again-in-white-house-protest.html.

41. Leo Hickman, "Climate Scientists Back Call for Sceptic Thinktank to Reveal Backers," *Guardian*, January 22, 2012, https://www.theguardian.com/environment/2012/jan/23/climate-sceptic-lawson-thinktank-funding.

42. Marc Morano, "Era of Constant Electricity at Home Is ending, Says Power Chief," *Daily Telegraph*, March 2, 2011, http://www.climatedepot.com/2014/06/11/flashback-2011-were-all-north-koreans-now-era-of-constant-electricity-at-home-is-ending-says-uk-power-chief-families-would-have-to-get-used-to-on-ly-usi-2/.

43. Steven Goddard, "Livechat at Noon Here! Can a Tax Stop Global Warming?" Real Science, May 7, 2013, https://stevengoddard.wordpress.com/2013/05/07/la-daily-news-carbon-tax-to-make-you-rich-and-stop-global-warming/.

44. Robert H. Frank, "Carbon Tax Silence, Overtaken by Events," *New York Times*, August 25, 2012, http://www.nytimes.com/2012/08/26/business/carbon-tax-would-have-many-benefits-economic-view.html?_r=1&smid=tw-share.

45. Henry M. Paulson Jr., "The Coming Climate Crash," *New York Times*, June 21, 2014, https://www.nytimes.com/2014/06/22/opinion/sunday/lessons-for-climate-change-in-the-2008-recession.html?_r=0.

46. Yoram Bauman and Shi-Ling Hsau, "The Most Sensible Tax of All," *New York Times*, July 4, 2012, http://www.nytimes.com/2012/07/05/opinion/a-carbon-tax-sensible-for-all.html?_r=0.

47. "The Carbon Tax (editorial)," *Washington Post*, November 10, 2012, https://www.washingtonpost.com/opinions/the-carbon-tax/2012/11/10/6c576bfa-29f5-11e2-b4e0-346287b7e56c_story.html.

48. Fred Palmer, "A Fool's Errand: Al Gore's $15 Trillion Carbon Tax," *Washington Examiner*, May 9, 2017, http://www.washingtonexaminer.com/a-fools-errand-al-gores-15-trillion-carbon-tax/article/2622479.

49. Marc Morano, "Flashback 1992: Gore on CO_2 Emissions: 'No Government Mandated Requirements Would Be Necessary of Any

Kind'—'Purely Voluntary Measures,'" Climate Depot, August 3, 2015, http://www.climatedepot.com/2015/08/03/flashback-1992-gore-on-co2-emissions-no-government-mandated-requirements-would-be-necessary-of-any-kind-purely-voluntary-measures/.

50. Clip of Peter Jennings "Reporting the Apocalypse and Al Gore" ABC News, April 11, 1998, in Marc Morano and Mick Curran, *Climate Hustle*, 2016, www.ClimateHustle.com; "CyberAlert—04/13/1998—Big Tobacco Should Go Out of Business," Media Research Center, April 13, 1998, http://www.mrc.org/biasalerts/cyberalert-04131998-big-tobacco-should-go-out-business#2.

51. Philip Webster, "G8 Leaders Claim Historic Breakthrough on New Deal to Tackle Global Warming," *Times*, July 9 2009, https://www.thetimes.co.uk/article/g8-leaders-claim-historic-breakthrough-on-new-deal-to-tackle-global-warming-nkbhhsjlkx2.

52. Philip Stott, "Global Warming Is Not a Crisis," Intelligence Squared, March 14, 2007, https://www.intelligencesquaredus.org/debates/global-warming-not-crisis.

53. Ian Johnston, "Donald Trump Is a 'Threat to the Planet', Says World-Leading Climate Change Scientist," *Independent*, October 5, 2016, http://www.independent.co.uk/environment/donald-trump-climate-change-global-warming-threat-to-the-planet-michael-mann-a7345836.html.

54. Steve Milloy, "Climategate 2.0: Jones Says 2-Degree C Limit 'Plucked Out of Thin Air,'" Junk Science, November 23, 2011, https://junkscience.com/2011/11/climategate-2-0-jones-says-2o-limit-plucked-out-of-thin-air/.

55. Marc Morano, Prof. Roger Pielke Jr. on Origins of 2 Degree Temp Target: 'Has Little Xcientific Basis,'" Climate Depot, September 20, 2017, http://www.climatedepot.com/2017/09/20/prof-roger-pielke-jr-on-origins-of-2-degree-temp-target-has-little-scientific-basis/.

56. Marc Morano, "Lomborg Blasts UN Paris Treaty's $100 Trillion Price Tag for No Temp Impact: 'You Won't Be Able to Measure It in 100 years,'" February 16, 2017, http://www.climatedepot.com/2017/02/16/lomborg-blasts-un-paris-treatys-100-trillion-price-tag-for-no-temp-impact-you-wont-be-able-to-measure-in-100-years/.

57. Marc Morano, "Climate Skeptics on 'Historic' UN treaty: 'Does This Mean We Never Have to Hear about "Solving" Global Warming Again!?'" Climate Depot, December 12, 2015, http://www.climatedepot.com/2015/12/12/climate-skeptics-on-historic-un-treaty-does-this-mean-we-never-have-to-hear-about-solving-global-warming-again/.

58. Václav Klaus, "Freedom, Not Climate, Is at Risk," *Financial Times*, June 13, 2007, https://www.ft.com/content/9deb730a-19ca-11dc-99c5-000b5df10621.

59. "The Climate Skeptic's Guide to Pope Francis' U.S. Visit: Talking Points about the Pope & Global Warming," Climate Depot, September 21, 2015, http://www.climatedepot.com/2015/09/21/the-climate-skeptics-guide-to-pope-francis-u-s-visit-talking-points-about-the-pope-global-warming/.

60. Marc Morano, "The Climate Skeptic's Guide to Pope Francis' U.S. Visit: Talking Points About the Pope & Global Warming," Climate Depot, September 21, 2015, http://www.climatedepot.com/2015/09/21/the-climate-skeptics-guide-to-pope-francis-u-s-visit-talking-points-about-the-pope-global-warming/.

61. Interview of Vaclav Klaus, Morano and Curran, *Climate Hustle*.

62. "UK Scientist Philip Stott: Global Warming 'Has Become the Grand Political Narrative of the Age, Replacing Marxism as a Dominant Force for Controlling Liberty and Human Choices,'" July 25, 2011, http://www.climatedepot.com/2011/07/25/uk-scientist-philip-stott-global-warming-has-become-the-grand-political-

narrative-of-the-age-replacing-marxism-as-a-
dominant-force-for-controlling-liberty-and-
human-choices/.

Chapter 15:
Climbing onto the Climate Change Gravy Train

1. Brent Baker, "NBC Nightly News Joins News-week in Smearing Global Warming 'Deniers,'" MRC News Busters, August 16, 2007, https://www.newsbusters.org/blogs/nb/brent-baker/2007/08/16/nbc-nightly-news-joins-news-week-smearing-global-warming-deniers#ixzz3656YCKng.

2. Andrew C. Revkin, "Exxon-Led Group Is Giving a Climate Grant to Stanford," *New York Times*, November 21, 2002, http://www.nytimes.com/2002/11/21/us/exxon-led-group-is-giving-a-climate-grant-to-stanford.html.

3. Shannon Hall, "Exxon Knew about Climate Change Almost 40 Years Ago," Scientific American, October 26, 2015, https://www.scientificamerican.com/article/exxon-knew-about-climate-change-almost-40-years-ago/.

4. Marc Morano, "'False': Mike Shellenberger Takes Down the 'The Exxon Climate Denial Myth'—Exxon in 'Many Cases Advocated for Climate Policy!'" Climate Depot, September 21, 2015, http://www.climatedepot.com/2015/09/21/false-mike-shellenberger-takes-down-the-the-exxon-climate-denial-myth-exxon-in-many-cases-advocated-for-climate-policy/.

5. "Scientists Sniffing Out Methane Emissions from Cows Sniffing at the Wrong End," Canada Free Press, June 28, 2007, http://canadafreepress.com/2007/cover071107.htm, https://www.epw.senate.gov/public/index.cfm/press-releases-all?ID=38d98c0a-802a-23ad-48ac-d9f7facb61a7.

6. Steven J. Allen, "A Climate Hysteric's Fake Enemies List," *Wall Street Journal*, April 27, 2017, https://www.wsj.com/articles/a-climate-hysterics-fake-enemies-list-1493330851.

7. Marc Morano, "Warmist Randy Olson Laments: Morano Is 'Beating the Hell Out of the Climate Movement'—Says It's 'Myth' Skeptics Are Better Funded," Climate Depot, August 9, 2017, http://www.climatedepot.com/2017/08/09/warmist-randy-olson-laments-morano-is-beating-the-hell-out-of-the-climate-movement-says-its-myth-skeptics-are-better-funded/.

8. Ken Haapala, "A New Paradigm?" Science and Environmental Policy Project, July 7, 2014, http://www.sepp.org/twtwfiles/2014/TWTW%207-12-14.pdf.

9. Joanne Nova, "The Money Trail," ABC News, September 28, 2010 http://www.abc.net.au/news/2010-03-04/33114.

10. "The Global Warming Science Machine: $79 Billion and Counting," Science and Public Policy Institute, August 11, 2009, http://scienceandpublicpolicy.org/commentaries-essays/commentaries/global-warming-science-machine.

11. Michael Bastasch, "Global Warming Gets Nearly Twice as Much Taxpayer Money as Border Security," The Daily Caller, October 28, 2013, http://dailycaller.com/2013/10/28/global-warming-gets-nearly-twice-as-much-taxpayer-money-as-border-security/#ixzz365EuGPLN.

12. Christopher Flavelle, "To Protect Climate Money, Obama Stashed It Where It's Hard to Find," Bloomberg, March 15, 2017, https://www.bloomberg.com/news/articles/2017-03-15/cutting-climate-spending-made-harder-by-obama-s-budget-tactics.

13. Ibid.

14. Marc Morano, "Say What?! U. S. Department of Transportation Asks: 'How Might Climate Change Increase the Risk of Fatal Crashes in a Community?'" Climate Depot, August 29, 2016, http://www.climatedepot.com/2016/08/29/say-what-u-s-department-of-transportation-asks-how-might-climate-change-increase-the-risk-of-fatal-crashes-in-a-community/.

15. Marc Morano, "Trump's OMB Dir. Mulvaney: No More 'Crazy' Climate Spending," Climate Depot, May 23, 2017, http://www.climatedepot.com/2017/05/23/trumps-omb-dir-mulvaney-no-more-crazy-climate-spending/.

16. Ron Arnold, "Mainstream Media Don't Know Big Green Has Deeper Pockets than Big Oil," *Washington Examiner*, May 13, 2014, http://www.washingtonexaminer.com/mainstream-media-dont-know-big-green-has-deeper-pockets-than-big-oil/article/2548405?utm_source=CFACT+Updates&utm_campaign=708b62c1f9-Big_Green_s_billions5_14_2014&utm_medium=email&utm_term=0_a28eaedb56-708b62c1f9-269738545.

17. Miles Weiss, "Gore Invests $35 Million for Hedge Funds with EBay Billionaire," Canadian Hedge Watch, March 6, 2008, http://www.canadianhedgewatch.com/content/news/general/?id=2675&printer=1.

18. Ellen McGirt, "Al Gore's $100 Million Makeover," Fast Company, July 1, 2007, https://www.fastcompany.com/60067/al-gores-100-million-makeover

19. "Video/Transcript: The O'Reilly Factor Features Climate Depot on Gore's Path to Become the First 'Carbon Billionaire,'" Climate Depot, May 2, 2009, http://www.climatedepot.com/2009/05/02/videotranscript-the-oreilly-factor-features-climate-depot-on-gores-path-to-become-the-first-carbon-billionaire/.

20. Mike Allen, "Gore Launches $300 Million Campaign," *Politico*, March 30, 2008, http://www.politico.com/story/2008/03/gore-launches-300-million-campaign-009268.

21. Carol D. Leonnig, "Al Gore Has Thrived as Green-Tech Investor," *Washington Post*, October 10, 2012, https://www.washingtonpost.com/politics/decision2012/al-gore-has-thrived-as-green-tech-investor/2012/10/10/1dfaa5b0-0b11-11e2-bd1a-b868e65d57eb_story.html?utm_term=.09d50c312e5f.

22. Ibid.

23. "Maybe Al Gore Should Be the Subject of a RICO Investigation," Sunshine Hours, May 19, 2016, https://sunshinehours.net/2016/05/19/maybe-al-gore-should-be-the-subject-of-a-rico-investigation/.

24. Marc Morano, "Warren Buffet's Vice Chair: Gore's 'Not Very Smart' & 'an Idiot,' but Became Filthy Rich Investing in 'Global Warming,'" Climate Depot, June 24, 2017, http://www.climatedepot.com/2017/06/24/buffets-vice-chair-tells-investors-gores-not-very-smart-but-became-filthy-rich-investing-in-global-warming-hes-an-idiot/.

25. Gaia Vince, "One Last Chance to Save Mankind," New Scientist, January 21, 2009, https://www.newscientist.com/article/mg20126921.500-one-last-chance-to-save-mankind/?full=true&print=true.

26. Marc Morano, "AlGorjeera—It's Official: Al Gore Is by Far the Most Lavishly Funded Fossil Fuel Player in the Global Warming Debate Today," Climate Depot, January 4, 2013, http://www.climatedepot.com/2013/01/04/algorjeera-its-official-al-gore-is-by-far-the-most-lavishly-funded-fossil-fuel-player-in-the-global-warming-debate-today/.

27. Greg Pollowitz, "Al Gore Sells Out to Big Oil and Gas," *National Review*, January 3, 2013, http://www.nationalreview.com/planet-gore/336823/al-gore-sells-out-big-oil-and-gas-greg-pollowitz.

28. Linda Stasi, "Current Situation: Staffers Talk about First Meeting with Al Jazeera," *New York Post*, January 8, 2013, http://nypost.com/2013/01/08/current-situation-staffers-talk-about-first-meeting-with-al-jazeera/.

29. Felicity Barringer, "Answering for Taking a Driller's Cash," *New York Times*, February 13, 2012, http://www.nytimes.com/2012/02/14/science/earth/after-disclosure-of-sierra-clubs-gifts-from-gas-driller-a-roiling-debate.html?_r=1&.

30. Marc Morano, "Watch Video Now: CNN Hosts Rare Live Contentious Global Warming Debate—Marc Morano vs. Sierra Club's Michael Brune &

Philippe Cousteau Jr.—Full Transcript—Morano: 'So Record Cold Is Now Evidence of Man-Made Global Warming?" Climate Depot, December 11, 2013, http://www.climatedepot.com/2013/12/11/cnn-hosts-rare-live-contentious-global-warming-debate-marc-morano-vs-sierra-clubs-michael-brune-philippe-cousteau-jr-full-transcript-video-coming-soon-morano-so-record-cold-is-now-evi/.

31. Interview of Patrick Moore, Marc Morano and Mick Curran, *Climate Hustle*, 2016, www.ClimateHustle.com.

32. IRS 990 forms for 2012, Guidestar.com.

33. "Sierra Club," Activist Facts (Center for Organizational Research and Education), 2017, https://www.activistfacts.com/organizations/194-sierra-club/.

34. Steven F. Haywards, "Climate Cultists," *Weekly Standard*, June 16, 2014, http://www.weeklystandard.com/climate-cultists/article/794401?page=3.

35. Burgess Everett, "Reid: Kochs Cause Climate Change," *Politico*, May 7, 2014, http://www.politico.com/story/2014/05/harry-reid-koch-brothers-climate-change-106441.

36. Glenn Kessler, "Harry Reid's Claim That the Koch Brothers Are 'One of the Main Causes' of Climate Change," *Washington Post*, May 8, 2014, https://www.washingtonpost.com/news/fact-checker/wp/2014/05/08/harry-reids-claim-that-the-koch-brothers-are-one-of-the-main-causes-of-climate-change/?utm_term=.05d1d5b411bf.

37. Center for Responsive Politics, "Top Organization Contributors (2016)," OpenSecrets, http://www.opensecrets.org/orgs/list.php.

38. "The Heinz Awards" (7th), Heinz Foundation, March 5, 2001, http://www.heinzawards.net/recipients/james-hansen.

39. "Brokaw's Objectivity Compromised in Global Warming Special," U.S. Senate Environment and Public Works Committee, July 11, 2006, https://www.epw.senate.gov/public/index.cfm/press-releases-all?ID=CB71A459-1F37-4792-AE25-541FCCED0466.

40. "Inhofe to Blast Global Warming Media Coverage in Speech Today on the Senate Floor," U.S. Senate Environment and Public Works Committee, September 25, 2006, https://www.epw.senate.gov/public/index.cfm/2006/9/post-7ce43ec4-41a8-4c1f-a881-b9c4e361cd89

41. "Inhofe Complains the Media Failed to Report Climate Change 'Ketchup Money' Grant," Greenie Watch, September 29, 2006, http://antigreen.blogspot.com/2006/09/tunguska-event-responsible-for-warming.html.

42. Amy Ridenour, "Media Double-Standard on Global Warming 'Censorship,'" Newsbusters, June 3, 2008, https://www.newsbusters.org/blogs/nb/amy-ridenour/2008/06/03/media-double-standard-global-warming-censorship.

43. Reuters and Joshua Gardner, "When Simply Walking Outside Can Kill: Study Reveals Rising Summer Heat and Humidity Will Make Any Outdoor Activity across America DEADLY for Tomorrow's Retirees," *Daily Mail*, June 26, 2014, http://www.dailymail.co.uk/news/article-2671334/When-simply-walking-outside-kill-Study-reveals-rising-summer-heat-humidity-make-outdoor-activity-America-DEADLY-tomorrows-retirees.html.

44. Climate State, "Global Warming, What You Need to Know, with Tom Brokaw" YouTube, March 26, 2017, https://www.youtube.com/watch?v=ptNL780RWLI.

45. "Michael Oppenheimer," Princeton University Program in Science, Technology, and Environmental Policy, July 5, 2017, https://www.princeton.edu/step/people/faculty/michael-oppenheimer/.

46. "Environmental Defense Fund from Source Watch," NGO Watch, https://thewrongkindofgreen.wordpress.com/the-group-of-ten/environmental-defense-fund/.

47. Marc Morano, "UN IPCC's Michael Oppenheimer: 'An Activist First—A Scientist a Distant Second'— Scientific Work 'Roundly Trashed' Even by Fellow Warmists!" Climate Depot, January 4, 2011, http://www.

climatedepot.com/2011/01/04/un-ipccs-michael-oppenheimer-an-activist-first-a-scientist-a-distant-second-scientific-work-roundly-trashed-even-by-fellow-warmists/.

48. Marc Morano, "Streisand v. Inhofe—Outrage over Inhofe Singling Out Singer's Role in Promoting 'Global Warming'—but Streisand Admitted: 'I, and Others Have Spent Countless Millions on This Issue'—She Funded UN IPCC Lead Author to Tune of $250,000," Climate Depot, December 3, 2014, http://www.climatedepot.com/2014/12/03/streisand-v-inhofe-hollywood-outraged-that-inhofe-singled-out-singers-role-in-promoting-global-warming-but-streisand-admits-i-and-others-have-spent-countless-millions-on-this-issue/.

49. "2010 Heinz Awards Announced," *USA Today*, Sep 21, 2010, http://content.usatoday.com/communities/sciencefair/post/2010/09/2010-heinz-awards-announced/1#.WdhSCeJSzrc; "Brokaw's Objectivity Compromised."

50. "Newsweek Editor Calls Mag's Global Warming 'Deniers' Article 'Highly Contrived,'" U.S. Senate Committee on Environment and Public Works, August 12, 2007, https://webcache.googleusercontent.com/search?q=cache:rv0aH1zbMtAJ:https://www.epw.senate.gov/public/index.cfm/press-releases-all%3FID%3D58659aa0-802a-23ad-49d7-3d18075e69c3+&cd=1&hl=en&ct=clnk&gl=us.

51. Bjørn Lomborg, "$7 Trillion to Fight Climate Change?" *Slate*, November 29 2013, http://www.slate.com/articles/health_and_science/project_syndicate/2013/11/climate_change_the_eu_wants_to_spend_7_trillion_on_projects_that_will_barely.html.

52. Tom Bawden, "Davos Call for $14trn 'Greening' of Global Economy," *Independent*, January 22, 2013, http://www.independent.co.uk/news/business/news/davos-call-for-14trn-greening-of-global-economy-8460994.html.

53. Michael Bastasch, "World Spends $1 BILLION a Day on Global Warming," The Daily Caller, October 23, 2013, http://dailycaller.com/2013/10/23/world-spends-1-billion-a-day-on-global-warming/.

54. Janine Puhak, "Mars Candy Company Promises $1 Billion to Combat Climate Change," Fox News, September 7, 2017, http://www.foxnews.com/food-drink/2017/09/07/mars-promises-1-billion-to-combat-climate-change.html.

55. Dan Gainor, "Even U.N. Admits That Going Green Will Cost $76 Trillion," Fox News, July 06, 2011, http://www.foxnews.com/opinion/2011/07/06/even-un-admits-that-going-green-will-cost-76-trillion.html#ixzz1RLVXSRUg.

56. "UN High Priest of Global Warming Christiana Figueres Describes Her Job as 'Sacred,'...". N, January 23, 2014, http://newnostradamusofthenorth.blogspot.com/2014/01/un-high-priest-of-global-warming.html.

57. Roy W. Spencer, "Science under President Trump: End the Bias in Government-Funded Research," Roy Spencer, Ph.D. (blog), December 21, 2016, http://www.drroyspencer.com/2016/12/science-under-president-trump-end-the-bias-in-government-funded-research/.

58. Roy W. Spencer, "The AMS Scolds Rick Perry for Believing the Oceans Are Stronger than Your SUV," Roy Spencer, Ph.D. (blog), June 22, 2017, http://www.drroyspencer.com/2017/06/the-ams-scolds-rick-perry-for-believing-the-oceans-are-stronger-than-your-suv/.

59. Marc Morano, "Prominent Scientist Dissents: Renowned Glaciologist Declares Global Warming Is 'Going to Be a Big Plus'—Fears 'Frightening' Cooling—Warns Scientists Are 'Prostituting Their Science,'" Climate Depot, January 8, 2015, http://www.climatedepot.com/2015/01/08/prominent-scientist-dissents-internationally-renowned-glaciologist-declares-global-warming-is-good-not-bad-its-going-to-be-a-big-plus/.

60. Marc Morano, "The 2 Month Press Release! 2047 Doomsday Author Prof. Mora Admits 'Press Release' for Study Took 'Two Months to Prepare'—Study Took Two Months to Write," Climate Depot, July 2, 2014, http://www.climatedepot.com/2014/07/02/the-2-month-press-release-2047-doomsday-author-prof-mora-admits-press-release-for-study-took-two-months-to-prepare-study-took-two-month-to-write/.

61. Duane Thresher, "Follow the Money," Real Climatologists, September 18, 2017, http://columbia-phd.org/RealClimatologists/Articles/2017/09/18/Follow_The_Money/.

62. Richard Lindzen, "Thoughts on the Public Discourse over Climate Change," Merion West, April 25, 2017, http://merionwest.com/2017/04/25/richard-lindzen-thoughts-on-the-public-discourse-over-climate-change/.

63. James Varney, "Skeptical Climate Scientists Coming in from the Cold," RealClearInvestigations, December 31, 2016, http://www.realclearinvestigations.com/articles/2016/12/31/skeptical_climate_scientists_coming_in_from_the_cold.html.

64. Ibid.

65. Interview of Caleb Rossiter, Morano and Curran, *Climate Hustle*.

66. Devin Henry "White House: Climate Funding Is 'a Waste of Your Money,'" *Hill*, March 16, 2017, http://thehill.com/policy/energy-environment/324358-white-house-says-climate-funding-is-a-waste-of-your-money.

Chapter 16: Hypocrisy on Parade

1. "Al Gore at the Oscars," YouTube, September 23, 2007, https://www.youtube.com/watch?v=6klON8Eklck.

2. Charles Krauthammer, "Limousine Liberal Hypocrisy," *Time*, March 16, 2007, http://content.time.com/time/magazine/article/0,9171,1599714,00.htm.

3. "Gore Refuses to Take Personal Energy Ethics Pledge," U.S. Senate Committee on Environment and Public Works, March 21, 2007, https://www.epw.senate.gov/public/index.cfm/press-releases-republican?ID=7616011F-802A-23AD-435E-887BAA7069CA.

4. Tim Hains, "Al Gore: I Don't Have a Private Jet; 'I Live A Carbon-Free Lifestyle,'" Real Clear Politics, June 4, 2017, https://www.realclearpolitics.com/video/2017/06/04/al_gore_i_dont_have_a_private_jet_i_live_a_carbon-free_lifestyle.html.

5. Marc Morano, "Gore's Two Tales: Claims 'No Memory of Airplane Encounter,'" April 12, 2009, Climate Depot, http://www.climatedepot.com/2009/04/12/gores-two-talesnbspclaims-no-memory-of-airplane-encounter/.

6. Marc Morano, "Face to Face: Morano Confronts Gore with 'Climate Hustle' DVD in Australia! Gore Refuses to Accept, Departs in SUV," Climate Depot, July 13, 2017, http://www.climatedepot.com/2017/07/13/face-to-face-morano-confronts-gore-with-climate-hustle-dvd-in-australia-gore-refuses-to-accept-departs-in-suv/.

7. Elizabeth Blair, "Laurie David: One Seriously 'Inconvenient' Woman," *All Things Considered*, NPR, May 7, 2007, http://www.npr.org/templates/story/story.php?storyId=9969008.

8. Peter Kiefer, "Hollywood Is Gearing Up for a Week of Science and Environmental Activism," *Hollywood Reporter*, April 21, 2017, https://www.yahoo.com/movies/hollywood-gearing-week-science-environmental-activism-230104029.html.

9. "On a Mission to the Stars," *Guardian*, November 18, 2006, https://www.theguardian.com/environment/2006/nov/18/weekendmagazine.usnews.

10. Drew Johnson, "EXCLUSIVE: Al Gore's Home Devours 34 Times as More Electricity than Average U.S. Household," The Daily Caller, August 2, 2017, http://dailycaller.

com/2017/08/02/exclusive-al-gores-home-devours-34-times-more-electricity-than-average-u-s-household/.

11. "Editorial: Al Gore's in the Big House—and It Sucks Up a Lot of Juice," *Richmond Times-Dispatch*, August 4, 2017, http://www.richmond.com/opinion/our-opinion/editorial-al-gore-s-in-the-big-house-and-it/article_6c7d0121-59fd-51b9-b2b3-7737c7e5258a.html.

12. Marc Morano, "'This Was Not Supposed to Happen': Gore's Sequel Comes in Dismal 15th at Box Office—Gore Fans Allege Film 'Sabotaged' by Paramount," Climate Depot, August 7, 2017, http://www.climatedepot.com/2017/08/07/under-performance-gores-sequel-comes-in-dismal-15th-at-box-office-gore-fans-allege-film-was-sabotaged-by-paramount/.

13. Alexander C. Kaufman, "Al Gore's Stupendous Wealth Complicates His Climate Message. That Can Change," HuffPost, August 18, 2017, https://www.huffingtonpost.com/entry/al-gore-wealth_us_599709f2e4b0e8cc855d5c09.

14. "Al Gore Takes Private Jet," *Hannity's America*, Fox News, September 9, 2007, https://www.youtube.com/watch?v=2VV309lbB8c.

15. Jim Treacher, "Leo DiCaprio: 'I Will Fly Around the World Doing Good for the Environment,'" The Daily Caller, January 21, 2013, http://dailycaller.com/2013/01/21/leo-dicaprio-i-will-fly-around-the-world-doing-good-for-the-environment/.

16. Gareth Davies, "Carbon Footprints of the Telethon Stars: The Hand in Hand Hurricane Fund-Raiser Started with a Lecture about Global Warming—Then Celebrities with Multiple Homes, Cars and Private Jets Starting Soliciting Much Needed Money," *Daily Mail*, September 13, 2017, http://www.dailymail.co.uk/news/article-4880930/Leo-Bieber-s-jet-setting-added-Irma.html.

17. Marc Morano, "Watch: Leonardo DiCaprio Tells UN Summit: Climate Action 'Requires All of Us to Make Real Changes in the Way We Live Our Lives,'" Climate Depot, December 4, 2015http://www.climatedepot.com/2015/12/04/dicaprio-tells-un-summit-climate-action-requires-all-of-us-to-make-real-changes-in-the-way-we-live-our-lives/.

18. P. Gosselin, "DiCaprio's Private Jet Junket Burned 30,000 Liters of Fuel…Enough For 10,000 Cars an Entire Day!" No Tricks Zone, May 23, 2016, http://notrickszone.com/2016/05/23/dicaprios-private-jet-junket-burned-30000-liters-of-fuel-enough-for-10000-cars-an-entire-day/#sthash.Az1yGqw4.0qf6lRX9.dpbs.

19. "DiCaprio Takes Private Jet Extra 8,000 Miles to Collect Environmental Award," *New York Post*, May 21, 2016, http://www.foxnews.com/entertainment/2016/05/21/dicaprio-takes-private-jet-extra-8000-miles-to-collect-environmental-award.html.

20. "With Five Private Jets, Travolta Still Lectures on Global Warming," *Evening Standard*, March 30, 2007, https://www.standard.co.uk/showbiz/with-five-private-jets-travolta-still-lectures-on-global-warming-7244988.html.

21. Marc Morano, "Flashback 2000: Actor Chevy Chase Says Socialism Works, Cuba Is a Prime Example!" Climate Depot, September 9, 2010, http://www.climatedepot.com/2010/09/09/flashback-2000-actor-chevy-chase-says-socialism-works-cuba-is-a-prime-example/.

22. Ibid.

23. Kavita Daswani, "My Favorite Room: Donna Mills Takes a Golden Opportunity to Add Magic to Her Dining Room," *Los Angeles Times*, July 12, 2017, http://www.latimes.com/business/realestate/hot-property/la-fi-hp-my-favorite-room-donna-mills-20170715-story.html.

24. Aly Nielsen, "Harrison Ford on Climate Change: 'There Won't Be Any Damn People,'" Newsbusters, December 11, 2015, https://www.newsbusters.org/blogs/business/alatheia-larsen/2015/12/11/harrison-ford-climate-change-there-wont-be-any-damn-people.

25. Evan Halper and Michael Rothfeld, "This Puts Your Commute to Shame," *Los Angeles Times*, March 07, 2008, http://articles.latimes.com/2008/mar/07/local/me-arnold7.

26. Marc Morano, "Schwarzenegger Touts 'Air-Drying Your Clothes for 6 Months to Save 700 pounds of Carbon Dioxide,'" Climate Depot, April 25, 2009, http://www.climatedepot.com/2009/04/25/schwarzenegger-touts-airdrying-your-clothes-for-6-months-to-save-700-pounds-of-carbon-dioxide/.

27. Emily Gosden, "Greenpeace Executive Flies 250 Miles to Work," *Telegraph*, June 23, 2014, http://www.telegraph.c.ouk/news/earth/earthnews/10920198/Greenpeace-executive-flies-250-miles-to-work.html.

28. Derek Hunter, "Hurricane Telethon Gets Political Right at The Start," The Daily Caller, September 12, 2017, http://dailycaller.com/2017/09/12/hurricane-telethon-gets-political-right-at-the-start-video/.

29. Davies, "Carbon Footprints of the Telethon Stars."

30. Marc Morano, "Celebs Ignore Death, Poverty on MTV Enviro Series," Climate Depot, May 5, 2005, http://www.climatedepot.com/2009/05/19/flashback-cameron-diaz-lauds-traditional-tribal-lifestyles-which-lack-running-water-electricity-on-mtv-enviro-series/.

31. Morano, "Celebs Ignore Death, Poverty."

32. Jeff Poor, "Alec Baldwin: Climate Change Denial 'A Form of Mental Illness,'" Breitbart, April 2016, http://www.breitbart.com/video/2016/04/22/alec-baldwin-climate-change-denial-form-mental-illness/.

33. John D. Sutter, "Why Alec Baldwin Wants You to listen to Donald Trump," CNN, December 11, 2015, http://www.cnn.com/2015/12/10/opinions/sutter-alec-baldwin-climate-cop21/.

34. Tom Nelson, Super-Rich Left-Winger Chuck Lorre Fantasizes about How Great Life Allegedly Was before Fossil Fuels, the Wheel, Books, Penicillin, Indoor Plumbing, Average Lifespans over 20 Years, Etc," Tom Nelson (blog), October 9, 2012, http://tomnelson.blogspot.com/2012/10/super-rich-left-winger-chuck-lorre.html.

35. Emily Goodin, "Robert Redford: I Will be 'Devastated' If EPA Riders Pass," *Hill*, April 11, 2011, http://thehill.com/policy/energy-environment/155143-robert-redford-i-will-be-devastated-if-epa-riders-pass.

36. Belinda Luscombe, "Leaders & Visionaries: Robert Redford," *Time*, October 17, 2007, http://content.time.com/time/specials/2007/article/0,28804,1663317_1663319_1669890,00.html.

37. "'The Heat Is On' Introduction Narrated by Robert Redford," YouTube, April 16, 2007, https://www.youtube.com/watch?v=PQNwh4nLxws.

38. Samuel R. Avro, "Robert Redford Criticized as Racist and 'Enemy of the Poor,'" Energy Trends Insider, January 20, 2009, http://www.energytrendsinsider.com/2009/01/20/robert-redford-criticized-as-racist-and-enemy-of-the-poor/.

39. Robert Redford, "Stop Public Handouts to Oil, Gas and Coal Companies, Now," HuffPost, June 9, 2012 http://www.huffingtonpost.com/robert-redford/fossil-fuel-subsidies_b_1605146.html?utm_hp_ref=green&ncid=edlinkusaolp00000008.

40. "Robert Redford: Natural Disasters Not Odd Coincidences," *USA Today*, June 10, 2013, https://www.usatoday.com/story/opinion/2013/06/10/president-obama-climate-change-column/2407783/, http://www.msnbc.com/msnbc/robert-redford-why-its-time-kill-keystone-good.

41. "The Irish Film-Maker Who's Stirring Up a Green Hornet's Nest in Hollywood," *Independent*, February 5, 2011, https://www.pressreader.com/ireland/irish-independent-weekend-review/20110205/282359741181227.

42. Robert Redford, "United Airlines 'Life' Commercial," YouTube, September 29, 2007, https://www.youtube.com/watch?v=qkmSzk_yWr4.

43. Marc Morano, "Director James Cameron Unleashed: Calls for Gun Fight with Global Warming Skeptics: 'I Want to Call Those Deniers Out into the Street at High Noon and Shoot It Out with those Boneheads,'" Climate Depot, March 24, 2010, http://www.climatedepot.com/2010/03/24/director-james-cameron-unleashed-calls-for-gun-fight-with-global-warming-skeptics-i-want-to-call-those-deniers-out-into-the-street-at-high-noon-and-shoot-it-out-with-those-boneheads/.

44. Marc Morano, "From King of the World to Chicken of the Sea: Director James Cameron Challenges Climate Skeptics to Debate and Then Bails Out at Last Minute," Climate Depot, August 23, 2010, http://www.climatedepot.com/2010/08/23/from-king-of-the-world-to-chicken-of-the-sea-director-james-cameron-challenges-climate-skeptics-to-debate-and-then-bails-out-at-last-minute/.

45. Marc Morano, "Cameron's Spokesman: 'Morano Is Not at Cameron's Level to Debate, and That's Why It Didn't Happen. Cameron Should Be Debating Someone Who Is Similar to His Stature in Our Society,'" Climate Depot, August 24, 2010, http://www.climatedepot.com/2010/08/24/camerons-spokesman-morano-is-not-at-camerons-level-to-debate-and-thats-why-it-didnt-happen-cameron-should-be-debating-someone-who-is-similar-to-his-stature-in-our-society/.

46. Aly Sujo, "Pols Pig Out as Africa Starves," *New York Post*, August 28, 2002, http://nypost.com/2002/08/28/pols-pig-out-as-africa-starves-2/.

47. Marco Sibaja and Bradley Brooks, "James Cameron: Brazilian Dam Dispute a Real-Life 'Avatar,'" Associated Press, April 12, 2010, https://usatoday30.usatoday.com/life/people/2010-04-12-cameron-brazil-avatar_N.htm.

48. Glenn Whipp, "Is 'Avatar' a Message Movie? Absolutely, Says James Cameron," *Los Angeles Times*, February 10, 2010, http://articles.latimes.com/2010/feb/10/news/la-en-cameron10-2010feb10.

49. Luke Buckmaster, "James Cameron's Deepsea Challenge: I'm Not Just Some Rich Guy on a Ride," *Guardian*, August 20, 2014, https://www.theguardian.com/film/australia-culture-blog/2014/aug/21/james-cameron-on-deepsea-challenge-im-not-just-some-rich-guy-on-a-ride.

50. Jen Chaney, "James Cameron on the Special Edition of 'Avatar,' Eco-Conscious DVDs and BP," *Washinton Post*, August 27, 2010, http://voices.washingtonpost.com/celebritology/2010/08/talking_with_james_cameron_abo.html.

51. Phelim McAleer, "James Cameron—Hypocrite," October 20, 2010, YouTube, https://www.youtube.com/watch?v=TKZ4RolQxec; Suzannah Hills, "Hollywood Director James Cameron Donates His Record-breaking Submarine to Science to Promote Marine Research," *Daily Mail*, March 26, 2013, http://www.dailymail.co.uk/sciencetech/article-2299422/Hollywood-director-James-Cameron-donates-submarine-science.html.

Chapter 17:
Child Propaganda

1. Justin Fishel, "John Kerry Signs Climate Deal with Granddaughter Seated in His Lap," ABC News, April 22, 2016, http://abcnews.go.com/International/john-kerry-signs-climate-deal-granddaughter-seated-lap/story?id=38600097.

2. "Inhofe Slams DiCaprio and Laurie David in Two-Hour Senate Floor Speech Debunking Climate Fears," U.S. Senate Committee on Environment and Public Works, October 26, 2007, https://www.epw.senate.gov/public/index.cfm/press-releases-all?ID=DDDC4451-802A-23AD-4000-A9B55ED9489A.

3. Scott Roxborough, "DiCaprio Sheds Light on '11th Hour,'" *Hollywood Reporter*, May 20, 2007, http://www.hollywoodreporter.com/news/dicaprio-sheds-light-11th-hour-136753.

4. Kyle Smith, "Leo DiCaprio's Parents Brainwashed Him into Becoming an Environmentalist Freak," *New York Post*, October 19, 2016, http://nypost.com/2016/10/19/leo-dicaprios-parents-brainwashed-him-into-becoming-an-environmentalist-freak/.

5. "Inhofe slams DiCaprio and Laurie David."

6. "Scientist Silenced by Politician for His Position on Global Warming," Hot Air, October 1, 2007, https://hotair.com/archives/2007/10/01/scientist-silenced-by-politician-for-his-position-on-global-warming/.

7. Marc Morano, "The Weather Channel Video Uses Young Kids to Promote 'Global Warming' Fears—'Dear Mom & Dad, Climate Change Could Be Very Catastrophic,'" Climate Depot, November 3, 2016, http://www.climatedepot.com/2016/11/03/the-weather-channel-uses-young-kids-to-promote-global-warming-fears-dear-mom-dad-climate-change-could-be-very-catastrophic/.

8. Ibid.

9. Marc Morano, "'We Don't Need No CO_2'—'Don't Need No Bath': Children 'Astronauts' Awarded by UN for 'Climate Song,'" Climate Depot, February 22, 2016, http://www.climatedepot.com/2016/02/22/we-dont-need-no-co2-dont-need-no-bath-children-astronauts-awarded-by-un-for-climate-song/.

10. Marc Morano, "Singer/Actress Katy Perry Does UN Video: 'Man-Made Climate Change Is Hurting Children around the World,'" Climate Depot, December 8, 2015, http://www.climatedepot.com/2015/12/08/singeractress-katy-perry-does-un-video-man-made-climate-change-is-hurting-children-around-the-world/.

11. "Bill Nye Wants Fox news to Get Real about Climate Change," Wired, September 16, 2017, https://www.wired.com/2017/09/geeks-guide-bill-nye/.

12. Marc Morano, "Watch: Kids Used to Promote Gore's Sequel: 11-Year Old: 'People Are Releasing Toxic Gases That Are Ruining the World,'" Climate Depot, August 15, 2017, http://www.climatedepot.com/2017/08/15/watch-kids-used-to-promote-gores-sequel-11-year-old-people-are-releasing-toxic-gases-that-are-ruining-the-world/.

13. 24 Hour News 8 Web Staff, "Boy Starts Site to Teach Other Kids about Global Warming," Wood TV, September 26, 2016, http://woodtv.com/2016/09/26/boy-starts-site-to-teach-other-kids-about-global-warming/.

14. Darragh Johnson, "Climate Change Scenarios Scare, and Motivate, Kids," *Washington Post*, April 16, 2007, http://www.washingtonpost.com/wp-dyn/content/article/2007/04/15/AR2007041501164_pf.html.

15. "Inhofe Slams DiCaprio and Laurie David in Two-Hour Senate Floor Speech Debunking Climate Fears," U.S. Senate Committee on Environment and Public Works, October 26, 2007, https://www.epw.senate.gov/public/index.cfm/press-releases-all?ID=DDDC4451-802A-23AD-4000-A9B55ED9489A.

16. Interview of Patrick Moore, Marc Morano and Mick Curran, *Climate Hustle*, 2016, www.ClimateHustle.com

17. Interview of Judith Curry, Morano and Curran, *Climate Hustle*.

18. "Labour 'Is Brainwashing Pupils with Al Gore Climate Change Film' Says Father in Court," *Daily Mail*, September 27, 2007, http://www.dailymail.co.uk/news/article-484249/Labour-brainwashing-pupils-Al-Gore-climate-change-film-says-father-court.html.

19. Marc Morano, "Special Report: Morano Testifies in WVA on Climate School Curriculum: 'We Must Not Tell Kids There Is No Debate and No Dissent Is Allowed,'" Climate Depot, January 15, 2015, http://www.climatedepot.com/2015/01/15/morano-testifies-in-wva-on-climate-school-curriculum-we-must-not-tell-kids-there-is-no-debate-and-no-dissent-is-allowed/.

20. Ibid.

21. Marc Morano, "Update: Victory for Skeptics! W.Va. School Board Votes 'to Allow Classroom Debate on Climate Change,'" Climate Depot, April 9, 2015, http://www.climatedepot.com/2015/04/09/update-victory-for-skeptics-w-va-school-board-votes-to-allow-classroom-debate-on-climate-change/.

22. Paul Bedard, "Liberals Seek K-12 Ban on Book Challenging Global Warming," *Washington Examiner*, April 4, 2017, http://www.washingtonexaminer.com/liberals-seek-k-12-ban-on-book-challenging-global-warming/article/2619341.

23. Shasta Kearns Moore, "Portland School Board Bans Climate Change–Denying Materials," *Portland Tribune*, May 19, 2016, http://portlandtribune.com/sl/307848-185832-portland-school-board-bans-climate-change-denying-materials.

24. Marc Morano, "Watch: Morano on Fox News in Heated Climate Debate over Schools Banning Books Doubting 'Global Warming,'" Climate Depot, June 1, 2016, http://www.climatedepot.com/2016/06/01/watch-morano-on-fox-news-in-heated-climate-debate-over-schools-banning-books-doubting-global-warming/.

25. Shen Lu and Katie Hunt, "China's One-Child Policy Goes but Heartache Remains," CNN, December 31, 2015, http://www.cnn.com/2015/12/31/asia/china-second-child-policy-in-effect/.

26. Marie-Louise Olson, "'No Children, Happy to Go Extinct,' Tweets Weatherman after Grim Climate-Change Report Made Him Cry (Now He's Considering a Vasectomy)," *Daily Mail*, September 28, 2013, http://www.dailymail.co.uk/news/article-2436551/A-weatherman-breaks-tears-vows-NEVER-fly-grim-climate-change-report.html.

27. William Teach, "Warmists Who Originally Promised Never To Have A Baby Has A Baby," Right Wing News, Mar,ch 6, 2015, http://rightwingnews.com/climate-change/warmists-who-originally-promised-never-to-have-a-baby-has-a-baby/.

28. John Nolte, "Science-Denier Bill Nye: Is It Time To 'Penalize People for Having Extra Kids?'" The Daily Wire, April 26, 2017, http://www.dailywire.com/news/15759/science-denier-bill-nye-it-time-penalize-people-john-nolte#exit-modal.

29. Marc Morano, "NPR: 'Should We Be Having Kids in the Age of Climate Change?' 'We Should Protect Our Kids by Not Having Them,'" August 18, 2016, http://www.climatedepot.com/2016/08/18/npr-should-we-be-having-kids-in-the-age-of-climate-change-we-should-protect-our-kids-by-not-having-them/; Jennifer Ludden, "Should We Be Having Kids in the Age of Climate Change?" NPR, August 18, 2016, http://www.npr.org/2016/08/18/479349760/should-we-be-having-kids-in-the-age-of-climate-change.

30. William M. Briggs, "Global Warming Alarmists Plead: Save the Children by Not Having Them," The Stream, August 23, 2016, https://stream.org/global-warming-alarmists-plead-save-children-by-not-having-them/.

31. Andrew Follett, "NPR: Should We Be Having Kids in Light of Global Warming?" August 18, 2016, The Daily Caller, http://dailycaller.com/2016/08/18/npr-should-we-be-having-kids-in-light-of-global-warming/.

32. Jeff Cox, "Contraception Key in Climate Change Fight: Gore and Gates," CNBC, January 24, 2014, https://www.cnbc.com/2014/01/24/contraception-key-in-climate-change-fight-gore-and-gates.html.

33. Lubos Motl, "It Is Immoral for Al Gore to Organize 'Fertility Management' for Other Nations," The Reference Frame, January 27, 2014, https://motls.blogspot.com/2014/01/it-is-immoral-for-al-gore-to-organize.html.

34. Eric Roston, "Climate Change Kills the Mood: Economists Warn of Less Sex on a Warmer Planet," Bloomberg, November 2, 2015, https://www.bloomberg.com/news/articles/2015-11-02/climate-change-kills-the-mood-economists-warn-of-less-sex-on-a-warmer-planet.

Chapter 18:
Bypassing Democracy to Impose Green Energy Mandates

1. Michael Bastasch, "Obama's Law School Professor: EPA Is 'Burning the Constitution,'" The Daily Caller, March 17, 2015, http://dailycaller.com/2015/03/17/obama-law-school-prof-epa-is-burning-the-constitution/.

2. Steven Mufson, "The Last Minute Obama-McCain Coal Debate," Newsweek, November 3, 2008, http://newsweek.washingtonpost.com/postglobal/energywire/2008/11/the_last_minute_obama-mccain_c.html.

3. Marc Morano, "The Chinafication of America': Reaction to Obama Bypassing the Senate with UN Climate Treaty: 'Obama is taking a page from China's government,'" Climate Depot, August 27, 2014, http://www.climatedepot.com/2014/08/27/the-chinafication-of-america-reaction-to-obama-bypassing-the-senate-with-un-climate-treaty-obama-is-taking-a-page-from-chinas-government/.

4. Ibid.

5. Marc Morano, "Left-wing Env. Scientist Bails Out of Global Warming Movement: Declares It a 'Corrupt Social Phenomenon...Strictly an Imaginary Problem of the 1st World Middleclass,'" Climate Depot, July 26, 2010, http://www.climatedepot.com/2010/07/26/leftwing-env-scientist-bails-out-of-global-warming-movement-declares-it-a-corrupt-social-phenomenonstrictly-an-imaginary-problem-of-the-1st-world-middleclass/.

6. Kimberley A. Strassel, "Scott Pruitt's Back-to-Basics Agenda for the EPA," Wall Street Journal, February 18, 2017, https://www.wsj.com/amp/articles/scott-pruitts-back-to-basics-agenda-for-the-epa-1487375872.

7. Justin Sink, "WH: Carbon Rules Needed to Combat Extreme Weather," Hill, July 1, 2014, http://thehill.com/video/energy-environment/211070-wh-carbon-rules-needed-to-combat-extreme-weather.

8. Roger Pielke Jr., "Some Perspective on the US EPA Carbon Regulations," Roger Pielke Jr.'s Blog, June 3, 2014, http://rogerpielkejr.blogspot.com/2014/06/some-perspective-on-us-epa-carbon.html.

9. Marc Morano, "EPA Chief Admits Obama Regs Have No Measurable Climate Impact: 'One One-Hundredth of a Degree?' EPA Chief McCarthy Defends Regs as 'Enormously Beneficial'—Symbolic Impact," Climate Depot, July 15, 2015, http://www.climatedepot.com/2015/07/15/epa-chief-admits-obama-regs-have-no-measurable-climate-impact-one-one-hundredth-of-a-degree-epa-chief-mccarthy-defends-regs-as-enormously-beneficial-symbolic-impact/.

10. Anthony Watts, "EPA's Jackson and Energy Sec. Chu on the Senate Hot Seat," Watts Up with That, July 7, 2009, https://wattsupwiththat.com/2009/07/07/epas-jackson-and-energy-sec-chu-on-the-senate-hot-seat/.

11. Marc Morano, "Former Obama Energy Chief Slams EPA Climate Regs: 'Falsely Sold as Impactful'—'All U.S. Annual Emissions Will be Offset by 3 Weeks of Chinese Emissions,'" Climate Depot, May 26, 2016, http://www.climatedepot.com/2016/05/26/former-obama-energy-chief-slams-epa-climate-

regs-falsely-sold-as-impactful-all-u-s-annual-
emissions-will-be-offset-by-3-weeks-of-chinese-
emissions/.

12. Paul Driessen, "CFACT Energy & Environment
Truth File: 2014," Committee for a Constructive
Tomorrow, 2014, https://www.cfact.org/
wp-content/uploads/2014/10/Energy__
Environment_Truth_File_100614_Online.pdf

13. Ibid.

14. Interview of Robert Giegengack, Marc Morano
and Mick Curran, *Climate Hustle*, 2016, www.
ClimateHustel.com.

15. James Powers, "Economist Blasts Paris Treaty
$100T Price Tag: 'Not a Good Deal,'" Fox
Business News "Varney & Company," Feburary
14, 2017, https://www.newsbusters.org/blogs/
business/2017/02/15/economist-blasts-paris-
treaty-100t-price-tag-not-good-deal.

16. "Lomborg: Trump Is Right to Reject Paris
Climate Deal: It's Likely to Be a Costly Failure,"
Daily Telegraph, June 1, 2017, https://www.
thegwpf.com/bjorn-lomborg-trump-is-right-to-
reject-paris-climate-deal-its-likely-to-be-a-costly-
failure/.

17. Bjørn Lomborg, "Trump's Climate Plan Might
Not Be So Bad after All," *Washington Post*,
November 21, 2016, https://www.
washingtonpost.com/opinions/trumps-climate-
plan-might-not-be-so-bad-after-all/2016/11/21/
f8c37aa8-acef-11e6-a31b-4b6397e625d0_story.
html.

18. Marc Morano, "Lomborg Blasts UN Paris Treaty's
$100 Trillion Price Tag for No Temp Impact: 'You
Won't Be Able to Measure It in 100 Years,'"
Climate Depot, February 16, 2017, http://www.
climatedepot.com/2017/02/16/lomborg-blasts-
un-paris-treatys-100-trillion-price-tag-for-no-
temp-impact-you-wont-be-able-to-measure-in-
100-years/.

19. Marc Morano, "Statistician: UN Climate Treaty
Will Cost $100 Trillion—To Have No Impact—
Postpone Warming by Less than Four Years by

2100," Climate Depot, January 17, 2017, http://
www.climatedepot.com/2017/05/31/alert-trump-
is-officially-pulling-u-s-out-of-un-paris-climate-
pact/.

20. Daniel Halper, "Claim: Germany Spends $110
Billion to Delay Global Warming by 37 Hours,"
Weekly Standard, March 30, 2013, http://www.
weeklystandard.com/claim-germany-spends-
110-billion-delay-global-warming-37-hours/
article/712223.

21. Bjørn Lomborg, "$7 Trillion to Fight Climate
Change?" Slate, November 29, 2013, http://www.
slate.com/articles/health_and_science/project_
syndicate/2013/11/climate_change_the_eu_
wants_to_spend_7_trillion_on_projects_that_
will_barely.html.

22. Oliver Milman, "James Hansen, Father of
Climate Change Awareness, Calls Paris Talks 'a
Fraud,'" *Guardian*, December 2, 2015, https://
www.theguardian.com/environment/2015/
dec/12/james-hansen-climate-change-paris-
talks-fraud.

23. Emily Atkin, "The Troubling Return of Al Gore,"
New Republic, July 24, 2017, https://newrepublic.
com/article/143966/troubling-return-al-gore-
profile-inconvenient-sequel.

24. Devin Henry, "EPA: US Greenhouse Gas
Emissions Declined in 2015," *Hill*, February 14,
2017, http://thehill.com/policy/energy-
environment/319537-epa-us-greenhouse-gas-
emissions-declined-in-2015.

25. Andrew Follett, "Chart: American CO_2
Emissions Are WAY Down Due to Fracking," The
Daily Caller, May 9, 2016, http://dailycaller.
com/2016/05/09/chart-american-co2-emissions-
are-way-down-due-to-fracking/.

26. "Take Back Al Gore's Nobel and Give It to the
Fracking Industry," Investor's Business Daily,
February 17, 2017, http://www.investors.com/
politics/editorials/take-back-al-gores-nobel-and-
give-it-to-the-fracking-industry/.

27. Driessen, "CFACT Energy & Environment Truth File."

28. Michael Bastasch, "CONSENSUS: Global Warming Isn't a Top Concern For 90% of Americans," The Daily Caller, July 17, 2017, http://dailycaller.com/2017/07/17/consensus-global-warming-isnt-a-top-concern-for-90-of-americans/.

29. Marc Morano, "Gallup: 'Global Warming' Ranks at Bottom of Americans' ENVIRONMENTAL Concerns—Concern Falls to 1989 Levels," Climate Depot, March 25, 2015, http://www.climatedepot.com/2015/03/25/gallup-global-warming-ranks-at-bottom-of-americans-environmental-concerns-concern-falls-to-1989-levels/.

30. Michael Bastasch, "Al Gore Calls Global Warming 'The Most Serious Challenge We Face,' but 90% of Americans Disagree," The Daily Caller, July 17, 2017, http://dailycaller.com/2017/07/17/consensus-global-warming-isnt-a-top-concern-for-90-of-americans/.

31. Andrew Follettt, "Poll: Environment Is Least Important Issue to Americans," The Daily Caller, December 21, 2016, http://dailycaller.com/2016/12/21/poll-environment-is-least-important-issue-to-americans/.

32. Jeff Dunetz, "Pew Research: Americans Don't Believe the '97% Consensus of Climate Scientists' Claim," The Lid, December 8, 2016, http://lidblog.com/bogus-97-study/.

33. Christopher Cadelago, "Jerry Brown Predicts 'Negative, and Very Powerful' Reaction If Donald Trump Halts Climate Change Action," Sacramento Bee, December 6, 2016, http://www.sacbee.com/news/politics-government/capitol-alert/article119292988.html.

34. Marc Morano, "Scientific Smackdown: Skeptics Voted the Clear Winners against Global Warming Believers in Heated NYC Debate," Climate Depot, March 16, 2007, http://www.climatedepot.com/2013/12/19/why-warmists-hate-debate-flashback-2007-scientific-smackdown-skeptics-voted-the-clear-winners-against-global-warming-believers-in-heated-nyc-debate-realclimate-orgs-gavin-schmidt-appeared-so-de/.

35. "Climatology's Nutcracker," Bishop Hill (blog), May 14, 2013, http://www.bishop-hill.net/blog/2013/5/14/climatologys-nutcracker.html.

36. James Delingpole, "UN Poll Shows Climate Change Is the Lowest of All Global Concerns," Breitbart, October 26, 2016, http://www.breitbart.com/london/2016/10/26/un-poll-puts-climate-change-lowest-global-concerns/.

37. "New Poll: Most Americans Want Government to Combat Climate Change, but Voters Deeply Divided along Party Lines on Paying for Solutions," Energy Policy Institute at the University of Chicago, Associated Press, September 14, 2016, http://www.apnorc.org/PDFs/EnergyClimate/Press%20Release_EPIC%20AP-NORC%20Energy%20Policy%20Poll_Final.pdf.

38. Mandy Mayfield, "Bill Nye: Older People Need to 'Die' Out Before Climate Science Can Advance," Washington Examiner, July 19, 2017, http://www.washingtonexaminer.com/bill-nye-older-people-need-to-die-out-before-climate-science-can-advance/article/2629163.

39. Marc Morano, "Watch: NASA's Chief Climate Scientist Gavin Schmidt Claims Texans Won't Listen to 'Liberal, Jewish Atheist from NYC' about 'Global Warming,'" Climate Depot, January 6, 2016, http://www.climatedepot.com/2016/01/06/watch-nasas-chief-climate-scientist-need-less-science-to-message-in-texas-more-cultural-understanding-on-youtube/.

Chapter 19: Green Colonialism

1. John Kerry, "Remarks at Boston College's 138th Commencement Ceremony," U.S. Department of State, May 19, 2014, https://2009-2017.state.gov/secretary/remarks/2014/05/226291.htm.

2. Marc Morano, "Flashback 2002: Jerry Brown Says 'It's Not Viable' for Poverty Stricken Developing World to Emulate Prosperity of U.S.," Climate Depot, August 20, 2009, http://www.climatedepot.com/2009/08/20/flashback-2002-jerry-brown-says-its-not-viable-for-poverty-stricken-developing-world-to-emulate-prosperity-of-us/.

3. Marc Morano, "Flashback 2000: Actor Chevy Chase Says Socialism Works, Cuba Is a Prime Example!" Climate Depot, September 9, 2010, http://www.climatedepot.com/2010/09/09/flashback-2000-actor-chevy-chase-says-socialism-works-cuba-is-a-prime-example/.

4. Megan Rowling, "One in Seven People Still Live without Electricity—World Bank—TRFN," Reuters, May 19, 2015, https://ca.news.yahoo.com/one-seven-people-still-live-without-electricity-world-073404900.html.

5. Marc Morano, "SPECIAL REPORT: More Than 1000 International Scientists Dissent over Man-Made Global Warming Claims—Challenge UN IPCC & Gore," Climate Depot, December 8, 2010, http://www.climatedepot.com/2010/12/08/special-report-more-than-1000-international-scientists-dissent-over-manmade-global-warming-claims-challenge-un-ipcc-gore-2/.

6. "New Report Shows How Africa's Electricity Providers Can Be Profitable and Still Make Electricity Affordable," World Bank, October 27, 2016, http://www.worldbank.org/en/news/press-release/2016/10/27/how-africas-electricity-providers-can-be-profitable-and-still-make-electricity-affordable.

7. Roger Pielke and Daniel Sarewitz, "Climate Policy Robs the World's Poor of Their Hopes," Financial Times, February 26, 2014, http://sciencepolicy.colorado.edu/admin/publication_files/2014.03.pdf.

8. Ibid.

9. Marlo Lewis, "CO$_2$ Emissions, Life Expectancy, per Capita GDP: The Real Hockey Stick," Global Warming, December 27, 2012, http://www.globalwarming.org/2012/12/27/the-real-hockey-stick/; https://www.thegwpf.org/content/uploads/2012/12/Goklany-Number-One-Threat.pdf.

10. Marlo Lewis, "CO$_2$ Emissions, Life Expectancy, Per Capita GDP: The Real Hockey Stick," GlobalWarming.org, December 27, 2012, http://www.globalwarming.org/2012/12/27/the-real-hockey-stick/.

11. Alex Epstein, "Epstein: We live on the cleanest, safest planet in history," Stossel, Fox Business Channel, January 24, 2014, http://video.foxbusiness.com/v/3099991586001/?#sp=show-clips.

12. Marc Morano, "Politically Left Scientist Dissents—Calls President Obama 'Delusional' on Global Warming," Climate Depot, September 23, 2014, http://www.climatedepot.com/2014/09/23/politically-left-scientist-dissents-calls-president-obama-delusional-on-global-warming/.

13. Interview of Caleb Rossiter, Marc Morano and Mick Curran, Climate Hustle, 2016, www.ClimateHustle.com.

14. Marc Morano, "Fired for 'Diverging' on Climate: Progressive Professor's Fellowship 'Terminated' after WSJ OpEd Calling Global Warming 'Unproved Science,'" Climate Depot, June 12, 2014, http://www.climatedepot.com/2014/06/12/fired-for-diverging-on-climate-progressive-professors-fellowship-terminated-after-wsj-oped-calling-global-warming-unproved-science/.

15. Jason Howerton, "Obama in Africa: 'The Planet Will Boil Over' If Everybody Has a Car, Air Conditioning and a Big House," The Blaze, July 1, 2013, http://www.theblaze.com/stories/2013/07/01/obama-in-africa-the-planet-will-boil-over-if-everybody-has-a-car-air-conditioning-and-a-big-house/.

16. Valerie Volcovici, "World Bank Plans to Limit Financing of Coal-Fired Power Plants," Reuters, June 26, 2013, http://in.reuters.com/article/

usa-climate-world-bank-coal-plants/world-bank-plants-to-limit-financing-of-coal-fired-power-plants-idINDEE95Q00E20130627.

17. Brad Plumer, "The U.S. Will Stop Financing Coal Plants Abroad. That's a Huge Shift," *Washington Post*, June 27, 2013, https://www.washingtonpost.com/news/wonk/wp/2013/06/27/the-u-s-will-stop-subsidizing-coal-plants-overseas-is-the-world-bank-next/?utm_term=.c927a6d3f39a.

18. Pielke and Sarewitz, "Climate Policy Robs the World's Poor."

19. Mike Pflanz, "Africa's Energy Consumption Growing Fastest in World," *Christian Science Monitor*, January 1, 2013, https://www.csmonitor.com/World/Africa/2013/0101/Africa-s-energy-consumption-growing-fastest-in-world.

20. Marc Morano, "Environmentalist Laments Introduction of Electricity," Free Republic, August 26, 2002, http://www.freerepublic.com/focus/news/739362/posts.

21. Ibid.

22. Willis Eschenbach, "We Have Met the 1%, and He Is Us," Watts Up WithT hat, January 13, 2013, https://wattsupwiththat.com/2013/01/13/we-have-met-the-1-and-he-is-us/.

23. Marc Morano, "Prof. Roger Pielke Jr. in Financial Times: Climate Activists 'Promote Green Imperialism That Helps Lock in Poverty'—'Climate Policy Robs the World's Poor of their Hopes,'" Climate Depot, February 28, 2014, http://www.climatedepot.com/2014/02/28/prof-roger-pielke-jr-in-financial-times-climate-activists-promote-green-imperialism-that-helps-lock-in-poverty-climate-policy-robs-the-worlds-poor-of-their-hopes/.

24. "Diplomat: Averting Climate Change Means Leaving Fossil Fuels in Ground," UPI, September 23, 2013, https://www.upi.com/Science_News/2013/09/23/Diplomat-Averting-climate-change-means-leaving-fossil-fuels-in-ground/UPI-68671379973018/?spt=hs&or=sn.

25. "Panic, Outcry at Government Charcoal Ban," IRIN News, January 16, 2009, http://www.irinnews.org/news/2009/01/16/panic-outcry-government-charcoal-ban.

26. Victoria Averill, "African Trade Fears Carbon Footprint Backlash," BBC, February 21, 2007, http://news.bbc.co.uk/2/hi/business/6383687.stm.

27. Rama Lakshmi, "India Rejects Calls For Emission Cuts," *Washington Post*, April 13, 2009, http://www.washingtonpost.com/wp-dyn/content/article/2009/04/12/AR2009041202452.html.

28. Marc Morano, "S. African Activist Slams UN's 'Green Climate Fund': 'Government to Govt Aid Is a Reward for Being Better than Anyone Else at Causing Poverty'—'It Enriches the People Who Cause Poverty,'" Climate Depot, December 11, 2011, http://www.climatedepot.com/2011/12/11/s-african-activist-slams-uns-green-climate-fund-government-to-govt-aid-is-a-reward-for-being-better-than-anyone-else-at-causing-poverty-it-enriches-the-people-who-cause-poverty/.

29. Ibid.

30. Ibid.

31. Emily Wax, "In Poorer Nations, Energy Needs Trump Climate Issues," *Washington Post*, September 8, 2009, http://www.washingtonpost.com/wp-dyn/content/article/2009/09/08/AR2009090804019.html?hpid=topnews.

32. Andrew C. Revkin, "Dispatches from Rio and Nepal: Knife Fights over Firewood, *New York Times*, June 22, 2012, AMhttps://dotearth.blogs.nytimes.com/2012/06/22/dispatches-from-rio-and-nepal-knife-fights-over-firewood/.

33. Bjørn Lomborg, "The Deadliest Environmental Threat (It's Not Global Warming)," *New York Post*, April 21, 2014, http://nypost.com/2014/04/21/the-deadliest-environmental-threat-its-not-global-warming/.

34. Bill McKibben, "The Fossil Fuel Resistance," *Rolling Stone*, April 11, 2013, http://www.

rollingstone.com/politics/news/the-fossil-fuel-resistance-20130411?print=true.

35. Steve Milloy, "NO…Poor Countries Can't Bypass Fossil Fuels Like They Bypassed Land Lines for Cell Phones," Junk Science, July 28, 2013, https://junkscience.com/2013/07/no-poor-countries-cant-bypass-fossil-fuels-like-they-bypassed-land-lines-for-cell-phones/.

36. Paul Driessen, "More Solar Jobs Is a Curse, Not a Blessing," CFACT, May 9, 2017, https://www.cfact.org/2017/05/09/more-solar-jobs-is-a-curse-not-a-blessing/.

37. Leo Hickman, "James Lovelock on Shale Gas and the Problem with 'Greens,'" Guardian, June 15, 2012, https://www.theguardian.com/environment/blog/2012/jun/15/james-lovelock-fracking-greens-climate?newsfeed=true.

38. James Lovelock, "Renewable Energy: Not All It's Cracked Up to Be?" Sustainable Technology Forum, July 23, 2010, https://sustainabletechnologyforum.com/renewable-energy-not-all-its-cracked-up-to-be_14776.html.

39. "Lovelock Recants," Bishop Hill, January 25, 2013, http://bishophill.squarespace.com/blog/2013/1/25/lovelock-recants.html.

40. John R. Christy, "Written Statement of John R. Christy," House Energy and Power Subcommittee, September 20, 2012, https://archives-energycommerce.house.gov/sites/republicans.energycommerce.house.gov/files/Hearings/EP/20120920/HHRG-112-IF03-WState-ChristyJ-20120920.pdf.

41. Bjørn Lomborg, "Lomborg: Obama Should Confront Climate Change Fantasies," USA Today, June 25, 2013, https://www.usatoday.com/story/opinion/2013/06/25/obama-climate-change-georgetown-column/2455723/.

42. "U.S. Energy Production, Consumption Has Changed Significantly since 1908," November 1, 2016, https://www.eia.gov/todayinenergy/detail.php?id=28592.

Chapter 20:
The Way Forward

1. "Endangerment and Cause or Contribute Findings for Greenhouse Gases under the Section 202(a) of the Clean Air Act," Environmental Protection Agency, July 11, 2017, https://www.epa.gov/ghgemissions/endangerment-and-cause-or-contribute-findings-greenhouse-gases-under-section-202a-clean.

2. Stephen Moore, "2016's Biggest Loser: Big Green," Washington Times, January 1, 2017, http://www.washingtontimes.com/news/2017/jan/1/big-green-biggest-loser-of-2016/.

3. Marc Morano, "Watch: Morano on Fox Debating EPA & Climate: 'This Is the End of Superstition in Washington,'" Climate Depot, February 22, 2017, http://www.climatedepot.com/2017/02/22/watch-morano-on-fox-debating-epa-climate/.

4. Marc Morano, "Terry O'Sullivan, Head of Laborers' Int. Union on Keystone pipeline Project: 'The [Obama] Admin. Chose to Support Environmentalists over Jobs—Job-killers Win, American Workers Lose,'" Climate Depot, November 15, 2011, http://www.climatedepot.com/2011/11/15/terry-osullivan-head-of-laborers-int-union-on-keystone-pipeline-project-the-obama-admin-chose-to-support-environmentalists-over-jobs-mdash-jobkillers-win-american-workers-lose/.

5. Marc Morano, "'Keystone Uncensored'—Labor Leader Terry O'Sullivan Who Twice Endorsed Obama, Now Calls Administration 'Gutless,' 'Dirty' and More," Climate Depot, April 22, 2014, http://www.climatedepot.com/2014/04/22/keystone-uncensored-labor-leader-terry-osullivan-who-twice-endorsed-obama-now-calls-calls-administration-gutless-dirty-and-more/.

6. Donald J. Trump, "Statement by President Trump on the Paris Climate Accord," Office of the Press Secretary, The White House, June 01, 2017, https://www.whitehouse.gov/the-press-office/2017/06/01/statement-president-trump-paris-climate-accord.

7. Marc Morano, "Watch: Morano in Contentious Al Jazeera TV Debate: 'Trump Waded into a Religious War by Going after Climate Pact,'" Climate Depot, June 5, 2017, http://www.climatedepot.com/2017/06/05/watch-morano-in-contentious-al-jazeera-tv-debate-trump-waded-into-a-religious-war-by-going-after-climate-pact/.

8. "Trump Cuts Show Paris Treaty Is a Paper Tiger: Bjorn Lomborg," *USA Today*, March 29, 2017, https://www.usatoday.com/story/opinion/2017/03/29/trump-order-paris-treaty-emissions-bjorn-lomborg-column/99737238/.

9. Valerie Richardson, "Not So Hot: Al Gore's 'Inconvenient Sequel' Meets with Skepticism Even from Left," *Washington Times*, July 27, 2017, http://www.washingtontimes.com/news/2017/jul/27/al-gores-inconvenient-sequel-meets-with-skepticism/.

10. David G. Victor, et al., "Prove Paris Was More than Paper Promises," *Nature*, August 1, 2017, http://www.nature.com/news/prove-paris-was-more-than-paper-promises-1.22378.

11. Marc Morano, "Bravo! Climate Skeptics Rejoice! Trump Echoes Climate Depot's Call to Dismantle & Defund UN/EPA Climate Agenda!" Climate Depot, May 26, 2016, http://www.climatedepot.com/2016/05/26/bravo-trump-echoes-climate-depots-morano-call-to-dismantle-unepa-climate-agenda/.

12. "Climate Skeptic Shreds Paris Agreement at UN 'Global Warming' Conference," Associated Press, November 16, 2016, http://www.climatedepot.com/2016/11/16/un-armed-security-shuts-down-skeptics-after-trump-event-shredded-un-climate-treaty-at-summit/.

13. Morano, "Bravo! Climate Skeptics Rejoice!"

14. Tim Marcin, "Trump's Paris Agreement Decision Takes Effect One Day after the 2020 Election," *Newsweek*, June 1, 2017, http://www.newsweek.com/trumps-paris-agreement-decision-takes-effect-one-day-after-2020-election-619326.

15. "A President Trump Could Wreck Progress on Global Warming," *Washington Post*, October 4, 2016, https://www.washingtonpost.com/opinions/a-president-trump-could-wreck-progress-on-global-warming/2016/10/04/1e6df606-85c3-11e6-ac72-a29979381495_story.html?utm_term=.699824f4a287.

16. Marc Morano, "Gore Admits Paris Pact Symbolic—Makes Incorrect Claims about Greenland, Sea Levels & Extreme Weather on Fox News" Climate Depot, June 4, 2017, http://www.climatedepot.com/2017/06/04/gore-makes-false-climate-claims-about-greenland-sea-levels-extreme-weather-on-fox-news/.

17. Eric Wolff, et al., "Ex-Obama team distressed As Trump Guts Climate Regs," *Politico*, March 28, 2017, http://www.politico.com/story/2017/03/trump-obama-climate-regulations-236611.

18. Alex Cramer, "Al Gore Says 'We're Going to Win' Battle against Global Warming," *Hollywood Reporter*, June 26, 2017, http://www.hollywoodreporter.com/news/al-gore-going-win-battle-global-warming-inconvenient-sequel-red-carpet-screening-1024543.

19. "US Congress Told 'Climate Change Is Not Real,'" ABC News, March 26, 2009, http://www.abc.net.au/news/2009-03-26/us-congress-told-climate-change-is-not-real/1631962.

20. Emily Tillett, "Al Gore Says fight for Paris Accord, Climate Will Continue despite Trump's Policies," CBS News, August 2, 2017, https://www.cbsnews.com/news/al-gore-says-fight-for-paris-accord-climate-will-continue-despite-trumps-policies/.

Appendix

1. "World Broadcast Premier of 'Amazon Rainforest: Clearcutting the Myths,'" American Investigator, http://web.archive.org/web/20001206222300/http://www.americaninvestigator.net/.

2. U.S. Senate Committee on Environment and Public Works, https://www.epw.senate.gov/public/index.cfm?FuseAction=Minority.

Blogs&ContentRecord_id=79541c08-802a-23ad-4789-56296e2d061b.

3. James Inhofe, "Skeptic's Guide to Debunking Global Warming Alarmism: Hot & Cold Media Spin Cycle: A Challenge to Journalists Who Cover Global Warming," 2006,

4. Marc Morano, "U.S. Senate Minority Report Update: More Than 700 (Previously 650) International Scientists Dissent over Man-Made Global Warming Claims: Scientists Continue to Debunk 'Consensus' in 2008: Update: December 10, 2010: More Than 1000 International Scientists Dissent over Man-Made Global Warming Claims," December 10, 2010, https://www.epw.senate.gov/public/index.cfm/press-releases-republican?ID=D6D95751-802A-23AD-4496-7EC7E1641F2F

5. Marc Morano, "Climate Depot Wins Another Award! Daily Caller: Climate Depot Shares '2010 Award for Political Incorrectness' with Senator Inhofe," Climate Depot, December 29, 2010, https://www.climatedepot.com/2010/12/29/climate-depot-wins-another-award-daily-caller-climate-depot-shares-2010-award-for-political-incorrectness-with-sen-inhofe/.

6. Ari Natter, "Climate Changed: Alexandria Ocasio-Cortez's Green New Deal Could Cost $93 Trillion, Group Says," Bloomberg News, February 25, 2019, https://www.bloomberg.com/news/articles/2019-02-25/group-sees-ocasio-cortez-s-green-new-deal-costing-93-trillion.

7. See Tom Elliott, "Ocasio-Cortez: Fixing 'Global Warming' Requires 'Massive Government Intervention,'" grabienews, February 7, 2019, https://news.grabien.com/story-ocasio-cortez-fixing-global-warming-requires-massive-governm; Michael Palicz, "Green New Deal: 'Air Travel Stops Becoming Necessary,'" Americans for Tax Reform, February 7, 2019, https://www.atr.org/green-new-deal-air-travel-stops-becoming-necessary.

8. According to "an overview circulated by [its] prononents." Jeff Cox, "Ocasio-Cortez's Green New Deal Offers 'Economic Security' for Those 'Unwilling to Work,'" CNBC, February 7, 2019, https://www.cnbc.com/2019/02/07/ocasio-cortezs-green-new-deal-offers-economic-security-for-those-unwilling-to-work.html; Michael Palicz, "Green New Deal: 'Air Travel Stops Becoming Necessary,'" Americans for Tax Reform, February 7, 2019, https://www.atr.org/green-new-deal-air-travel-stops-becoming-necessary.

9. Tom Nelson, "Yipes: UN Climate Hoax Chief Talks Openly about a [Bad-Weather-Preventing 'Centralized Transformation'That Is 'Going to Make the Life of Everyone on the Planet Very Different," Tom Nelson (blog), November 21, 2012, http://tomnelson.blogspot.com/2012/11/yipes-un-climate-hoax-chief-talks.html.

10. Michael Bastasch, "Liberal Campaigner Calls 'Green New Deal' a Plan to 'Redistribute Power' from Rich to Poor," Daily Caller, February 5, 2019, https://dailycaller.com/2019/02/05/ocasio-cortez-green-deal-redistribute-wealth/.

11. Joanne Nova, "Flashback: IPCC Official Admits UN Climate Meetings Redistribute Wealth in One of the "Largest Economic Conferences since World War II," JoNova, September 7, 2015, http://joannenova.com.au/2015/09/flashback-ipcc-official-admits-un-climate-meetings-redistribute-wealth-largest-economic-conference-since-wwii/.

12. Yaron Steinbuch, "AOC Explains Why 'Farting Cows' Were Considered in the Green New Deal,"

13. Bjørn Lomborg, "Where's the Beef? Ask Green Campaigners," Shine, November 30, 2018, https://www.shine.cn/opinion/1811305941/?mc_cid=d873354042&mc_eid=bcd216d9bf.

14. Steve Milloy, "Green New Deal revealed as communism" (tweet), Twitter, February 7, 2019, https://twitter.com/JunkScience/status/1093504373014040579.

15. Leo Marx, "American Institutions and Ecological Ideals," *Science* vol. 170 (November 27, 1970), 945–52, http://nomadism.org/pdf/ecology.pdf.

16. Al Gore, PBS NEWS HOUR, October 12, 2018, video embedded at Marc Morano, "Al Gore Admits UN IPCC Report Was 'Torqued Up' to Promote Political Action—'How Else Do They Get the Attention of Policy-Makers around the World," Climate Depot, October 23, 2018, https://www.climatedepot.com/2018/10/23/watch-gore-admits-un-ipcc-report-was-torqued-up-to-promote-politcal-action-how-else-do-they-get-the-attention-of-policy-makers-around-the-world/.

17. Joseph Hall, "Facing Climate Change Head-On Means Changing Capitalism: Naomi Klein," Star, September 13, 2014, https://www.thestar.com/news/insight/2014/09/13/facing_climate_change_headon_means_changing_capitalism_naomi_klein.html.

18. "The Green New Deal Is Better Than Our Climate Nightmare" (editorial), *New York Times*, February 23, 2019, https://www.nytimes.com/2019/02/23/opinion/green-new-deal-climate-democrats.html.

19. "Want a Green New Deal? Here's a Better One" (editorial) *Washington Post*, February 24, 2019, https://www.washingtonpost.com/opinions/want-a-green-new-deal-heres-a-better-one/2019/02/24/2d7e491c-36d2-11e9-af5b-b51b7ff322e9_story.html?noredirect=on&utm_term=.f1afa62ba85c.

20. Bryan *Walsh, Bryan, "Heroes of the Environment 2008: Ted Norhaus and Michael Shellenberger," Time, September 24, 2008, http://content.time.com/time/specials/packages/article/0,28804,1841778_1841779_1841804,00.html*

21. Michael Shellenberger, "I'm sorry, but I am calling bullshit " (thread), Twitter, February 7, 2019, https://twitter.com/ShellenbergerMD/status/1093591397364424704.

22. "LIUNA Endorses President Obama for Second Term as President," LIUNA, March 8, 2012, https://www.liuna.org/news/story/liuna-endorses-president-obama-for-second-term-as-president; "LiUNA Endorses Secretary Clinton for President," LIUNA, November 24, 2015, https://www.npmhu.org/media/news/liuna-endorses-secretary-clinton-for-president.

23. LIUNA, "Statement of #TerryOSullivan, General President of the Laborers' International Union of North America, On the 'Green New Deal'" (tweet and meme), Twitter, February 8, 2019, https://twitter.com/LIUNA/status/1093882327577567232/photo/1?ref_src=twsrc%5Etfw%7Ctwcamp%5Etweetembed%7Ctwterm%5E1093882327577567232&ref_url=https%3A%2F%2Fwelovetrump.com%2F2019%2F02%2F10%2Fmajor-union-boss-unloads-on-aocs-new-green-deal-will-destroy-workers-livelihoods%2F.

Index

K

L

M